Springer Series in Computational Mathematics

17

A. R. Conn N. I. M. Gould Ph. L. Toint

LANCELOT

A Fortran Package for Large-Scale Nonlinear Optimization (Release A)

With 38 Figures and 24 Tables

Springer-Verlag
Berlin Heidelberg New York
London Paris Tokyo
Hong Kong Barcelona
Budapest

Dr. A. R. Conn
Mathematical Sciences Department
IBM T. J. Watson Research Center
P. O. Box 218
Yorktown Heights, NY 10598, USA

Dr. N. I. M. Gould
Central Computing Department
Rutherford Appleton Laboratory
Chilton, Oxfordshire OX11 0QX, United Kingdom

Prof. Dr. Ph. L. Toint
Department of Mathematics
Facultés Universitaires ND de la Paix
61, rue de Bruxelles
B-5000 Namur, Belgium

Mathematics Subject Classification (1991): 90C30, 49M37, 90C06, 65Y15, 90C90, 65K05, 65K10, 49M27, 49M29

ISBN 3-540-55470-X Springer-Verlag Berlin Heidelberg New York
ISBN 0-387-55470-X Springer-Verlag New York Berlin Heidelberg

Printed in the United States of America

Typesetting: Camera ready by author
41/3140 - 5 4 3 2 1 0 - Printed on acid-free paper

To

Tony and Cyril Conn
Elizabeth and Ray Gould
Claire

with love and thanks

LANCELOT

Large And Nonlinear Constrained Extended Lagrangian Optimization Techniques

Preface

LANCELOT is a software package for solving large-scale nonlinear optimization problems. This book is our attempt to provide a coherent overview of the package and its use. This includes details of how one might present examples to the package, how the algorithm tries to solve these examples and various technical issues which may be useful to implementors of the software.

We hope this book will be of use to both researchers and practitioners in nonlinear programming. Although the book is primarily concerned with a specific optimization package, the issues discussed have much wider implications for the design and implementation of large-scale optimization algorithms. In particular, the book contains a proposal for a standard input format for large-scale optimization problems. This proposal is at the heart of the interface between a user's problem and the LANCELOT optimization package. Furthermore, a large collection of over five hundred test examples has already been written in this format and will shortly be available to those who wish to use them.

We would like to thank the many people and organizations who supported us in our enterprise. We first acknowledge the support provided by our employers, namely the the Facultés Universitaires Notre-Dame de la Paix (Namur, Belgium), Harwell Laboratory (UK), IBM Corporation (USA), Rutherford Appleton Laboratory (UK) and the University of Waterloo (Canada). We are grateful for the support we obtained from NSERC (Canada), NATO and AMOCO (UK).

Amongst the wonderful people and dogs whose presence and encouragement made LANCELOT possible, it is a pleasure to thank the Beachcombers, Michel Bierlaire, Ingrid Bongartz, Didier Burton, Jean-Marie Collin, Barbara Conn, Iain Duff, Johnny Engels, David Jensen, Helene Kellett, Penny King, Lucie Leclercq, Marc Lescrenier, Claire Manil, Jorge Moré, Jorge Nocedal, Michael Powell, John Reid, Annick Sartenaer, Bobby Schnabel, Jennifer Scott, Michael Saunders, Tina Stoecklin, Jacques Toint, Daniel Tuyttens, Margaret Wright and Tisa.

The list of our supporters would be incomplete if we didn't recognize the encouragements provided (albeit quite indirectly) by J. S. Bach, the Bell (Aldworth), J. Brahms, Captain Beefheart, D. Chostakovitch, the Chhokar Nepalese restaurant (Didcot), Côte d'Or, M. Davis, ECM Records, J. Garbarek, H. Hartung, Imaginary Records, Janet Lynns (Waterloo), the Jazz Butcher, Ch. Mingus, C. Monteverdi, Speyside and Islay, R. Towner, and the White Hart (Fyfield).

Spring 1992 — Andrew Conn, Nick Gould and Philippe Toint

Contents

List of Figures

List of Tables

Chapter 1. Introduction

1.1 Welcome to LANCELOT

LANCELOT, an acronym for Large And Nonlinear Constrained Extended Lagrangian Optimization Techniques, is a package of standard Fortran subroutines and utilities for solving large-scale nonlinearly constrained optimization problems. It has been designed for problems where the objective function is a smooth function of many real variables and where the value of these variables may be restricted by a finite set of smooth constraints. The package is intended to be especially effective when the number of variables is large.

At this point, it is probably worthwhile elaborating on what we mean by large. Firstly, this notion is clearly computer dependent. What is large on an IBM PS/2 or an Apple Macintosh is significantly different from what is large on a Thinking Machines CM2, an IBM 3090 or a Cray 2. The former machines have a substantially smaller memory and store than the latter, and therefore have more difficulty handling problems involving a large amount of data. Moreover, a highly nonlinear problem in five hundred variables could be considered large, whereas in linear programming it is possible to solve problems in more than five million variables.[1] The notion of size is thus also problem dependent. Furthermore, it depends upon the structure of the problem. Many large-scale nonlinear problems arise from the modelling of very complicated systems that may be subdivided into loosely connected subsystems. This structure may often be reflected in the mathematical formulation of the problem and exploiting it is often crucial if one wants to obtain an answer efficiently. The complexity of this structure is often a key factor in assessing the size of a problem. Lastly, the notion of a large problem depends upon the frequency with which one expects to solve a particular instance or closely related problem. This is because, in the case when one anticipates solving the same class of problems many times, one can afford to expend a significant amount of energy analyzing and exploiting the underlying problem structure.

By contrast to the highly successful package MINOS, [48], which is particularly appropriate for large problems where the number of nonlinear degrees

[1] This figure is undoubtedly already out of date.

of freedom is modest, the emphasis in LANCELOT is on problems which are significantly nonlinear, in the sense that they involve a large number of nonlinear degrees of freedom. In particular, LANCELOT may be rather inefficient for linearly constrained problems, especially in the presence of degeneracy. LANCELOT exploits a very pervasive type of structure using what we call *group partial separability.* This exploitation often leads to economies of storage which in turn enables one to solve quite large problems on small machines. It is the purpose of this book to introduce the necessary concepts and to detail how this solution process actually works. Fortunately, the fact that LANCELOT is effective for large-scale applications does not imply that it is unsuitable for solving either problems that only involve a few nonlinear degrees of freedom or problems that would not be considered large in the sense described above (although it might be less efficient than specialized packages).

LANCELOT was created out of necessity. There is an increasing demand for such software as the size and nonlinearity of the problems that practitioners are interested in solving steadily grows. This trend is caused by the more widespread awareness that both nature and society optimize and the universe is not linear. Therefore, *accurate* modelling of physical and scientific phenomena frequently leads to larger and larger nonlinear optimization problems and the same evolution is observed in data fitting, econometrics and operations research models. In particular, it is perhaps worth listing

- Discretisations of variational calculations and optimal control problems involving both state and control variables.

  These arise, for example, in quantum physics, tidal flow analysis, design (of aircraft, journal bearings and other mechanical devices), structural optimization and ceramics manufacturing.

- Nonlinear equations arising, both in their own right and in the solution of ordinary and partial differential equations.

  These occur, for instance, in elasticity, semiconductor simulation, chemical reaction modelling and radiative transfer.

- Nonlinear least squares or regression.

  Some examples include fluid dynamic calculations, tomography (both seismic and medical), combustion, isomerization and metal coating thickness assessment.

- Nonlinear approximation.

  These include antenna design, power transmission, maximum likelihood and robust regression.

- Nonlinear networks.

  Examples occur in traffic modelling, energy and water distribution/management systems, and neural networks. These problems can have hundreds of thousands of variables but they are tractable because of their very special structure.

- Other interesting problems occur in macro- and micro-economics, equilibrium calculations, production planning, energy scenario appraisal and portfolio analysis. These problems often give rise to quadratic programs, particularly in portfolio analysis.

The increasing interest in the solution of large-scale nonlinear optimization problems are also related to the realization by users that today's advances in computer technology are making the solution of such problems possible. Unfortunately, *efficient* algorithms for small-scale problems do not necessarily translate into efficient algorithms for large-scale problems. Perhaps the main reason for this is that in order to be able to handle large problems the algorithms have to necessarily be as simple as possible. Consequently, relative to many of the more successful algorithms for small, dense problems, the amount of information available at any given iteration may be severely restricted. This makes designing algorithms that are scale invariant (in the sense that, assuming infinite precision, quasi-Newton methods for unconstrained optimization are invariant under linear transformations) more difficult for large-scale problems. A second reason is that the *structure* of large-scale problems is of paramount importance. In small-scale problems, structure may be usefully exploited but this is not essential. As a consequence, it is just not adequate to take existing optimization software for small problems and to apply it on large ones, hoping that the increased capacity in computing will take care of the growth in problem size.

All this was already in the authors' mind, when, around 1986 after much debate and some experience, we agreed on the statement that it was indeed very difficult to solve large nonlinear optimization problems. Both a proven methodology and suitable software were clearly badly needed. The subsequent discussion and collaboration led first to the design of a robust trust region method for nonlinear minimization problems with simple bounds [8], followed by a more sophisticated method capable of handling the general nonlinear programming problem [12]. The implementation that went along with these theoretical developments always had as a priority the capacity to solve large-scale problems, using the structure of partially separable [34], and later that of group partially separable functions [10].

Once the first prototype of the much-discussed large-scale nonlinear programming package LANCELOT was available, it soon became apparent that the very statement of these problems (they were clearly needed for testing the code!) was in itself a rather formidable challenge. In particular, the amount of information present in the structure of a large-scale optimization problem, although crucial for the acceptable performance of the algorithms, was also very difficult to specify in a complete and understandable format. Further discussion with other researchers in the area soon convinced the authors that one possible approach would be to use a "standard input", that is a simple formal language in which these structural concepts could be expressed unambiguously. Such a form of input has been rather well-established and succesful in the more restricted domain of linear programming, where the MPS format [14], despite many known shortcomings, has, and still is playing, a considerable role. It was therefore decided that a Standard Input Format (SIF) for large nonlinear problems was the next step. This of course required more discussions and tests, but a proposal eventually materialized (see Chapter 7). The proposal was also soon turned into software to 'decode' the problems specified in this format into structures[2] that LANCELOT could actually use. This SIF interface to LANCELOT is currently our preferred method of communicating with the package.

Today, more than five hundred problems have been specified using the SIF input and most of them have subsequently been solved by LANCELOT. This collection features a wide variety of problem types: large and small, academic and practical, simple and difficult. It has been our experience that, once understood, the SIF approach is in fact quite efficient for problem specification. We started to be really enthusiastic when somebody came in to one of our offices with a highly structured constrained nonlinear programming problem arising from his research area (energy scenario simulation) where the number of variables was in the thousands. In just one hour, the problem was specified, validated (and even solved!) using the SIF and LANCELOT. Although this favourable scenario will (alas) probably remain exceptional, it indicates the potential effectiveness of the approach.

Of course, the SIF is not restricted to its use in tandem with LANCELOT (see Chapter 7 for a discussion). We hope that we will be able to convince you that it provides a practical and effective way to state and solve your nonlinear programming problems.

The present guide is intended to help you to install and use the LANCELOT optimization package. It is organized as follows.

- The next section provides a first introduction to the structure of nonlinear optimization problems that we set out to exploit. This clarifies a number of the concepts that will be used throughout the book. Because these notions are re-explained in later sections, it is possible for the non-technically oriented user to skip this section at first reading.

- Chapter 2 gives a more gentle introduction to the ways one might specify

[2]Numerical data and Fortran subprograms, as a matter of fact.

optimization problems for LANCELOT, using the Standard Input Format (SIF). We accomplish this by leading the reader through a hierarchy of steadily more difficult examples. Our intention is that by the end of this chapter the user will be capable of specifying a broad range of realistic problems.

- Chapter 3 contains an in-depth description of the numerical optimization algorithms available within the package. This chapter is mainly intended for optimization specialists. The novice should be reassured that sensible default settings are provided with the package as distributed. However, the more sophisticated the users understanding of the underlying numerical algorithms, the more they have at hand to guide them, in the specification of their problems, in the options available when solving the problems and in the assessment of the results.

- Chapter 4 describes how the specification of various options is handled. These options enable the user to, for instance, check function derivatives, control the level and frequency of output and specify a variety of useful algorithmic options.

- Chapter 5 describes the way in which the various components of the LANCELOT package work together. The main purpose of this chapter is to assist those who wish to install LANCELOT on an environment not accomodated by the various default installation procedures provided. The reader should be aware that as LANCELOT evolves the number of different installations will grow. Consequently, they should consult the current LANCELOT distribution for information regarding the range of machines for which installation procedures are available.

- Chapter 6 describes the installation procedures for several well-known operating systems. The operating systems for which installation procedures are provided are Unix and its derivatives (AIX, Ultrix and UNICOS), CMS and VMS.

- Chapter 7 is the technical reference document for the Standard Input Format.

- Chapter 8 contains details of the Fortran argument lists for the main LANCELOT procedures so that they can be called as stand-alone subroutines in other applications.

- Some perspectives are proposed in the Coda.

- Finally, Appendix A describes the conditions under which the use of the package is permitted. Users who intend to use LANCELOT should be aware of and abide by these conditions.

We hope that you will find LANCELOT both useful and reasonably friendly. We welcome any comments that might lead to improvements in future releases.

1.2 An Introduction to Nonlinear Optimization Problem Structure

As we have already mentioned, structure is an integral and significant aspect of large-scale problems. Structure is often equated with sparsity; indeed the two are closely linked when the problem is linear. However, sparsity is not the most important phenomenon associated with a nonlinear function; that role is played by invariant subspaces. The *invariant subspace* of a function $f(x)$ is the set of all vectors w for which $f(x+w) = f(x)$ for all possible vectors x. This phenomenon encompasses function sparsity. For instance, the function

$$f(x_1, x_2, \cdots, x_{1000}) = x_{500}^2$$

has a gradient and Hessian matrix each with a single nonzero, has an invariant subspace of dimension 999, and is, by almost any criterion, sparse. However the function

$$f(x_1, x_2, \cdots, x_{1000}) = (x_1 + \cdots + x_{1000})^2$$

has a completely dense Hessian matrix but still has an invariant subspace of dimension 999, the set of all vectors orthogonal to a vector of ones. The importance of invariant subspaces is that nonlinear information is not required for a function in this subspace. We are particularly interested in functions which have large (as a percentage of the overall number of variables) invariant subspaces. This allows for efficient storage and calculation of derivative information. The penalty is, of course, the need to provide information about the subspace to an optimization procedure.

A particular objective function $F(x)$ is unlikely to have a large invariant subspace itself. However, many reasonably behaved functions may be expressed as a sum of *element* functions, each of which does have a large invariant subspace. This is certainly true if the function is sufficiently differentiable and has a sparse Hessian matrix ([34]). Thus, rather than storing a function as itself, it pays to store it as the sum of its elements. The elemental representation of a particular function is by no means unique and there may be specific reasons for selecting a particular representation. Specifying Hessian sparsity is also supported in the present proposal, but we believe that it is more efficient and also much easier to specify the invariant subspaces directly.

LANCELOT considers the problem of minimizing or maximizing an objective function of the form

$$F(x) = \sum_{i \in I_O} g_i \left(\sum_{j \in J_i} w_{i,j} f_j(\bar{x}_j) + a_i^T x - b_i \right), \quad x = (x_1, x_2, \cdots, x_n), \qquad (1.2.1)$$

within the "box" region

$$l_i \le x_i \le u_i, \quad l \le i \le n \tag{1.2.2}$$

(where either bound on each variable may be infinite), and where the variables are required to satisfy the extra conditions

$$g_i\left(\sum_{j \in J_i} w_{i,j} f_j(\bar{x}_j) + a_i^T x - b_i\right) = 0 \; (i \in I_E) \tag{1.2.3}$$

and

$$0 \left\{ \begin{matrix} \le \\ \ge \end{matrix} \right\} g_i\left(\sum_{j \in J_i} w_{i,j} f_j(\bar{x}_j) + a_i^T x - b_i\right) \left\{ \begin{matrix} \le \\ \ge \end{matrix} \right\} r_i, \; (i \in I_I) \tag{1.2.4}$$

for some index sets I_0, I_E and I_I and (possibly infinite) values r_i. The univariate functions g_i are known as *group functions.* The argument

$$\sum_{j \in J_i} w_{i,j} f_j(\bar{x}_j) + a_i^T x - b_i$$

is known as the i-th *group.* The functions $f_j, j = 1, \cdots, n_e$, are called *nonlinear* element functions. They are functions of the problem variables $\bar{x}_j$, where the $\bar{x}_j$ are either small subsets of x or such that f_j has a large invariant subspace for some other reason. The constants $w_{i,j}$ are known as *weights.* Finally, the function $a_i^T x - b_i$ is known as the *linear* element for the i-th group.

It is more common to call the group functions in (1.2.3) equality constraint functions, those in (1.2.4) inequality constraint functions and the sum of those in (1.2.1) the objective function.

When stating a structured nonlinear optimization problem of the form (1.2.1)–(1.2.4), we need to specify the group functions, linear and nonlinear elements and the way that they all fit together.

1.2.1 Problem, Elemental and Internal Variables

A nonlinear element function f_j is assumed to be a function of the problem variables $\bar{x}_j$, a subset of the overall variables x. Suppose that $\bar{x}_j$ has n_j components. Then one can consider the nonlinear element function to be of the *structural* form $f_j(v_1, \cdots, v_{n_j})$, where we assign $v_1 = \bar{x}_{j1}, \cdots, v_{n_j} = \bar{x}_{jn_j}$. The *elemental* variables for the element function f_j are the variables v and, while we need to associate the particular values $\bar{x}_j$ with v, it is the elemental variables which are important in defining the character of the nonlinear element functions.

As an example, the first nonlinear element function for a particular problem might be

$$(x_{29} + x_3 - 2x_{17})e^{x_{29}-x_{17}} \tag{1.2.5}$$

which has the structural form

$$f_1(v_1, v_2, v_3) = (v_1 + v_2 - 2v_3)e^{v_1 - v_3}, \tag{1.2.6}$$

where we need to assign $v_1 = x_{29}, v_2 = x_3$ and $v_3 = x_{17}$. For this example, there are three elemental variables.

The example may be used to illustrate a further point. Although f_1 is a function of three variables, the function itself is really only composed of *two* independent parts; the product of $v_1 + v_2 - 2v_3$ with $e^{v_1 - v_3}$, or, if we write $u_1 = v_1 + v_2 - 2v_3$ and $u_2 = v_1 - v_3$, the product of u_1 with e^{u_2}. The variables u_1 and u_2 are known as *internal* variables for the element function. They are obtained as *linear combinations* of the elemental variables. The important feature as far as an optimization procedure is concerned is that each nonlinear function involves as few internal variables as possible, as this allows for compact storage and more efficient derivative approximation.

It frequently happens, however, that a function does not have useful internal variables. For instance, another element function might have structural form

$$f_2(v_1, v_2) = v_1 \sin v_2, \tag{1.2.7}$$

where for example $v_1 = x_6$ and $v_2 = x_{12}$. Here, we have broken f_2 down into as few pieces as possible. Although there are internal variables, $u_1 = v_1$ and $u_2 = v_2$, they are the same in this case as the elemental variables and there is no virtue in exploiting them. Moreover it can happen that although there are special internal variables, there are just as many internal as elemental variables and it therefore doesn't particularly help to exploit them. For instance, if

$$f_3(v_1, v_2) = (v_1 + v_2) \log(v_1 - v_2), \tag{1.2.8}$$

where, for example, $v_1 = x_{12}$ and $v_2 = x_2$, the function can be formed as $u_1 \log u_2$, where $u_1 = v_1 + v_2$ and $u_2 = v_1 - v_2$. But as there are just as many internal variables as elementals, it will not normally be advantageous to use this internal representation. Finally, although an element function may have useful internal variables, the user may decide not to exploit them. The optimization procedure should still work but at the expense of extra storage and computational effort.

In general, there will be a linear transformation from the elemental variables to the internal ones. For example in (1.2.6), we have

$$\begin{pmatrix} u_1 \\ u_2 \end{pmatrix} = \begin{pmatrix} 1 & 1 & -2 \\ 1 & 0 & -1 \end{pmatrix} \begin{pmatrix} v_1 \\ v_2 \\ v_3 \end{pmatrix} \tag{1.2.9}$$

while in (1.2.7), we have

$$\begin{pmatrix} u_1 \\ u_2 \end{pmatrix} = \begin{pmatrix} 1 & 0 \\ 0 & 1 \end{pmatrix} \begin{pmatrix} v_1 \\ v_2 \end{pmatrix} \tag{1.2.10}$$

In general the transformation will be of the form

$$u = Wv \tag{1.2.11}$$

and this transformation is *useful* if the matrix W has fewer rows than columns.

1.2.2 Element and Group Types

It is quite common for large nonlinear programming problems to be defined in terms of many nonlinear elements. It is also common that these elements, although using different problem variables, are structurally the same as each other. For instance, the function

$$\sum_{i=1}^{n-1} (x_i x_{i+1})^i \tag{1.2.12}$$

naturally decomposes into the sum of $n-1$ group functions, $\alpha, \alpha^2, \cdots, \alpha^{n-1}$. Each group is a nonlinear element function $v_1 v_2$ of the two elemental variables v_1 and v_2 evaluated for different pairs of problem variables. More commonly, the elements may be arranged into a few classes; the elements within each class are structurally the same. For example, the function

$$\sum_{i=1}^{n-1} (x_i x_{i+1} + x_1/x_i)^i \tag{1.2.13}$$

naturally decomposes into the sum of the same $n-1$ group functions. Each group is the sum of two nonlinear elements $v_1 v_2$ (where $v_1 = x_i$ and $v_2 = x_{i+1}$) and v_1/v_2 (where $v_1 = x_1$ and $v_2 = x_i$). A further common occurrence is the presence of elements which have the same structure, but which differ in using different problem variables *and other auxiliary parameters*. For instance, the function

$$\sum_{i=1}^{n-1} (i x_i x_{i+1})^i \tag{1.2.14}$$

naturally decomposes into the sum of the same $n-1$ group functions. Each group is a nonlinear element $p_1 v_1 v_2$ of the single parameter p_1 and two elemental variables v_1 and v_2 evaluated for different values of the parameter and pairs of problem variables. Any two elements which are structurally the same are said to be of the same *type*. Thus examples (1.2.12) and (1.2.14) use a single element type, where as (1.2.13) uses two types. When defining the data for problems of the form (1.2.1)–(1.2.4), it is unnecessary to define each nonlinear element in detail. All that is actually needed is to specify the characteristics of the element types and then to identify each f_j by its type and the indices of its problem variables and (possibly) auxiliary parameters.

The same principal may be applied to group functions. For example, the group functions that make up

$$\sum_{i=1}^{n-1}(x_i x_{i+1})^2 \tag{1.2.15}$$

have different arguments but are structurally all the same, each being of the form $g_i(\alpha) = \alpha^2$. As a slightly more general example, the group functions for

$$\sum_{i=1}^{n-1} i(x_i x_{i+1})^2 \tag{1.2.16}$$

have different arguments and depend upon different values of a parameter but are still structurally all the same, each being of the form $g(\alpha) = p_1\alpha^2$ for some parameter p_1. Any two group functions which are structurally the same are said to be of the same *type;* the structural function is known as the *group type* and its argument is the *group-type variable.* Once again, using group types makes the task of specifying the characteristics of individual group functions more straightforward. The group type $g(\alpha) = \alpha$ is known as the *trivial* type. Trivial groups occur very frequently and are considered to be the default type. It is then only necessary to specify non-trivial group types.

1.2.3 An Example

We now consider the small example problem,

$$\text{minimize} \quad F(x_1, x_2, x_3) \equiv x_1^2 + x_2^4 x_3^4 + x_2 \sin(x_1 + x_3) + x_1 x_3 + x_2$$

subject to the bounds $-1 \leq x_2 \leq 1$ and $1 \leq x_3 \leq 2$. There are a number of ways of casting this problem in the form (1.2.1). Here, we consider partitioning F into groups as

$$\underset{\substack{\uparrow \\ \text{group 1}}}{(x_1)^2} \quad + \quad \underset{\substack{\uparrow \\ \text{group 2}}}{(x_2x_3)^4} \quad + \quad \underset{\substack{\uparrow \\ \text{group 3}}}{(x_2 \sin(x_1 + x_3) + x_1x_3 + x_2)}$$

Notice the following:

1. group 1 uses the non-trivial group function $g_1(\alpha) = \alpha^2$. The group contains a single *linear* element; the element function is x_1.

2. group 2 uses the non-trivial group function $g_2(\alpha) = \alpha^4$. The group contains a single *nonlinear* element; this element function is x_2x_3. The element function has *two* elemental variables, v_1 and v_2, say, (with $v_1 = x_2$ and $v_2 = x_3$) but there is no useful transformation to internal variables.

3. group 3 uses the trivial group function $g_3(\alpha) = \alpha$. The group contains two *nonlinear* elements and a single *linear* element x_2. The first nonlinear element function is $x_2 \sin(x_1 + x_3)$. This function has *three* elemental variables, v_1, v_2 and v_3, say, (with $v_1 = x_2$, $v_2 = x_1$ and $v_3 = x_3$), but may be expressed in terms of *two* internal variables u_1 and u_2, say, where $u_1 = v_1$ and $u_2 = v_2 + v_3$. The second nonlinear element function is $x_1 x_3$, which has two elemental variables v_1 and v_2 (with $v_1 = x_1$ and $v_2 = x_3$) and is of the same type as the nonlinear element in group 2.

Thus we see that we can consider our objective function to be made up of three groups; the first and second are non-trivial (and of different types) so we will have to provide our optimization procedure with function and derivative values for these at some stage. There are three nonlinear elements, one from group two and two more from group three. Again this means that we shall have to provide function and derivative values for these. The first and third nonlinear element are of the same type, while the second element is a different type. Finally one of these element types, the second, has a useful transformation from elemental to internal variables so this transformation will need to be set up.

1.2.4 A Second Example

We now consider a different sort of example, the unconstrained problem,

$$\text{minimize} \quad F(x_1, \cdots, x_{1000}) \equiv \sum_{i=1}^{999} \sin(x_i^2 + x_{1000}^2 + x_1 - 1) + \tfrac{1}{2} \sin(x_{1000}^2). \tag{1.2.17}$$

Once again, there are a number of ways of casting this problem in the form (1.2.1), but the most natural is to consider the argument of each sine function as a group — the group function is then $g_i(\alpha) = p_1 \sin \alpha$, $1 \le i \le 1000$, for various values of the parameter p_1. Each group but the last has two nonlinear elements, x_{1000}^2 and x_i^2 $1 \le i \le 999$ and a single linear element $x_1 - 1$. The last has no linear element and a single nonlinear element, x_{1000}^2. A single element type, v_1^2, of the elemental variable, v_1, covers all of the nonlinear elements.

Thus we see that we can consider our objective function to be made up of 1000 nontrivial groups, all of the same type, so we will have to provide our optimization procedure with function and derivative values for these at some stage. There are 1999 nonlinear elements, two from each group except the last, but all of the same type and again we shall have to provide function and derivative values for these. As there is so much structure to this problem, it would be inefficient to pass the data group-by-group and element-by-element. Clearly, one would like to specify such repetitious structures using a convenient shorthand.

1.2.5 A Final Example

As a third example, consider the constrained problem in the variables $x_1, \cdots, x_{100}$ and y

$$\text{minimize} \quad \tfrac{1}{2}((x_1 - x_{100})x_2 + y)^2 \tag{1.2.18}$$

subject to the constraints

$$x_1 x_{i+1} + (1 + \tfrac{2}{i})x_i x_{100} + y \leq 0 \quad (1 \leq i \leq 99), \tag{1.2.19}$$

$$0 \leq (\sin x_i)^2 \leq \tfrac{1}{2} \quad (1 \leq i \leq 100), \tag{1.2.20}$$

$$(x_1 + x_{100})^2 = 1 \tag{1.2.21}$$

and the simple bounds

$$-1 \leq x_i \leq i \quad (1 \leq i \leq 100). \tag{1.2.22}$$

As before, there are a number of ways of casting this problem in the form (1.2.1)–(1.2.4). We chose to decompose the problem as follows:

1. the objective function group uses the non-trivial group function $g(\alpha) = \frac{1}{2}\alpha^2$. The group contains a single *linear* element; the element function is y. There is also a nonlinear element $(x_1 - x_{100})x_2$. This element function has three elemental variables, v_1, v_2 and v_3, say (with $v_1 = x_1$, $v_2 = x_{100}$ and $v_3 = x_2$); there is a useful transformation from elemental to internal variables of the form $u_1 = v_1 - v_2$ and $u_2 = v_3$ and the element function may then be represented as $u_1 u_2$.

2. The next set of groups, inequality constraints, $x_1 x_{i+1} + (1+2/i)x_i x_{100} + y \leq 0$ for $1 \leq i \leq 99$ are of the form (1.2.4) with no lower bounds. Each uses the trivial group function $g(\alpha) = \alpha$ and contains a single *linear* element, y, and two *nonlinear* elements $x_i x_{i+1}$ and $(1+2/i)x_i x_{100}$. Both nonlinear elements are of the same type, $p_1 v_1 v_2$, for appropriate variables v_1 and v_2 and parameter p_1, and there is no useful transformation to internal variables.

3. The following set of groups, again inequality constraints, $0 \leq (\sin x_i)^2 \leq \frac{1}{2}$ for $1 \leq i \leq 100$, are of the form (1.2.4) with both lower and upper bounds. Each uses the non-trivial group function $g(\alpha) = \alpha^2$ and contains a single *nonlinear* element of the type $sin v_1$ for an appropriate variable v_1. Notice that the group types for these groups and for the objective function group are both of the form $g(\alpha) = p_1\alpha^2$, for some parameter p_1, and it may prove more convenient to use this form to cover both sets of groups.

4. The last group, an equality constraint, $(x_1 + x_{100})^2 - 1 = 0$, is of the form (1.2.3). Again, this group uses the trivial group function $g(\alpha) = \alpha$ and

contains a single *linear* element, -1, and a single *nonlinear* element of the type $(v_1 + v_2)^2$ for appropriate elemental variables v_1 and v_2. Once more, a single internal variable, $u_1 = v_1 + v_2$ can be used and the element is then represented as u_1^2.

Thus we see that we can consider our problem to be made up of 201 groups of two different types so we will have to provide our optimization procedure with function and derivative values for these at some stage. There are 200 nonlinear elements of four different types and again this means that we shall have to provide function and derivative values for these. As for the previous example, there is so much structure to this problem that it would be inefficient to pass the data group-by-group and element-by-element. Again, we will introduce ways to specify this repetitious structure using a convenient shorthand.

Chapter 2. A SIF/LANCELOT Primer

2.1 About this Chapter

If you read this line, this is probably because you are interested in solving some optimization problem(s), and that you hope that the LANCELOT package might help you. The purpose of this chapter is to help this hope materialize: starting from simple examples, we intend to tell you

- how to specify the problem you are interested in, and
- how to use LANCELOT to solve it.

We will do both at once, because this process is actually quite simple. Of course, specifying a problem and solving it are formally two different tasks, each requiring a different set of tools. This will force us to distinguish these tasks in our examples as well, but we will always try to remember that specifying the problem is only a step towards its practical solution.

The general approach that we intend to use throughout this chapter is to learn some basic problem specification techniques first and to see how to solve the resulting simple problems, gradually building on these to tackle more and more complex cases. This manual is therefore a learning tool, and should not be considered as a reference guide for expert users. Such a reference text is presented in Chapter 7.

2.2 The Heritage of MPS

As we have already stated in the book's introduction, the very idea of defining a *Standard Input Format* (SIF) was forced on the authors by the sheer difficulty they experienced in specifying large problems. It was also motivated by the successful record of MPS, a standard input format for linear programming (LP). As the purpose of the SIF is to cover the general mathematical programming problem, of which linear programs are specialized examples, it is natural to require that the MPS syntax constitute a subset of the SIF syntax. This option has the further advantage that the sizeable collection of LP test problems already available in MPS format can also be used within the SIF context *without modification.*

However, preserving MPS compatibility has disadvantages. Maybe the worst is the requirement of using a fixed format (that is of aligning keywords and data in well defined columns). MPS also has a (somewhat outdated) "style" for the keywords themselves that we had to adopt in the SIF syntax definition. The authors are well aware that a completely new design would have avoided those old fashioned "features", but the price to pay was to give up MPS compatibility. That seemed, to us, to be an even worse choice.

One must also stress that a Standard Input Format by no means replaces a true modelling language. Such languages have already been developed in the context of linear programs (OMNI [37], for instance) and also in a more general context (GAMS [2] or AMPL [24]). However, none of these actually provides a way of specifying the nonlinear structure as does the SIF. Moreover, they do not belong to the public domain, which puts additional restrictions on their widespread use by the scientific community.

Although the previous paragraph sounds a little bit like a disclaimer (all right, it is one), we strongly believe that using the SIF is now a viable and efficient technique, and we hope that you will share this belief when you know more about its innards.

2.3 Getting Started with the SIF: Linear Programs

2.3.1 The Fields in a SDIF Statement

Before we actually begin to specify our first mathematical programming problem, we have to briefly discuss some prerequisites about the way words and numbers are understood within the SIF syntax.

When you specify a problem in SIF, you do so by writing one or more *files*. We start by considering the first file, the Standard *Data* Input Format (or SDIF for short). We will return to the subsequent files after gaining some knowledge of the first.

The SDIF contains a number of ordered *sections* (just as in MPS) using five kinds of objects: keywords, codes, numbers, names and Fortran names.

- The *keywords* are the headers (titles) of the different sections that together constitute the problem SDIF. We will meet these sections (and their titles) in due course in this introduction.

- *Codes* consist of one or two upper case letters. We will also learn about them as we go along. Their purpose is to specify, within the sections, various kinds of information on the problem at hand.

- *Numbers* consist of at most 12 characters which may include a decimal point and an optional sign (a positive number is assumed unless a - is given). The value may be followed by a decimal exponent, written as an `E` or `D`, followed by a signed or unsigned one or two digit integer.

Blanks are not allowed within numbers. Hence "3.14", "-0.001", "10" and "+4.23D-06" are numbers but "2.0D 5" is not. Observe that the decimal point is optional, and that it is therefore possible to specify integer numbers as "10" or "-3".

- *Names* (that you will want to attach to various parts of the problem specification, as variables, constraints...) are made up of *valid characters.* The valid characters are all characters whose ASCII decimal code lies in the range 32 to 126 (binary 0100000 to 11111110, hex 20 to 7E). These include lower and upper case roman alphabetic characters, the digits 0 to 9, the blank character and other mathematical and grammatical symbols. The allowed length for a name varies between 6 and 10, as we will discover later. The only excluded names are "'SCALE'", "'DEFAULT'", "'MARKER'", "'INTEGER'", "'ZERO-ONE'"[1] and the name consisting entirely of blanks.

  For instance, "NAME", "lower name" , "3.141592" and "@&#+-*/_!!" are valid names. Of course, you would usually choose names that mean something to you, because that helps in understanding the relationships between the parts of the problems, so you might want to use "COST", "km/h", "PI" or "Heavens!" instead of the above.

- The SIF/LANCELOT interface produces Fortran subroutines that are subsequently used by the LANCELOT optimizers. These routines contain variables that are named within the SIF problem specification. Because they appear not only in the SIF file, but also in the Fortran subroutines, these names are called *Fortran names.* They are restricted to contain at most 6 characters (upper-case alphabetic letters and digits only) and must not start with a digit.

In addition to these three categories of objects, you may also want to insert comments into your problem description. This is obtained by starting a line with *, and anything that follows on that line is your comment, as in

```
* This useful comment was written by me today (version 1).
```

As alluded to above, codes, names and numbers must occur in well defined *fields* within the line. The definition for the 6 fields that can occur[2] on a line is as follows

[1]The 'INTEGER' and 'ZERO-ONE' names are excluded, not because they have any special meaning for LANCELOT, but because future extensions of the SIF are likely to use them for special purposes ... guess what!

[2]We will also see later that, in two particular cases, the fields are defined somewhat differently (see Sections 2.6.1 and 2.5.1).

```
F1 <---F2---><---F3---><----F4---->    <---F5---><----F6---->
|  |         |         |          |    |         |          |
2  5         15        25         36   40        50         61
```

where the numbers below indicate the columns in which the respective fields start. Note that the fields (4 and 6) are 12 characters wide: field 4 ends in column 36 while field 6 ends in column 61. The fields contain different types of information:

- field 1 contains a code,
- fields 2, 3 and 5 contain names,
- fields 4 and 6 contain numbers.

Schematically, a typical SDIF line is thus organized as

```
F1 <---F2---><---F3---><----F4---->    <---F5---><----F6---->
cd <--name--><--name--><--number-->    <--name--><--number-->
```

Of course, such an organization is very rigid and has a number of drawbacks, but it also has the advantage of simplicity and readability. It is also absolutely necessary because blank characters are significant in MPS names, and can therefore not be used as separators. Section 2.8.10 will indicate how the rigidity of this format may be alleviated, but we will use it until then.

A word of caution might be appropriate here. Notice that the blank characters are significant within names, which implies that "`A `" and "` A`" are different names! We therefore recommend that you flush all names to the left of the field where they appear, in order to avoid mistakes that are quite hard to detect...

Finally, we said that the SDIF sections must appear in a certain order, as is the case in MPS. This order is implicitly given in all the examples we will see (watch out!), but is also formalized in Section 2.8.7 if you have any doubt.

2.3.2 The Simplest Example

We are now ready to specify our first problem. Obviously, we will start with a very simple example. In fact, it is the simplest example that we could think of. The problem is to find the smallest real nonnegative number, or, in its mathematical form, the following linear program:

$$\min_{x \in \Re} x \tag{2.3.1}$$

subject to the constraint

$$x \geq 0. \tag{2.3.2}$$

As you can see, it is very hard to find a simpler one. Let us now see how we can specify it in the SDIF.

The first thing we have to do is to define a name for this problem, "SIMPLEST" for instance[3]. We then give it that name by writing:

```
NAME          SIMPLEST
```

In this line, the keyword **NAME** starts in column 1 and the name itself is in field 3.

We then have to specify the variable of the problem, x. The task of specifying the problem's variables and naming them is performed in the **VARIABLES** section. As indicated above, the SDIF is divided into a number of such "sections", just as in the MPS format. Each one has its well defined purpose, that we will discover little by little below. After specifying the **VARIABLES** section, our problem description then becomes

```
NAME          SIMPLEST

*   The simplest linear program

VARIABLES
    x
```

Notice that we have also added a comment to describe the problem. This comment is surrounded by two blank lines: blank lines are totally ignored in SDIF: they can thus be used to "space out" the content of the specification and make it more readable. Then we have started the **VARIABLES** "section" by using the associated keyword, starting in column 1. All keywords (marking the beginning of sections) start in column 1, just as **VARIABLES** and **NAME**. Finally, we have declared that **x** is the (only) variable for the problem.

Once the problem is named and its variable(s) known, we must of course define the objective function. Ours is quite simple and is just equal to $1.0 \times x$. This is what we mean by completing our specification:

```
NAME          SIMPLEST

*   The simplest linear program

VARIABLES
    x
GROUPS
 XN Object    x         1.0
ENDATA
```

[3]This is one of the few cases where there is an extra restriction on the length of a name. As many users may wish to use the name of the problem as the name of the file containing the problem description, it was necessary to restrict the length of the problem's name to what can be accepted as a filename within most popular operating systems. It turns out that some unfriendly systems still limit the length of filenames to 8 characters, so this is the limit in SDIF as well.

As above, we first started a section: this one is called GROUPS and its purpose is to state the objective function (and, later, the constraints) of the problem as well as the linear contribution of the variables to these functions. The objective is declared by using the code XN. Observe that XN is a code and thus starts in column 2. We then say that variable x appears in the objective function (that we call Object) linearly with a coefficient equal to 1.0. Notice that 1.0 appears in field 4, one of the two "numerical" fields (the other being field 6). As a matter of fact, it can be written anywhere within this field.

Then we have added the keyword ENDATA, which says that this is the end of the SDIF (and, in this simple case, of the SIF) specification! Quite a short description, isn't it? We now want to use LANCELOT to solve this problem. Assuming we just saved our problem description in a file called SIMPLEST.SIF[4], we simply issue the basic command to run LANCELOT (double precision version) on a SIF problem description, i.e.

```
sdlan SIMPLEST
```

and watch what happens. Your faithful computer then answers:

```
PROBLEM NAME: SIMPLEST

Double precision version will be formed.

The objective function uses      1 linear group


There is       1 variable  bounded only from below


objective function value =    0.00000000000000D+00

 x                       0.00000000000000D+00
```

Well, this is quite obviously a brief summary of our problem, followed by an equally brief statement of its solution!

But you will immediately object: "How come LANCELOT understands that the variable x must be nonnegative, although we did not specify it anywhere?" In fact, we did not forget it, but specified it implicitly. We used the convention that SDIF (as MPS) assumes that *all variables are restricted to be nonnegative, unless stated otherwise.* This type of information appears in another section surprisingly called BOUNDS, so that our problem can equally be specified by

```
NAME          SIMPLEST
```

[4] At least on a VMS or Unix system. On a CMS system the file might be called SIMPLEST SIF. See Chapter 6 for more details.

```
*   The simplest linear program (version 2)

VARIABLES
    x
GROUPS
 XN Object    x            1.0
BOUNDS
 XL SIMPLEST  x            0.0
ENDATA
```

We added a BOUNDS section, in which we specified that variable x has a lower bound of 0, using the code XL. (Rerunning LANCELOT with this more explicit SIF description of course still produces the same answer.) The repetition of the problem's name after the XL code is a consequence of our decision to be MPS compatible: MPS indeed requires that bounds "sets" (that is the set of bound specifications corresponding to one instance of the problem) be explicitly named.

Specifying upper bounds is just as easy: if we wish to impose the additional constraint

$$x \leq 5, \tag{2.3.3}$$

we simply modify our problem statement to

```
NAME          SIMPLEST

*   The simplest linear program (version 3)

VARIABLES
    x
GROUPS
 XN Object    x            1.0
BOUNDS
 XU SIMPLEST  x            5.0
ENDATA
```

remembering now that the line

```
 XL           x            0.0
```

is implicitly present. LANCELOT now says:

```
 PROBLEM NAME: SIMPLEST

 Double precision version will be formed.

 The objective function uses      1 linear group


 There is      1 variable  bounded from below and above
```

```
objective function value =     0.00000000000000D+00

 x                         0.00000000000000D+00
```

The only difference from what we saw above is in the summary line that now states that there is 1 variable (the only one here) that is bounded from below (by 0) and from above (by 5). Other type of bounds can of course be imposed on a variable; some of the usual relevant codes and their meanings are given in Table 2.1. They may appear in any order, the last one used taking precedence.

Codes	Meaning
XL	Bounded below by the value...(Lower)
XU	Bounded above by the value...(Upper)
XP	Unbounded above (Plus infinity)
XM	Unbounded below (Minus infinity)
XR	Unbounded above and below (fRee)
XX	Bounded above and below by the common value...(fiXed)

Table 2.1: Codes for bounding variables (in the BOUNDS section).

Note that a variable can be fixed (its lower and upper bounds are equal), which in effect means that it is not allowed to vary! It therefore acts as parameter in the problem definition, sometimes a very useful trick.

There is another part of the problem that we avoided entirely above: the specification of a starting point for the minimization. Of course, this is not so important for linear programs (as SIMPLEST), but may be vital for nonlinear problems. *If nothing is explicitly said, the SDIF assumes that the starting point is the origin.* So in fact we started solving SIMPLEST from its solution! Is LANCELOT capable of solving the problem when started from somewhere else? Fortunately yes, as you can quickly check by introducing a different starting value for x (using the code XV, where V stands for Variable, in a section obviously called START POINT).

```
NAME          SIMPLEST

*   The simplest linear program (version 4)

VARIABLES
    x
GROUPS
 XN Object    x         1.0
START POINT
 XV SIMPLEST  x         3.141592
ENDATA
```

(The starting points are also given the problem's name, just as the bound sets.) Running LANCELOT again (just in case) now dispells your last doubts.

2.3.3 Complicating our First Example

Equipped with our fresh knowledge, we now enter the realm of problems with more complex constraints. As a cautious first step, we (barely) complicate our initial problem (2.3.1)–(2.3.2) and consider

$$\min_{x,y\in\Re} x+1 \tag{2.3.4}$$

subject to the constraints

$$x \geq 0 \tag{2.3.5}$$

and

$$x+2y=2, \tag{2.3.6}$$

where we have added one variable (y) and the constraint (2.3.6). This is easily translated into

```
NAME          EXTRASIM

*   An extremely simple linear program

VARIABLES
    x
    y
GROUPS
 XN Object    x         1.0
 XE Cautious  x         1.0          y         2.0
CONSTANTS
    EXTRASIM  Object    -1.0
    EXTRASIM  Cautious  2.0
BOUNDS
 XR           y
ENDATA
```

The introduction of y and its declaration as a free variable doesn't call on anything new. What is new is the code XE employed in the GROUPS section to specify Equality constraints. It is followed by the name of the constraint (Cautious will do nicely) and a list of the variables occurring linearly in the constraint (x and y) together with their respective nonzero coefficients (1.0 and 2.0 here). We also specified the right-hand-side (constant) part of the new constraint in the CONSTANTS section in a rather obvious way, specifying that the constant value associated with the constraint Cautious. Just notice that the name of the relevant constraint now appears in field 3. In fact, *the difference between field 2 and fields 3 and 5 is that the former is used to indicate names of the object being* defined *by the current statement, while the latters contain names of already defined objects*: the statement

```
 XE Cautious  x          1.0             y          2.0
```

defines the constraint `Cautious`, while

```
    EXTRASIM  Cautious  2.0
```

defines the set of constants named `EXTRASIM` and merely assigns a constant value to the already defined constraint. Another novelty in `EXTRASIM` is that we have added the constant 1 to the objective function. This constant must therefore be specified in the `CONSTANTS` section together with the right-hand-side of the constraints, which we have done by writing

```
    EXTRASIM Object    -1.0
```

Notice the we have written "`-1.0`" instead of the expected "`1.0`", because the objective has no proper right-hand-side, and thus the constant should be put in the left-hand-side, which implies it changes sign!

After saving the specification in `EXTRASIM.SIF`, this yields

```
sdlan EXTRASIM

 PROBLEM NAME: EXTRASIM

 Double precision version will be formed.

 The objective function uses       1 linear group

 There  is      1 linear equality constraint

 There  is      1 free variable
 There  is      1 variable  bounded only from below


 objective function value =    1.00000000000000D+00

                x              0.00000000000000D+00
                y              1.00000000000000D+00
```

which is indeed what we expect: the problem summary has been updated and the new solution calculated by LANCELOT is correct.

As the first step was not too difficult, let us be a bit more daring and add two more constraints to our problem. We will still consider (2.3.4)–(2.3.6), but subject to the additional requirements that

$$2x + y \geq 2 \tag{2.3.7}$$

and

$$y - x \leq 1. \tag{2.3.8}$$

Our problem specification is soon completed and becomes

```
NAME          SUPERSIM

*   A super-simple linear program

VARIABLES
    x
    y
GROUPS
 XN Object    x          1.0
 XE Cautious  x          1.0            y          2.0
 XG Daring    x          2.0            y          1.0
 XL Foolish   x          -1.0           y          1.0
CONSTANTS
    SUPERSIM  Object    -1.0
    SUPERSIM  Cautious  2.0
    SUPERSIM  Daring    2.0
    SUPERSIM  Foolish   1.0
BOUNDS
 XR           y
ENDATA
```

The code XL is used to specify constraints where a Less than or equal signs separates the variables from the constant. Similarly, XG specifies "Greater than or equal" constraints. This is summarized in Table 2.2.

Codes	Meaning		
XN	Objective function – constant		
XL	Constraint	$\leq$	constant
XE	Constraint	$=$	constant
XG	Constraint	$\geq$	constant

Table 2.2: Codes for declaring constraint types (in the GROUPS section).

We can now return to LANCELOT with the new problem file SUPERSIM.SIF, and type

```
sdlan SUPERSIM

 PROBLEM NAME: SUPERSIM

 Double precision version will be formed.

 The objective function uses      1 linear group

 There  is      1 linear equality constraint
 There are      2 linear inequality constraints

 There  is      1 free variable
```

```
There  is       1 variable  bounded only from below
There are       2 slack variables



objective function value =    1.66666666666667D+00

           x                  6.66666666666667D-01
           y                  6.66666666666666D-01
           Daring             0.00000000000000D+00
           Foolish            1.00000000000000D+00
```

A quick glance at the result shows the solution is determined by both the `Daring` and `Cautious`[5] constraints, but that the `Foolish` one was unnecessary[6] and does not play any role in the final outcome[7].

2.3.4 Summary

In our first contact with LANCELOT, we learned some basic concepts about the objects that are used in an SDIF problem description: *keywords* to delimit sections, *codes* to indicate all kinds of informations that are relevant to a given section, *names* which can be given to variables, the objective function and constraints, and finally *numbers*.

We also introduced a number of sections.

`NAME` : defines the name of the problem.

`VARIABLES` : starts the section where the problem variables are given a name.

`GROUPS` : starts the section where the objective function (code `XN`) and constraints (codes `XL`, `XE` or `XG`) are given a name, and where the linear contribution of each variable to these is specified. The meaning of these codes is summarized in Table 2.2.

`CONSTANTS` : starts the section where the constant terms of the groups are named and defined. These are the constant term in the objective function (if any) and the right-hand-sides of the constraints (a zero right-hand-side is assumed unless otherwise stated).

`BOUNDS` : starts the section where the bounds on the variables are named and defined (a lower bound of 0 is assumed on the variables unless stated otherwise). The codes for specifying bounds are summarized in Table 2.1.

[5] Equality constraints are always active at the solution!

[6] As is often the case for such constraints in real life.

[7] At variance with what usually happens in real life.

START POINT : starts the section where the proposed starting point for the problem is named and specified using the code XV (the value 0 is assigned to all variables unless stated otherwise).

ENDATA : declares the end of the SDIF file for the current problem.

Finally note that empty CONSTANTS, BOUNDS or START POINT sections, i.e. sections where the default specification is appropriate for all objects covered in the section (constants, bounds or starting point components respectively), may safely be omitted, as is the case in our SIMPLEST example, where all three sections are missing. These sections require a name to be given to the specification that they contain. It is of course necessary to choose the same name for all the sections: in the above examples, we have taken the convention to use the problem's name in all cases.

2.4 SDIF Acquires a New Skill: Using the Named Parameters

2.4.1 What are the Named Parameters?

Let us return for a moment to our SUPERSIM problem, as described in section 2.3.3 and remove the unnecessary Foolish constraint from its definition. We are now interested in testing how the problem solution varies when we change the value of the constant appearing in both the Daring and Cautious constraints (a typical "sensitivity" study). One way of exploring this issue is to change the two corresponding lines in the CONSTANTS section (setting a common value for both constants that is different from 2.0), and then to run LANCELOT again. We might then change those two lines once more in order to try a third value, etc. However, the SDIF provides a way to change only one line instead of two for each value we want to try. The idea is to introduce a *parameter* to which we assign the common constant used in both our constraints and then to use this parameter in the definition of the constraints. Such a parameter will have a name as do all objects within the SDIF. If Level is the name we wish to give to this parameter, the description of problem SUPERSIM becomes:

```
NAME SUPERSIM

*   A super-simple linear program (version 2)

VARIABLES
    x
    y
GROUPS
 XN Object    x         1.0
 XE Cautious  x         1.0            y          2.0
```

```
 XG Daring    x          2.0            y         1.0
CONSTANTS
 RE Level                2.0
    SUPERSIM  Object     -1.0
 Z  SUPERSIM  Cautious                  Level
 Z  SUPERSIM  Daring                    Level
BOUNDS
 XR SUPERSIM  y
ENDATA
```

We have introduced here two new codes, namely `RE` and `Z`. The first (`RE`) is used to assign the real value appearing in field 4 to the real parameter `Level` (the mnemonic for the code comes from the interpretation that `Level` Equals 2.0). The second (`Z`) merely says that the value to assign to the constants in the `Cautious` and `Daring` constraints is no longer to be found in the numerical field 4, but that the value of the parameter whose name appears in field 5 (`Level`) is to be used instead.

If we now wish to create a variant of the `SUPERSIM` problem, it is easy to modify only the value given in field 4 of the definition of `Level`, for instance by rewriting this line as

```
 RE Level                3.141565
```

Notice that we have assigned a value to the parameter `Level` just before it is needed for the first time, but we could also introduce it anywhere in the specification *before its value is actually needed.* For instance, the specification

```
NAME          SUPERSIM

*   A super-simple linear program (version 3)

 RE Level                2.0
VARIABLES
    x
    y
GROUPS
 XN Object    x          1.0
 XE Cautious  x          1.0            y         2.0
 XG Daring    x          2.0            y         1.0
CONSTANTS
 Z  SUPERSIM  Cautious                  Level
 Z  SUPERSIM  Daring                    Level
BOUNDS
 XR SUPERSIM  y
ENDATA
```

is equally valid: we have placed the definition of `Level` in the 0-th section of the SDIF. This is often done for problem dependent parameters, because it makes them easy to find for later modifications.

2.4.2 The X vs. Z Codes

Our last example is probably an excellent occasion to introduce a difference in codes that will appear quite often in the SDIF syntax: the difference between codes starting with an `X` and codes starting with a `Z`. Observe that in the line

```
Z  SUPERSIM  Cautious                          Level
```

one has to decide somehow that the value of the constant for the `Cautious` constraint is not given by a number in the numerical field 4, but is instead given by the actual value that the parameter `Level` currently happens to have. We will use the `Z` code to indicate that we expect the value to be defined by that of an already defined named parameter.

In general, *the SDIF will use codes starting with* `X` *whenever the value associated with the action specified by the code is to be found in one (or both) of the "numerical fields" 4 and 6. Codes starting with* `Z` *will be used whenever this value is that of the parameter whose name appears in the "name field" 5.* To complicate things a bit further, the code consisting of the single character `X` can sometimes[8] be omitted, as above.

To illustrate this point, just observe that

```
   SUPERSIM  Cautious  2.0
   SUPERSIM  Daring    2.0
```

is equivalent to

```
X  SUPERSIM  Cautious  2.0
X  SUPERSIM  Daring    2.0
```

which, in turn, is equivalent to

```
RE Level               2.0
Z  SUPERSIM  Cautious                          Level
Z  SUPERSIM  Daring                            Level
```

This equivalence holds in the `CONSTANTS` section, as in our example, but also anywhere else in the SDIF problem specification! In fact, a number of the codes we have seen so far also have a `Z` form. For instance, the `GROUPS` section of the `SUPERSIM` problem could be written as

```
RE one                 1.0
RE two                 2.0
ZN Object    x                                 one
ZE Cautious  x                                 one
ZE Cautious  y                                 two
ZG Daring    x                                 two
ZG Daring    y                                 one
```

[8]But not always: see Section 2.7.1.

although this is somewhat verbose in this case. Note that we have continued the description of the linear variables in groups `Cautious` and `Daring` on more than one line because we now have to use one line per variable appearing in the group. One can even write

```
RE 1.0                   1.0
RE 2.0                   2.0
ZN Object    x                          1.0
ZE Cautious  x                          1.0
ZE Cautious  y                          2.0
ZG Daring    x                          2.0
ZG Daring    y                          1.0
```

where we have defined two parameters called "`1.0`" and "`2.0`" (both are valid names), to which we have assigned the numerical values 1.0 and 2.0. We then have used the parameters *by their names*, hence the `Z`-type codes! This trick sometimes helps, but be careful not to define a variable called "`1.0`" whose numerical value is 0.0 or any other value different from 1.0: this can generate appreciable confusion and be very hard to read or correct at a later stage...

2.4.3 Arithmetic with Real Named Parameters

It does not take much playing with the parameters we just introduced before one wishes to perform simple arithmetic operations involving their values. Fortunately, these parameter operations are possible: assignment, the four basic arithmetic operations and a few basic numerical functions are indeed available in the SDIF.

We already saw the most elementary case above in the line

```
RE Level                 2.0
```

This is the simple assignment of a real numerical value to a named parameter. One may wish to add a numerical value to that of an already defined parameter: this is done by

```
RA Higher     Level      1.4
```

where the constant 1.4 is added to the value of `Level` to obtain the new real parameter `Higher`, i.e.

$$\texttt{Higher} := 1.4 + \texttt{Level}. \tag{2.4.9}$$

The code is `RA` because it is a Real `A`ddition. A Real `S`ubtraction is also available in the form

```
RS Rest       Higher     7.0
```

where we have specified the operation

$$\texttt{Rest} := 7.0 - \texttt{Higher}, \tag{2.4.10}$$

and not the reverse[9]! Similarly, the Real Multiplication is specified as

```
RM Plenty    Rest       27.3
```

which just means

$$\texttt{Plenty} := 27.3 \times \texttt{Rest} \tag{2.4.11}$$

while the Real Division is written as

```
RD Pi        Plenty     308,75572599
```

to compute

$$\texttt{Left} := 308,75572599/\texttt{Plenty}. \tag{2.4.12}$$

Note that the first argument divides the second, and not the opposite. Also note that the value of the parameter Plenty in field 3 must be different from 0![10] One can finally take the square root of a number and assign the result to a parameter, as in

```
RF Pi        SQRT       9.869604
```

where the code RF stands for Real Function, and where the name of the considered function is indicated by its name in field 3. The square root's name is SQRT, but other functions are also available: they are given in Table 2.3. One can write,

```
RF Zero      SIN        3.14159265
```

or

```
RF Zero      ARCCOS     1.0
```

for instance.

The next step is, of course, to perform arithmetic operations involving parameters only, and no longer explicitly given numerical values. This is achieved in a way very similar to what we just saw. In fact, the now familiar sequence of operations

```
RE Level                2.0
RA Higher    Level      1.4
RS Rest      Higher     7.0
RM Plenty    Rest       27.3
RD Pi        Plenty     308,75572599
RF Pi        SQRT       9.869604
```

---

[9]This somewhat unexpected definition turns out to be the right one in many cases. Also, it would be extremely easy to add -7.0 with the RA operation if the reverse operation is wanted.

[10]This is *not* 0 factorial nor 0 factorial to the tenth power!

Name	Meaning		Restriction
SQRT	Square root	$f(x) = \sqrt{x}$	$x \geq 0$
ABS	Absolute value	$f(x) = \|x\|$	
EXP	Exponential	$f(x) = e^x$	
LOG	Natural logarithm	$f(x) = \log(x)$	$x > 0$
LOG10	Logarithm in base 10	$f(x) = \log_{10}(x)$	$x > 0$
SIN	Sine	$f(x) = \sin(x)$	
COS	Cosine	$f(x) = \cos(x)$	
TAN	Tangent	$f(x) = \tan(x)$	
ARCSIN	Inverse sine	$f(x) = \sin^{-1}(x)$	$\|x\| \leq 1$
ARCCOS	Inverse cosine	$f(x) = \cos^{-1}(x)$	$\|x\| \leq 1$
ARCTAN	Inverse tangent	$f(x) = \tan^{-1}(x)$	
HYPSIN	Hyperbolic sine	$f(x) = \sinh(x)$	
HYPCOS	Hyperbolic cosine	$f(x) = \cosh(x)$	
HYPTAN	Hyperbolic tangent	$f(x) = \tanh(x)$	

Table 2.3: Available real functions in SDIF

can equally be rewritten as

```
RE Level                 2.0
RE A                     1.4
RE B                     7.0
RE C                     27.3
RE D                     308,75572599
RE E                     9.869604

R+ Higher   Level                     A
R- Rest     B                         Higher
R* Plenty   Rest                      C
R/ Pi       D                         Plenty
R( Pi       SQRT                      E
```

where we have used the parameters A, B, C, D and E to hold the numerical values explicitly given before. The codes R+, R-, R*, R/[11] and R(have replaced RA, RS, RM, RD and RF respectively. Also note that the order of the operands for the subtraction (R-) and division (R/) operations has been reversed[12]...

2.4.4 Integer Parameters

In Section 2.3.1, we have indicated that parameters come in two kinds: reals and integers. The arithmetic operations we have defined up to now only handle

[11] As for RD, the value of the parameter whose name appears in field 5 must be different from zero.

[12] Yes, we think this is confusing too.

real parameters; it is thus natural to introduce their integer counterparts. This is done most easily: the codes used for handling integer parameters differ from those used for real ones only in their first letter, where the R of Real is simply replaced by the I of Integer! The following lines are thus perfectly valid:

```
IE Two                  2
IE One                  1
IA Three      Two       1
IS Four       Three     7
ID Five       Four      20
IM MinusOne   One       -1
I+ Nine       Four                      Five
I- Seven      Nine                      Two
I* MinusTwo   Two                       MinusOne
I/ OneAgain   Three                     Two
```

There is no integer equivalent of RF and R(, of course, because the functions of Table 2.3 take a real argument and return a real value. Observe also that ID and I/ stand for integer division, i.e. the real result of the division is truncated to an integer.

Finally, it is also possible to truncate a real parameter to obtain an integer one, and also to promote an integer parameter to real, using the codes IR and RI respectively. This is illustrated in the following lines:

```
RE Pi                   3.1415926536
IR Nearpi     Pi
RI Pi-approx  Nearpi
```

which specifies that the integer parameter **Nearpi** has the value 3 and that the real parameter **Pi-approx** has the value 3.0.

2.4.5 Summary

In this section, we have introduced the real and integer named parameters, as well as the operations that the SDIF is capable of performing with them. The type of a parameter (real or integer) is specified by its first definition in the problem specification[13].

Real parameters are defined and operated upon in lines with codes starting with the letter R, while integer parameters are defined and modified in lines with codes starting with the letter I.

The second letter of the code indicates what operation is to be performed with the content of fields 3 to 5 in order to define the value of the parameter named in field 2. Table 2.5 summarizes these operations. In this table the symbols f_i stands for the content of the i-th field. The table also indicates whether or not the order in which the operands appear is important.

[13] It should also be its first occurrence in the file!

Code first letter	Type of the defined parameter
R	Real parameters and numbers
I	Integer parameters and numbers

Table 2.4: The meaning of the first letter in parameter related codes

Code second letter	Operations	Order
E	$f_2 = f_4$	
A	$f_2 = f_4 + f_3$	
S	$f_2 = f_4 - f_3$	•
M	$f_2 = f_4 \times f_3$	
D	$f_2 = f_4/f_3$	•
=	$f_2 = f_3$	
+	$f_2 = f_3 + f_5$	
-	$f_2 = f_3 - f_5$	•
*	$f_2 = f_3 \times f_5$	
/	$f_2 = f_3/f_5$	•

Table 2.5: The meaning of the second letter in codes for parameter arithmetic

We have also seen that simple real functions can be applied to real parameters to define new real parameters, using the syntax given in Table 2.6 where f_i again stands for the content of the i-th field and where $F(\cdot)$ is a real function whose name is chosen in Table 2.3 and appears in field 3.

Code	Operation
RF	$f_2 = F(f_4)$
R(	$f_2 = F(f_5)$

Table 2.6: Syntax for applying simple functions to real parameters

Finally, conversion to Real from Integer and to Integer from Real are supplied by the codes RI and IR respectively.

2.5 Groups, Linear Least-squares and Unary Operators

After exploring the civilized domain of simple linear programs with LANCELOT and his faithful S(D)IF, we now venture in a wilder territory: that of nonlinear functions. In this section, we will however be cautious and only consider how to specify the tamest of all nonlinear problems, the linear least-squares problem.

2.5.1 Group Functions and Unary Operators

Let us first consider the very simple nonlinear problem of minimizing the square of the difference between two nonnegative variables x and y subject to the constraint that x and y sum to 1.0. Formally, the problem can be expressed as

$$\min_{x,y\in\Re}(x-y)^2 \tag{2.5.13}$$

subject to the constraints that

$$x+y=1.0 \tag{2.5.14}$$

and also that

$$x\geq 0 \text{ and } y\geq 0. \tag{2.5.15}$$

Assume, for an instant, that the square is not present in the objective definition. Using our knowledge of SDIF for linear programs, we could write the corresponding specification as follows.

```
 NAME           WRONG

*   Not quite a simple constrained linear least-squares problem

VARIABLES
    x
    y
GROUPS
 XN Object    x         1.0          y         -1.0
 XE Constr    x         1.0          y          1.0
CONSTANTS
 X  WRONG     Constr    1.0
ENDATA
```

This would be quite simple indeed, but the square in (2.5.14) can unfortunately not be forgotten. The idea is now to specify a nonlinear transformation of the objective group (in our case, raising it to the power two). Such transformations are called *group transformations* or *group functions* in SIF. Each such transformation is said to be of a specific *group type*, and all group types are introduced in a special purpose section cunningly called `GROUP TYPE`. For our problem, we need a `GROUP TYPE` section of the form

```
GROUP TYPE
 GV SQUARE    ALPHA
```

where we first used the keyword `GROUP TYPE` to indicate the beginning of the new section, and then the code `GV` to specify that the new group type, named

SQUARE, involves a *Group Variable*, called ALPHA, that is the variable needed to specify the group transformation (here, the transformation is α^2).

Of course, we still have to say that the objective group is of type SQUARE (that is involves the raising to the square) while the linear constraint Constr is of the usual type, called the trivial type. In fact, the trivial type uses the identity (or trivial) transformation of its implicit group variable, hence its name. Assigning types to groups is one of the purposes of yet another new section, called GROUP USES, which can be written as follows.

```
GROUP USES
 XT Object    SQUARE
```

Again we used the appropriate keyword to start the section, and specified that the objective group Object is of Type SQUARE, using the code XT. Observe that we did not assign any type to the group Constr, *in which case the type* trivial *is assumed.*

One may then think that we have specified everything, but this is not quite the case: we still have to define what we mean when we say that the group type SQUARE of group variable ALPHA is equal (in our simple example) to the square of that variable. We also have to specify the derivatives of this associated function in a form that can be converted to a useful (Fortran) subroutine at a later stage. Specifying the actual functional form and derivatives of the nonlinear group functions is done in another file of our problem file: the SGIF (Standard *Group* Input Format). In fact, the SDIF and SGIF parts of our problem specification are conceptually quite different and it is even possible to keep them in two different files! We will however avoid this complication here, and simply append the SGIF for our problem to its SDIF definition. The SGIF starts with the keyword GROUPS followed by the problem name (in field 3), i.e.

```
GROUPS        TAME
```

We next specify the individual nontrivial groups of our problem (just one, in our example) in a section of the SGIF called INDIVIDUALS as follows:

```
INDIVIDUALS
 T  SQUARE
 F                      ALPHA**2
 G                      2.0 * ALPHA
 H                      2.0
ENDATA
```

The code T indicates the type of group we are about to describe: this is the SQUARE type in our case. The functional expression for the group (as a function of its group variable(s)) is then specified in a line starting with the code F (as in Function). Its gradient with respect to its (unique) variable is then specified in the line starting with the code G (as in Gradient). Its second derivative is next specified in the line starting with the code `H` (as in `Hessian`). Finally, and as for the SDIF, the SGIF is ended with the `ENDATA` keyword.

Gathering all this new material, we obtain the following specification file.

```
NAME          TAME

*   A simple constrained linear least-squares problem

VARIABLES
    x
    y
GROUPS
 XN Object    x          1.0            y         -1.0
 XE Constr    x          1.0            y         1.0
CONSTANTS
 X  TAME      Constr     1.0
GROUP TYPE
 GV SQUARE    ALPHA
GROUP USES
 XT Object    SQUARE
ENDATA

GROUPS        TAME
INDIVIDUALS
 T  SQUARE
 F                       ALPHA**2
 G                       2.0 * ALPHA
 H                       2.0
ENDATA
```

After saving these lines in the file `TAME.SIF`, we can once more verify that LANCELOT can solve the problem for us by simply issuing the command

```
sdlan TAME
```

and verifying that our faithful knight has once more done his job.

The observant reader surely noticed that we have broken one more of our conventions of the first section: in the line

```
 G                       2.0 * ALPHA
```

for instance, we see that field 4 does not contain a number, but instead seems to contain an arithmetic expression involving the group variable `ALPHA`! In fact, the line format for the functions' specification within the `INDIVIDUALS` section of the SGIF is different from that in the SDIF, and is now of the form

```
F1 <---F2---><---F3---><------------------F7------------------>
cd <--name--><--name--><--------- Fortran expression ---------->
```

where the new field 7 is 41 characters long, starts in column 25 (as field 4) and ends in column 65. Field 7 may contain any Fortran expression for defining function, gradient or Hessian values. Observe that this restriction also imposes that the (arbitrary) name given to the group variables (ALPHA, for instance) must be valid Fortran names (they cannot start by a digit, for instance).

These restrictions do not apply however to the SGIF lines specifying the type of element being defined: the element type names can be as long as 10 characters and are not restricted by the more stringent rules applying to valid Fortran names: we could choose "Alpha**2" instead of "SQUARE" without a problem.

At this point, it is interesting to note that the introduction of group functions does, in fact, provides the necessary tool for specifying *unary* problems. Unary problems have their objective and constraint functions sharing a specific structure: each of these functions is a one-to-one twice differentiable real transformation of a linear expression involving the problem variables. In our example, we used the trivial and square transformations, but the mechanism is ready for handling more complex ones, as in the problem

$$\min_{x,y\in\Re} e^{(x-2y)} \tag{2.5.16}$$

subject to the constraint

$$\sin(y-x-1)=0 \tag{2.5.17}$$

and the bounds

$$-2 \le x \le 2 \quad \text{and} \quad -1.5 \le y \le 1.5. \tag{2.5.18}$$

Using the group technique, this can now be simply specified as

```
NAME          ALSOTAME

*   Another simple constrained  problem

VARIABLES
    x
    y
GROUPS
 XN Object    x          1.0          y          -2.0
 XE Constr    x          -1.0         y          1.0
CONSTANTS
 X  ALSOTAME  Constr     1.0
BOUNDS
 XL ALSOTAME  x          -2.0
 XU ALSOTAME  x          2.0
 XL ALSOTAME  y          -1.5
 XU ALSOTAME  y          1.5
GROUP TYPE
 GV EXPN      ALPHA
 GV SINE      ALPHA
GROUP USES
 XT Object    EXPN
```

```
 XT Constr    SINE
ENDATA

GROUPS        ALSOTAME
TEMPORARIES
 R  EXPA
 R  SINA
 M  EXP
 M  SIN
 M  COS
INDIVIDUALS
 T  EXPN
 A  EXPA               EXP( ALPHA )
 F                     EXPA
 G                     EXPA
 H                     EXPA

 T  SINE
 A  SINA               SIN( ALPHA )
 F                     SINA
 G                     COS( ALPHA )
 H                     - SINA
ENDATA
```

Notice the additional TEMPORARIES. section that we have introduced in the SGIF. This section is used to declare the type of temporary variables and/or functions that are used within the Fortran expressions of the INDIVIDUALS section. In our example, we use the variables EXPA and SINA to avoid additional calls to the exponential or sine intrinsic functions. These variables are declared to be Real by using the code R. Similarly, the exponential and sine are supplied by the Machine, hence the code M. (If they were needed, temporary Integer variableswould be defined by using the code I.)

The last new feature introduced in our example is the code A in the INDIVIDUALS section, which is used to Assign a value to a temporary variable (EXPA and SINA in our case). All assignment lines (i.e. lines starting with code A) appear before the function line (with code F). It is also required that all statements associated with a single group type are kept together under the relevant type declaration, as in the example above. The INDIVIDUALS section

```
INDIVIDUALS
*   An example of incorrect mixing of group types
 T  EXPN
 A  SINA               SIN( ALPHA )
 A  EXPA               EXP( ALPHA )
 F                     EXPA
 G                     EXPA
 H                     EXPA

 T  SINE
```

```
 F                      SINA
 G                      COS( ALPHA )
 H                      - SINA
```

is therefore *incorrect*, because the variable SINA is assigned a value in the part of the section corresponding to the group type EXPN instead of the part corresponding to the SINE group type.

2.5.2 Using more than a Single Group in the Objective Function

Let us consider now a somewhat less (?) trivial linear least-squares problem. The problem is to fit a straight line of the form $ax+b$ to a set of five measurements at x_1, x_2, x_3, x_4 and x_5 with respectives values y_1, y_2, y_3, y_4 and y_5. We also want the slope of the straight line to be nonnegative. Furthermore, we require that the affine function $ax+b$ corresponding to this line has a value lower than a given constant c at $x = 1$. Mathematically, this can expressed as

$$\min_{a,b\in\Re} \tfrac{1}{2}\sum_{i=1}^{5}(ax_i+b-y_i)^2 \tag{2.5.19}$$

subject to the bound

$$a \geq 0 \tag{2.5.20}$$

and the additional constraint

$$a+b \leq c. \tag{2.5.21}$$

The technique is now to *use several groups to describe the objective function.* We will choose the groups to correspond to each one of the five terms in our objective (2.5.19), which gives us the following specification

```
NAME          LSQFIT

*   An elementary constrained linear least-squares fit

*   Data points
 RE X1                    0.1
 RE X2                    0.3
 RE X3                    0.5
 RE X4                    0.7
 RE X5                    0.9
*   Observed values
 RE Y1                    0.25
 RE Y2                    0.3
 RE Y3                    0.625
 RE Y4                    0.701
 RE Y5                    1.0
*   Upper bound at 1.0
 RE C                     0.85
```

```
VARIABLES
    a
    b
GROUPS
*   Objective function groups
 ZN Obj1      a                        X1
 XN Obj1      b         1.0
 XN Obj1      'SCALE'   2.0
 ZN Obj2      a                        X2
 XN Obj2      b         1.0
 XN Obj2      'SCALE'   2.0
 ZN Obj3      a                        X3
 XN Obj3      b         1.0
 XN Obj3      'SCALE'   2.0
 ZN Obj4      a                        X4
 XN Obj4      b         1.0
 XN Obj4      'SCALE'   2.0
 ZN Obj5      a                        X5
 XN Obj5      b         1.0
 XN Obj5      'SCALE'   2.0
*   Constraint group
 XL Cons      a         1.0            b         1.0
CONSTANTS
 Z  LSQFIT    Obj1                     Y1
 Z  LSQFIT    Obj2                     Y2
 Z  LSQFIT    Obj3                     Y3
 Z  LSQFIT    Obj4                     Y4
 Z  LSQFIT    Obj5                     Y5
 Z  LSQFIT    Cons                     C
BOUNDS
 XR LSQFIT    b
GROUP TYPE
 GV SQUARE    ALPHA
GROUP USES
 XT Obj1      SQUARE
 XT Obj2      SQUARE
 XT Obj3      SQUARE
 XT Obj4      SQUARE
 XT Obj5      SQUARE
ENDATA

GROUPS        LSQFIT
INDIVIDUALS
 T  SQUARE
 F                      ALPHA * ALPHA
 G                      ALPHA + ALPHA
 H                      2.0
ENDATA
```

We added one further new construct in this specification: the scaling factors for the groups. You have noticed the line

```
XN Obj1        'SCALE'   2.0
```

and other similar ones in the `GROUPS` section. This line indicates that the group `Obj1` is to be *divided* ("`SCALEd`") by the factor 2.0 (that is multiplied by 0.5!) before it is incorporated in the objective function. The string `'SCALE'` is a reserved name[14] to indicate this use of the value that is specified further in the line. Such group scaling factors can also be specified via a named parameter, as would be the case if we had written

```
RE two                   2.0
ZN Obj1        'SCALE'                  two
```

using an `Z`-type code instead of an `X`-type one. Such a scaling may of course be specified for any group, including constraint groups, if any. We could of course incorporate the factor $\frac{1}{2}$ (common to all groups) in the definition of the group function itself, but we would then miss the opportunity to illustrate an interesting construct...

Also note that we just defined more than one objective group (as announced), using the codes `ZN` and `XN` for each of the five groups `Obj1...Obj5`. We have interspersed `X` and `Z` codes, just to refresh your mind about their differences (`X` for numerical values, `Z` for values via parameters). Also observe that, although there are five nonlinear groups, we only had to define the `SQUARE` group type once, a very handy feature when the number of observed values increases. We did not declare any type for the constraint group in the `GROUP USES` section because this group is of trivial type, and therefore needs no explicit typing. Finally, we have defined all the problem dependent data (data points and observed values) at the beginning of the file, which makes later modifications easy.

2.5.3 Robust Regression and Conditional Statements in the SGIF

The `SQUARE` group type is probably the most common in practical applications, just because least-square calculations occur so frequently in many fields of science. It is interesting to note that robust regression techniques (see [39] for more details) can also be accommodated quite easily in this framework. For instance, consider the `LSQFIT` problem with the Huber loss function

$$h_k(\alpha) = \begin{cases} \frac{1}{2}\alpha^2 & \text{for } |\alpha| \le k \\ k|\alpha| - \frac{1}{2}k^2 & \text{for } |\alpha| > k \end{cases} \qquad (2.5.22)$$

[14] Defined in MPS.

instead of the ℓ_2 measure of (2.5.19). The problem is now expressed as

$$\min_{a,b\in\Re} \sum_{i=1}^{5} h_k(ax_i + b - y_i) \tag{2.5.23}$$

(where h is defined in (2.5.22)) subject to the bound (2.5.20).

In fact, we just have to introduce (2.5.22) as a group type. This raises the apparent difficulty that the Huber function involves a conditional statement, which we have not yet mentioned in the SGIF syntax. You will be relieved to learn that such a statement nevertheless exists. The idea is to define a few additional temporary parameters in the `TEMPORARIES` section, of which one (`OUT`, say) is a *logical parameter*, and then assign "true" or "false" to it (using a Fortran statement) in the `INDIVIDUALS` section, depending whether or not the condition

$$|\alpha| > k \tag{2.5.24}$$

holds. We then write

```
A  HUBERK                 1.5
A  ABSA                   ABS( ALPHA )
A  OUT                    ABSA .GT. HUBERK
```

We may then compute the value of the function and derivatives depending on the value of the logical parameter `OUT`. This is what is specified on the following lines

```
I  OUT        FF          HUBERK * ABSA - 0.5 * HUBERK * HUBERK
E  OUT        FF          0.5 * ABSA * ABSA
```

In other words, If `OUT` is true, then `FF` is set to the value of the Huber function when (2.5.24) holds, Else (i.e. if `OUT` is false), `FF` is set to the alternate value. The principle is identical for the first derivative, except that we have the additional problem that the sign of α may now be important when (2.5.24) holds. We therefore write

```
A  NEGOUT                 OUT .AND. ( ALPHA .LT. 0.0 )
A  POSOUT                 OUT .AND. ( ALPHA .GE. 0.0 )
I  POSOUT     GG          HUBERK
I  NEGOUT     GG          - HUBERK
E  OUT        GG          ALPHA
```

which puts the first derivative value in the temporary variable `GG`. Finally, the second derivative is obtained in `HH` much in the same way

```
I  OUT        HH          0.0
E  OUT        HH          1.0
```

Of course, we must declare all these temporary variables in the `TEMPORARIES` section:

```
TEMPORARIES
 L  OUT
 L  NEGOUT
 L  POSOUT
 R  ABSA
 R  HUBERK
 R  FF
 R  GG
 R  HH
 M  ABS
```

where the code L is used to declare that OUT, NEGOUT and POSOUT are logical parameters. Putting the complete problem together, we finally obtain

```
NAME          HUBFIT

*   An elementary fit using the Huber loss function
...
(problem data, VARIABLES, GROUPS, CONSTANTS and BOUNDS sections
 as above)
...
GROUP TYPE
 GV HUBER     ALPHA
GROUP USES
 XT Obj1      HUBER
 XT Obj2      HUBER
 XT Obj3      HUBER
 XT Obj4      HUBER
 XT Obj5      HUBER
ENDATA

GROUPS        HUBFIT
TEMPORARIES
 L  OUT
 L  NEGOUT
 L  POSOUT
 R  ABSA
 R  HUBERK
 R  FF
 R  GG
 R  HH
 M  ABS
INDIVIDUALS
 T  HUBER
 A  HUBERK                 1.5
 A  ABSA                   ABS( ALPHA )
 A  OUT                    ABSA .GT. HUBERK
*  Case 1: alpha is outside [-k, k]
 I  OUT       FF           HUBERK * ABSA - 0.5 * HUBERK * HUBERK
 A  NEGOUT                 OUT .AND. ( ALPHA .LT. 0.0 )
```

```
 A  POSOUT               OUT .AND. ( ALPHA .GE. 0.0 )
 I  POSOUT    GG         HUBERK
 I  NEGOUT    GG         - HUBERK
 I  OUT       HH         0.0
*  Case 2: alpha is inside [-k, k]
 E  OUT       FF         0.5 * ABSA * ABSA
 E  OUT       GG         ALPHA
 E  OUT       HH         1.0
*  Assign the final values
 F                       FF
 G                       GG
 H                       HH
ENDATA
```

We have simply reordered the statements to group the cases and added a few comments.

Of course, the group function has discontinuous second derivatives at $-k$ and k, but fortunately LANCELOT can still cope with this problem as trying out quickly reveals...

Two final remarks on the SGIF part of the specification file:

- The SGIF in fact contains one more (optional) section, called GLOBALS. This section is intended to contain the assignment of *global variables* whose value does not depend on any particular type of group. We could use this section in our example, because there is only one group type, and thus the parameter HUBERK does not depend on the group type and could be assigned a value in the GLOBALS section. If we make this (unnecessarily verbose) choice, the new section is simply

  ```
  GLOBALS
   A  HUBERK               1.5
  ```

  We can then remove the statement assigning 1.5 to HUBERK from the INDIVIDUALS section.

  Note that the GLOBALS section uses the same format as the INDIVIDUALS one and that any variable appearing in field 2 must have been declared in the TEMPORARIES section. Interestingly, the conditional assignment lines with codes I and E can also be used in the GLOBALS section.

  The practical use of this technique is of doubtful interest in our very simple case, but can bring advantages when there are several group types within the same problem that involve complicated common subexpressions.

- Although it did not happen in either of our simple examples, it may well be the case that the Fortran expression for the group value or for one of its derivatives is too long to fit in field 7 on a single line. The solution is obviously to write more than one line, all "continuation lines" starting with the codes `I+`, `E+`, `A+`, `F+`, `G+`, or `H+`, depending on the type of expression they provide continuation for. For instance, a peculiar but correct way to write

```
I  OUT       FF          HUBERK * ABSA - 0.5 * HUBERK * HUBERK
```

could be

```
I  OUT       FF          HUBERK * ABSA
I+                                     - 0.5 * HUBERK * HUBERK
```

There may be up to 19 such continuation lines[15].

2.5.4 Summary

In this section, we have introduced the groups and their associated group functions. We saw that the SIF is able to represent the problem of minimizing a nonlinear objective function containing one or more groups subject to bounds on the variables and to nonlinear constraints, each of which corresponds to a separate group. Mathematically speaking the problem is of the form

$$\min_{x\in\Re^n} \sum_{i=1}^{p} \frac{1}{s_i} g_i(a_i^T x - b_i) \tag{2.5.25}$$

subject to the constraints

$$\frac{1}{s_i} g_i(a_i^T x - b_i) \left\{ \begin{array}{c} \leq \\ = \\ \geq \end{array} \right\} 0 \quad (i = p+1, \ldots, m) \tag{2.5.26}$$

and to the bounds

$$l_j \leq x_j \leq u_j \quad (j = 1, \ldots, n). \tag{2.5.27}$$

The specification of the group functions $g_i(\cdot)$ is made in two steps: group function types are first declared in the SDIF file and each group assigned its type; the group functions associated to each type are subsequently described (i.e. expressions for their value and derivatives are specified) in the SGIF file for the problem file. This latter part provides special sections

- to declare temporary variables and user or machine supplied functions (the `TEMPORARIES` section),
- to assign values to global temporary variables (the `GLOBALS` section),
- to assign the temporary variables, group function value, first and second derivative using Fortran assignment statements (the `INDIVIDUALS` section).

[15] This limit is due to Fortran.

Furthermore, the `GLOBALS` and `INDIVIDUALS` sections also provide a mechanism allowing for the conditional assignment of variables, which in turn allows the specification of piecewise group functions. These conditional assignments use logical parameters and "if-then-else" codes.

2.6 Introducing more Complex Nonlinearities: the Element Functions

At last, we are ready to introduce the not-so-gentle (perhaps, even vicious) problems, that is problems involving more serious nonlinearities. In general, the use of the group functions alone will not be sufficient to cover the more complex cases we will examine in this section. Instead, or often besides, we will need additional ways of specifying nonlinear components of our objective and/or constraint functions.

2.6.1 A Classic: The Rosenbrock "Banana" Unconstrained Problem

As before, we will introduce the necessary concepts with the help of an example. In this case, we will choose a classic[16]: the famous Rosenbrock unconstrained minimization problem in two variables. Mathematically, the problem is the following:

$$\min_{x,y\in\Re} 100(y-x^2)^2+(x-1)^2. \tag{2.6.28}$$

It is sometimes called the "banana valley" problem because drawing the level curves of (2.6.28) reveals a narrow curved valley (induced by the first term in the expression) whose bottom has itself a comparatively gentle curvature (the effect of the second term), the resulting level curves taking the form of nested "bananas".

Like a large percentage of nonlinear problems, this example has the least-squares structure, but the terms inside the squares are no longer all linear: the first one indeed involves the square of the variable x. Therefore, we will want to use the least-squares group function, but we will also need additional constructs to take the more complicated structure into account. Proceeding cautiously, we may first want to specify as much of the problem as we can with the techniques that we have learned already. This will give a file of the form

```
NAME          ROSENBR

*   A first attempt to specify the famous Rosenbrock problem
```

[16] To nonlinear programmers, of course.

```
VARIABLES
    x
    y
GROUPS
 XN Obj1      y            1.0
 XN Obj1      'SCALE'      0.01
 XN Obj2      x            1.0
CONSTANTS
 X  ROSENBR   Obj2         1.0
BOUNDS
 XR ROSENBR   x
 XR ROSENBR   y
GROUP TYPE
 T  SQUARE    ALPHA
GROUP USES
 T  Obj1      SQUARE
 T  Obj2      SQUARE
ENDATA

GROUPS        ROSENBR
INDIVIDUALS
 T  SQUARE
 F                         ALPHA * ALPHA
 G                         ALPHA + ALPHA
 H                         2.0
ENDATA
```

We now would like to indicate that group Obj1 is not as simple and that it contains a nonlinear *element function*, as we call such troublesome terms. You will not be surprised to hear that the task is achieved by specifying *element types* in a section called ELEMENT TYPE, and then assigning these types to the actual element functions that appear in the different groups[17]. More precisely, we write the following

```
ELEMENT TYPE
 EV SQ        V
```

for the ELEMENT TYPE section, thereby specifying that we consider a (single) element type called SQ[18] depending on a single *elemental variable* V. At this stage, this elemental variable V is floating around freely, and we now must define the particular problem variable (x or y) that will be assigned to V. This is the task of the ELEMENT USES section, which, in our case, reads as follows

```
ELEMENT USES
 T  XSQ       SQ
```

[17] The parallel with the groups is hopefully obvious.

[18] We used SQ instead of SQUARE just to avoid the confusion between the group and element types, but using the same name is perfectly legal.

```
 ZV XSQ       V                          x
```

The line starting with the code T assigns the element type SQ to the element named XSQ, while the line starting with the code ZV assigns the problem variable x to the the elemental variable V *for the particular element* XSQ.

The next task is then to say that group Obj1 involves the element XSQ, which is done in the GROUP USES section. This latter section thus becomes

```
GROUP USES
 T  Obj1      SQUARE
 XE Obj1      XSQ       -1.0
 T  Obj2      SQUARE
ENDATA
```

where the code XE indicates that we assign Element functions to a group (in our case XSQ to the group Obj1), after multiplication by the *weighting factor* -1.0.

Finally, we still have to specify the actual nonlinear behaviour that we have hidden behind the definition of the element types. As was the case for the group functions, this is done in another file of the problem specification file, called the Standard *Element* Input Format (SEIF), whose syntax is very similar to that of the SGIF. For the Rosenbrock function, this SEIF is as follows

```
ELEMENTS      ROSENBR
INDIVIDUALS
 T  SQ
 F                        V**2
 G  V                     2.0 * V
 H  V         V           2.0
ENDATA
```

The only difference (beyond the different names associated with the type and involved variable) between the SEIF and the the least-squares SGIF is the fact that an element type might involve several variables (in more complicated cases, see below), and therefore that it is necessary to indicate with respect to which variable differentiation is performed. Here, it is of course with respect to V for the gradient, and with respect to V and V for the Hessian. At variance with the group function specification, the gradient and Hessian specification are optional in the SEIF. However, it is our experience that it is very often helpful to provide expressions for gradient and Hessian entries: it usually pays off in faster and more robust convergence of LANCELOT[19]. *We therefore recommend that expressions for gradient and Hessian components be supplied whenever possible.*

Assembling the pieces of the puzzle, we thus obtain the complete SIF specification for the Rosenbrock problem which is given below.

---

[19] And also of all algorithms that can exploit user-supplied exact derivative information.

```
NAME          ROSENBR

*   The famous Rosenbrock problem

VARIABLES
    x
    y
GROUPS
 XN Obj1      y          1.0
 XN Obj1      'SCALE'    0.01
 XN Obj2      x          1.0
CONSTANTS
 X  ROSENBR   Obj2       1.0
BOUNDS
 XR ROSENBR   x
 XR ROSENBR   y
START POINT
 XV ROSENBR   x          -1.2
 XV ROSENBR   y          1.0
ELEMENT TYPE
 EV SQ        V
ELEMENT USES
 T  XSQ       SQ
 ZV XSQ       V                         x
GROUP TYPE
 T  SQUARE    ALPHA
GROUP USES
 T  Obj1      SQUARE
 XE Obj1      XSQ        -1.0
 T  Obj2      SQUARE
ENDATA

ELEMENTS      ROSENBR
INDIVIDUALS
 T  SQ
 F                       V**2
 G  V                    2.0 * V
 H  V         V          2.0
ENDATA

GROUPS        ROSENBR
INDIVIDUALS
 T  SQUARE
 F                       ALPHA**2
 G                       2.0 * ALPHA
 H                       2.0
ENDATA
```

(We have also specified the classical starting point for this problem). Note that the ELEMENT TYPE and ELEMENT USES section immediately precede the

GROUP TYPE and GROUP USES sections, and also that, by convention, the SEIF is written between the SDIF and the SGIF.

Clearly, the complexity of this specification may appear formidable for such a small example: comparing equation (2.6.28) and the above file for conciseness is not to the advantage of the latter. However, this simple case has been enough for us to become acquainted with a very general way of analyzing and describing a problem that will pay off many times when dealing with larger and more complex optimization problems.

2.6.2 The Internal Dimension and Internal Variables

There is one more structure to analyze before we move to the techniques allowing the specification of really large problems, namely the structure of the *internal variables* and *internal dimension.* As always, this structure is best introduced by an example.

Let us consider the (again very simple) problem of finding the pair of symmetric 2 by 2 matrices whose Frobenius distance is minimum and such that the first is positive semidefinite while the second is negative semidefinite. Once more we express the problem in its mathematical form, which is, in this case,

$$\min_{x_{i,j},y_{i,j}\in\Re} (x_{11}-y_{11})^2+2(x_{12}-y_{12})^2+(x_{22}-y_{22})^2 \tag{2.6.29}$$

subject to the constraints

$$x_{11}x_{22}-x_{12}^2 \geq 0, \tag{2.6.30}$$

$$y_{11}y_{22}-y_{12}^2 \leq 0 \tag{2.6.31}$$

and to the bounds

$$x_{11}\geq 0, \quad x_{22}\geq 0 \tag{2.6.32}$$

and

$$y_{11}\leq 0, \quad y_{22}\leq 0. \tag{2.6.33}$$

Note that the bounds on the diagonal entries of X (resp. Y) together with the determinant constraint (2.6.30) (resp. (2.6.31)) are sufficient to ensure the positive (resp. negative) semidefiniteness of this matrix[20]. Although the mathematical solution of this problem is immediate, its formulation in the above form is not so trivial (it indeed involves a nonlinear objective function and two nonlinear constraints), and will illustrate quite well how the different SIF constructs can contribute in defining a none too simple nonlinear minimization problem.

We first perform some now familiar tasks: we give a name to the problem, declare the problem variables, specify the nature of the three groups, state the bounds and define a starting point:

[20]The old Sylvester theorem.

```
NAME          MATRIX2

*   A small nonlinear matrix problem

VARIABLES
    X11
    X12
    X22
    Y11
    Y12
    Y22
GROUPS
 XN Frobdist
 XG Xposdef
 XL Ynegdef
BOUNDS
 XR MATRIX2   X12
 XM MATRIX2   Y11
 XR MATRIX2   Y12
 XM MATRIX2   Y22
START POINT
 XV MATRIX2   X11       1.0
 XV MATRIX2   X12       1.0
 XV MATRIX2   X22       1.0
 XV MATRIX2   Y11       1.0
 XV MATRIX2   Y12       1.0
 XV MATRIX2   Y22       1.0
```

We then examine the objective function (2.6.29), which constitutes its own group of trivial type. This implies that the group value is composed of three nonlinear element functions, each one of the form $(v - w)^2$. Examining this last formula, we quickly see that it consists of a nonlinear function (the square) of a linear combination of the variables appearing in the element[21]. In fact, we can write the simple transformation

$$(v - w)^2 = u^2 \tag{2.6.34}$$

where we define the *internal variable* u by

$$u = v - w. \tag{2.6.35}$$

This of course means that the derivatives of the element function $(v - w)^2$ have some additional structure that we want to exploit. More precisely, it is sufficient to specify the derivatives of u^2 with respect to the internal variable

[21] The similarity with the groups containing linear combinations of the variables must be striking to the attentive reader. The only difference is that groups also contain nonlinear elements besides their linear part.

u and (2.6.35) instead of specifying the derivatives of $(v - w)^2$ with respect to v and w. The number of internal variables of an element type is called its *internal dimension*, as opposed to the number of elemental variables, which is its *elemental dimension*. In the ELEMENT TYPE section, we therefore declare that our element function type has two Elemental Variables, named V and W, and also that it has one Internal Variable named U. This section therefore consists of

```
ELEMENT TYPE
 EV ISQ       V                        W
 IV ISQ       U
 EV SQ        X
 EV 2PR       X                        Y
```

where we have added the declaration of an element type called SQ for the SQuare appearing in the second terms of (2.6.30) and (2.6.31), and an element type called 2PR for the products appearing as first terms of the same constraint equations.

Each of the elements (three for the objective function and two for each constraint) must then be named explicitly and their elemental variables assigned the correct problem variables. This is done in the ELEMENT USES section.

```
ELEMENT USES
*   Elements of the objective
 XT XY11SQ    ISQ
 ZV XY11SQ    V                        X11
 ZV XY11SQ    W                        Y11
 XT XY12SQ    ISQ
 ZV XY12SQ    V                        X12
 ZV XY12SQ    W                        Y12
 XT XY22SQ    ISQ
 ZV XY22SQ    V                        X22
 ZV XY22SQ    W                        Y22
*   Constraint that det(X) >= 0
 XT X1122     2PR
 ZV X1122     X                        X11
 ZV X1122     Y                        X22
 XT X12SQ     SQ
 ZV X12SQ     X                        X12
*   Constraint that det(Y) <= 0
 XT Y1122     2PR
 ZV Y1122     X                        Y11
 ZV Y1122     Y                        Y22
 XT Y12SQ     SQ
 ZV Y12SQ     X                        Y12
```

Notice that we only needed three element types for seven different element functions.

We then assign all these elements to their groups:

```
GROUP USES
*   Objective function
 XE Frobdist  XY11SQ                    XY22SQ
 XE Frobdist  XY12SQ    2.0
*   det(X) >= 0
 XE Xposdef   X1122                     X12SQ     -1.0
*   det(Y) <= 0
 XE Ynegdef   Y1122                     Y12SQ     -1.0
```

Observe that we omitted the weighting factors whose value is equal to 1.0. This is allowed because *all unspecified element weighting factors are assumed to be 1.0.*

Finally, we specify the (very simple) nonlinear nature of the elements and groups, and obtain

```
NAME          MATRIX2
VARIABLES
    X11
    X12
    X22
    Y11
    Y12
    Y22
GROUPS
 XN Frobdist
 XG Xposdef
 XL Ynegdef
BOUNDS
 XR MATRIX2   X12
 XM MATRIX2   Y11
 XR MATRIX2   Y12
 XM MATRIX2   Y22
START POINT
 XV MATRIX2   X11       1.0
 XV MATRIX2   X12       1.0
 XV MATRIX2   X22       1.0
 XV MATRIX2   Y11       1.0
 XV MATRIX2   Y12       1.0
 XV MATRIX2   Y22       1.0
ELEMENT TYPE
 EV ISQ       V                         W
 IV ISQ       U
 EV SQ        X
 EV 2PR       X                         Y
ELEMENT USES
*   Elements from the objective
 XT XY11SQ    ISQ
 ZV XY11SQ    V                         X11
 ZV XY11SQ    W                         Y11
 XT XY12SQ    ISQ
```

```
 ZV XY12SQ    V                        X12
 ZV XY12SQ    W                        Y12
 XT XY22SQ    ISQ
 ZV XY22SQ    V                        X22
 ZV XY22SQ    W                        Y22
*   det(X) >= 0
 XT X1122     2PR
 ZV X1122     X                        X11
 ZV X1122     Y                        X22
 XT X12SQ     SQ
 ZV X12SQ     X                        X12
*   det(Y) <= 0
 XT Y1122     2PR
 ZV Y1122     X                        Y11
 ZV Y1122     Y                        Y22
 XT Y12SQ     SQ
 ZV Y12SQ     X                        Y12
GROUP USES
*   Objective function
 XE Frobdist  XY11SQ                   XY22SQ
 XE Frobdist  XY12SQ    2.0
*   det(X) >= 0
 XE Xposdef   X1122                    X12SQ     -1.0
*   det(Y) <= 0
 XE Ynegdef   Y1122                    Y12SQ     -1.0
ENDATA

ELEMENTS      MATRIX2
INDIVIDUALS
*   Square of an internal variable equal to the difference
*   between two elemental variables
 T  ISQ
 R  U         V         1.0            W         -1.0
 F                      U * U
 G  U                   U + U
 H  U         U         2.0
*   Square of an elemental variable
 T  SQ
 F                      X * X
 G  X                   X + X
 H  X         X         2.0
*   Product of 2 elemental variables
 T  2PR
 F                      X * Y
 G  X                   Y
 G  Y                   X
 H  X         Y         1.0
ENDATA
```

The only new feature appears in the INDIVIDUALS section of the SEIF, where we find the line

```
 R  U         V         1.0            W         -1.0
```

within the specification of the element type ISQ. As you would expect from the names, this lines just specifies the relation (2.6.35), that is the particular linear combination of the elemental variables V and W that gives the internal variable U. The code R is because this transformation is often called a Range transformation (see [34]).

Notice finally that we have not specified the four entries of the 2 by 2 Hessian matrix of the element type 2PR. The rules that govern the specification of derivatives in the SEIF are as follows.

1. The gradient or Hessian of the function associated with an element type is assumed to be unavailable unless at least one of its components is specified. The explicit specification of a single entry in a gradient or Hessian declares this gradient or Hessian to be available.
2. Either the i, j-th or the j, i-th entry of a Hessian matrix may be specified.
3. All entries of an available gradient or Hessian that are not explicitly specified are assumed to be identically zero.

Note that the last rule allows a description of the sparsity pattern of derivatives. However, we recommend that this technique is not used to completely specify problem structure — rather, one should use the more powerful analysis of the problem in terms of groups and elements.

2.6.3 Summary

We have introduced the nonlinear elements and their related concepts and constructs. This extends the class of problems that SIF/LANCELOT can handle to

$$\min_{x \in \Re^n} \sum_{i=1}^{p} \frac{1}{s_i} g_i \left(a_i^T x - b_i + \sum_{j \in J_i} w_j f_j(x) \right) \tag{2.6.36}$$

subject to the constraints

$$\frac{1}{s_i} g_i \left(a_i^T x - b_i + \sum_{j \in J_i} w_j f_j(x) \right) \left\{ \begin{array}{c} \leq \\ = \\ \geq \end{array} \right\} 0 \quad (i = p+1, \ldots, m) \tag{2.6.37}$$

and to the bounds

$$l_j \leq x_j \leq u_j \quad (j = 1, \ldots, n). \tag{2.6.38}$$

The $f_j(x)$ are the nonlinear element functions with their associated weighting factors w_j.

In order to specify the nonlinear element functions, six steps should be followed:

1. First specify suitable element types with their elemental and internal variables (in the `ELEMENT TYPE` section),

2. Assign a type to each element function appearing in the problem (in the `ELEMENT USES` section),

3. Assign the relevant problem variables to the elemental variables of the corresponding element type (in the `ELEMENT USES` section),

4. Assign the elements to their group(s)[22], together with their associated weighting factors (in the `GROUP USES` section),

5. State the particular linear combinations of the elemental variables that define the internal variables associated with the element types, if any (in the `INDIVIDUALS` section of the SEIF),

6. Specify expressions for calculating the value (and derivatives) of the nonlinear functions associated with the element types (in the `INDIVIDUALS` section of the SEIF).

2.7 Tools of the Trade: Loops, Indexing and the Default Values

Although specifying small problems using the constructs detailed above is acceptably concise, the sheer size of the specification soon becomes unbearable with larger ones. The specification of problem variables (one per line) is, for instance, unacceptably painful if there are many variables. The purpose of this section is to present the tools available within the SDIF that allow a much more flexible statement of problems involving many variables and/or many constraints. These tools essentially consist of facilities to index SDIF names (for variables, constraints, bounds, constants, elements, groups,...), constructs to specify loops on these indices, and codes that allow the setting of default values for many SDIF objects at once.

To be consistent with our past approach, we motivate the introduction of these features via an example: we consider here a classical problem in nonlinear variational calculations: the minimum surface problem. This problem is one of finding the surface of minimum area that interpolates a given continuous function on the boundary of the unit square. The unit square itself is discretized uniformly into p^2 smaller squares, as shown in Figure 2.1 with its associated labelling.

The variables of the problem will be chosen as the "height" of the unknown surface above the $(p+1)^2$ vertices of the p^2 smaller squares. The area of

[22]Yes, an element may be assigned to more than one group!

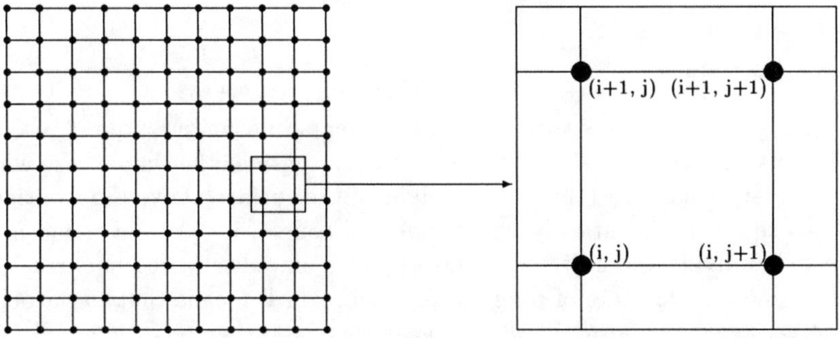

Figure 2.1: Discretisation of the unit square

the surface above the (i,j)-th discretized square is then approximated by the formula

$$s_{i,j}(x_{i,j}, x_{i+1,j}, x_{i,j+1}, x_{i+1,j+1}) = \frac{1}{p^2}\sqrt{1+\frac{p^2}{2}[(x_{i,j}-x_{i+1,j+1})^2+(x_{i,j+1}-x_{i+1,j})^2]}, \qquad (2.7.39)$$

where the squares and variables are indexed as shown in Figure 2.1. The variables in the sets

$$\{x_{1,j}\}_{j=1}^{p+1}, \quad \{x_{i,p+1}\}_{i=1}^{p+1}, \quad \{x_{p+1,j}\}_{j=1}^{p+1}, \text{ and } \{x_{i,1}\}_{i=1}^{p+1}, \qquad (2.7.40)$$

correspond to the boundary conditions and are assigned given values, while the others are left free. The complete problem is then to minimize the objective function

$$f(x) = \sum_{i,j=1}^{p} s_{i,j}(x_{i,j}, x_{i+1,j}, x_{i,j+1}, x_{i+1,j+1}) \qquad (2.7.41)$$

over all the free variables[23].

If we now examine the function $s_{i,j}$ in more detail, we easily see that we can decompose it in a nonlinear group function of the form

$$g_{i,j}(y) = \frac{1}{p^2}\sqrt{y}, \qquad (2.7.42)$$

[23] Various obstacle problems can also be obtained by specifying suitable bounds on the variables.

where y (the group variable) is given by

$$y = 1 + \frac{p^2}{2}(u_{i,j}^2 + v_{i,j}^2) \tag{2.7.43}$$

with

$$u_{i,j} \stackrel{\text{def}}{=} x_{i,j} - x_{i+1,j+1} \text{ and } v_{i,j} \stackrel{\text{def}}{=} x_{i,j+1} - x_{i+1,j}. \tag{2.7.44}$$

Each group thus contains two nonlinear elements of a unique type.

To illustrate the concepts of this section, we consider the simple case where the boundary condition (that is the height of the surface above the frontier of the unit square) is identically equal to 1. Of course, it does not make much sense to subdivide the unit square into four or nine discretized squares. The smallest reasonable value of p might be seven, which means that the problem has 64 variables appearing in 98 nonlinear elements, themselves merged in 49 groups.

2.7.1 Indexed Variables and Loops

When attempting to specify the minimum surface problem, it immediately appears that one would like to use variables with (double) indices, $x_{i,j}$ say. This is the case in fact for most of the large nonlinear problems arising from the discretization of an infinite dimensional variational calculation. Given the importance of this area, indexed variables are therefore a must, and have consequently been allowed within the SDIF syntax. Since it wasn't much more difficult, the SDIF also allows the possibility of indexing not only variable names, but most names that appear in the SDIF syntax.

In our minimum surface example, we would like to use a two dimensional array of variables of the form `X(i,j)`, which would nicely correspond to our mathematical statement. Furthermore, we also need to handle this array inside loops that would save us from writing down every single case. This is exactly what we do when we write the first lines of our problem specification as follows:

```
NAME          MINSURF
*    Discretization parameter
 IE P                     7
*    Other parameters
 IE 1                     1
 IA P+1         P         1
VARIABLES
 DO i           1                        P+1
 DO j           1                        P+1
 X  X(i,j)
 OD j
 OD i
```

Although the syntax above might seem reasonably self explanatory[24], it nevertheless requires some explanation.

[24] At least to Fortranolics!

The first point is that we just introduced two *loops*, one on the integer parameter `i` and the other on the integer parameter `j`. Such loops may appear anywhere within **VARIABLES**, **GROUPS**, **CONSTANTS**, **BOUNDS**, **START POINT**, **ELEMENT USES** and **GROUP USES** sections. Starting a loop is easy. The line starting with the `DO` code, i.e.

```
DO i         1                      P+1
```

initiates a loop on the integer parameter `i`, ranging from the value of the integer parameter whose name appears in field 3 to that of the integer parameter whose name appears in field 5. In this case, the parameter `i` will therefore take the values 1, 2, 3, 4, 5, 6, 7 and 8 successively. The loop is terminated further down by the line starting with the code `OD`, namely

```
OD i
```

which simply indicates the end of the loop using the integer parameter `i`. All statements between theses two lines constitute the "body" of the loop.

As is clear from the example, loops can be *nested*[25] up to a depth of at most three. It is even possible to specify a loop increment different from 1, by using the code `DI` as in

```
DO iloop     istart                 iend
DI i         incr
(body of the loop)
OD i
```

However, loops must not pass across section boundaries. The loop index[26] (`iloop`) will now take the values

$$\texttt{iloop} = \texttt{istart} + i \times \texttt{incr} \tag{2.7.45}$$

for all positive i for which `iloop` lies between (and including) `istart` and `iend`. The body of the loop is skipped if no such i exists. Note that a negative value of the *loop increment* `incr` is allowed.

Finally, we declared our two dimensional array of variables in the line

```
X  X(i,j)
```

Remember that we merely *give names* to our variables in this section. So how are the *names* of the indexed variables[27] built? The rule is very simple: if the integer parameters `i` and `j` have the values 2 and 3, say, then the string `X(i,j)` is equivalent to the string `X2,3`. *The last string* (sometimes called the

[25] It is of course forbidden to overlap loops in any other way.

[26] Also called *loop parameter*

[27] Also called *indexed arrays*, or simply *arrays*.

realization of the first one for the given values of the parameters) *is obtained by replacing the name of the involved parameter(s) by the string containing their value as integers and by dropping the parenthesis* (*and*). This procedure is called the "expansion" of the name. The loop

```
DO i          1                P
X  Z(i,j)
OD i
```

is thus entirely equivalent (when P has the value 7 and j the value 11) to

```
X  Z1,11
X  Z2,11
X  Z3,11
X  Z4,11
X  Z5,11
X  Z6,11
X  Z7,11
```

which, in the `VARIABLES` section, would define a hypothetical set of variables `Z1,11`, `Z2,11`, `Z3,11`, `Z4,11`, `Z5,11`, `Z6,11` and `Z7,11`. As the length of the names is limited to ten characters at most and since `X100,100,100` is already over this limit, the SDIF imposes a limit of three possible levels of indexing, corresponding to the three levels of loop nesting.

One additional point on indexed variables. You certainly noticed the use of the code `X` in the line

```
X  X(i,j)
```

although we did not use any code to declare variables in our previous examples. Of course, you wondered at the reason of this sudden introduction. The reason is that the SDIF interpreter needs a special code to realize that `X(i,j)` is to be expanded to `X12,17`, say, and is not just a fancy name with parenthesis and a comma (which is allowed in the MPS syntax). Hence, the following important rule: *every SDIF line containing an array name must start with a code whose first letter is either* `X` *or* `Z`, depending on where the associated data is to be read (see Section 2.4.2). The fact that we can use `X`- and `Z`-codes for ordinary (non array) names is because they are considered to be a special case of indexed names, a very handy convention. Observe that the `ELEMENT TYPE` and `GROUP TYPE` sections do not contain lines with such codes: array names are therefore not admissible in these sections[28].

Of course, we can also use indexed names for the groups (there is one group per small square, which we choose to index by the coordinates of its bottom-left corner), constants and bounds of our problem:

---

[28] They also exclude the use of loops.

```
GROUPS
 RI RP        P
 R* RPSQ      RP                        RP
 DO i         1                         P
 DO j         1                         P
 XN S(i,j)
 ZN S(i,j)                              RPSQ
 OD j
 OD i
CONSTANTS
 DO i         1                         P
 DO j         1                         P
 X  MINSURF   S(i,j)    -1.0
 OD j
 OD i
BOUNDS
*   Free variables inside the unit square
 IE 2                   2
 DO i         2                         P
 DO j         2                         P
 XR MINSURF   X(i,j)
 OD j
 OD i
*   Boundaries
 DO i         1                         P+1
 XX MINSURF   X(1,i)    1.0
 XX MINSURF   X(P+1,i)  1.0
 XX MINSURF   X(i,1)    1.0
 XX MINSURF   X(i,P+1)  1.0
 OD i
```

Note that the above lines stay valid if we change the value of P in the beginning of the file, a crucial advantage! Also note that we had to assign the value 2 to the integer parameter named 2 before using it as starting index for a loop.

Loops cannot be used (and are of debatable interest) in the ELEMENT TYPE section, which is then easily written as

```
ELEMENT TYPE
 EV ISQ       V                         W
 IV ISQ       U
```

where we specified the *unique* type of element present in our problem, that is the square of the difference between two variables.

Loops are however of great interest in the ELEMENT USES section:

```
ELEMENT USES
 DO i         1                         P
 IA i+1       i         1
 DO j         1                         P
 IA j+1       j         1
```

```
*   Diagonal
 XT A(i,j)    ISQ
 ZV A(i,j)    V                        X(i,j)
 ZV A(i,j)    W                        X(i+1,j+1)
*   Antidiagonal
 XT B(i,j)    ISQ
 ZV B(i,j)    V                        X(i,j+1)
 ZV B(i,j)    W                        X(i+1,j)
 OD j
 OD i
```

We have chosen to define A-elements for those involving the variables at the bottom-left and top-right corners of the discretized squares (the diagonal), and B-elements for those involving the other two corners (the antidiagonal). Notice the technique of defining i+1 by incrementing i within the loop where it is defined.

The groups of our problem are again of a unique type (square root):

```
GROUP TYPE
 GV SQROOT    ALPHA
```

and we can use the loop mechanism again to declare the type of our groups and assign the nonlinear elements to them:

```
GROUP USES
 RM WEIGHT    RPSQ      0.5
 DO i         1                        P
 DO j         1                        P
 XT S(i,j)    SQROOT
 ZE S(i,j)    A(i,j)                   WEIGHT
 ZE S(i,j)    B(i,j)                   WEIGHT
 OD j
 OD i
```

We started by computing the weighting factor ($p^2/2$) before looping over the groups.

The only thing left is to write the SEIF and SGIF parts of the specification file, and append them to what we have written already. The complete file is thus as follows:

```
NAME          MINSURF

*   A version of the minimum surface problem
*   on the unit square with simple boundary conditions.

*   Discretization parameter
 IE P                   7
*   Other parameters
 IE 1                   1
 IA P+1       P         1
```

```
 RI RP        P
 R* RPSQ      RP                         RP
VARIABLES
 DO i         1                          P+1
 DO j         1                          P+1
 X  X(i,j)
 OD j
 OD i
GROUPS
 DO i         1                          P
 DO j         1                          P
 XN S(i,j)
 ZN S(i,j)    'SCALE'                    RPSQ
 OD j
 OD i
CONSTANTS
 DO i         1                          P
 DO j         1                          P
 X  MINSURF   S(i,j)    -1.0
 OD j
 OD i
BOUNDS
*   Free variables inside the unit square
 IE 2                   2
 DO i         2                          P
 DO j         2                          P
 XR MINSURF   X(i,j)
 OD j
 OD i
*   Boundaries
 DO i         1                          P+1
 XX MINSURF   X(1,i)    1.0
 XX MINSURF   X(P+1,i)  1.0
 XX MINSURF   X(i,1)    1.0
 XX MINSURF   X(i,P+1)  1.0
 OD i
ELEMENT TYPE
 EV ISQ       V                          W
 IV ISQ       U
ELEMENT USES
 DO i         1                          P
 IA i+1       i         1
 DO j         1                          P
 IA j+1       j         1
*   Diagonal
 XT A(i,j)    ISQ
 ZV A(i,j)    V                          X(i,j)
 ZV A(i,j)    W                          X(i+1,j+1)
*   Antidiagonal
```

```
 XT B(i,j)     ISQ
 ZV B(i,j)     V                         X(i,j+1)
 ZV B(i,j)     W                         X(i+1,j)
 OD j
 OD i
GROUP TYPE
 GV SQROOT     ALPHA
GROUP USES
 RM WEIGHT     RPSQ      0.5
 DO i          1                         P
 DO j          1                         P
 XT S(i,j)     SQROOT
 ZE S(i,j)     A(i,j)                    WEIGHT
 ZE S(i,j)     B(i,j)                    WEIGHT
 OD j
 OD i
ENDATA

ELEMENTS       MINSURF
INDIVIDUALS
 T  ISQ
 R  U          V         1.0             W         -1.0
 F                       U * U
 G  U                    U + U
 H  U          U         2.0
ENDATA

GROUPS         MINSURF
TEMPORARIES
 R  SQRAL
 M  SQRT
INDIVIDUALS
 T  SQROOT
 A  SQRAL                SQRT( ALPHA )
 F                       SQRAL
 G                       0.5 / SQRAL
 H                       -0.25 / ( ALPHA * SQRAL )
ENDATA
```

In this file, we have omitted the START POINT section, implying that the value 0 should be used for all problem variables: it is up to the minimization algorithm to first set the fixed variables to their assigned values before the minimization proceeds. LANCELOT does this, of course.

At this point, it is probably a good idea to reread the last problem specification and make sure you understand every single line of it, before we move on to further complications. It is, in particular, important to note that changing the discretization mesh is a trivial exercise with this specification (how?), and that it is simple to modify the problem to incorporate more complicated boundary conditions and/or obstacles (that is constraints on the height of the surface at certain locations within the unit square).

2.7.2 Setting Default Values

Using loops to define values or names for array components is reasonably efficient, but the SDIF offer a useful alternative. This is the possibility of defining *defaults*. This notion is again best understood via examples. For instance, the BOUNDS section of the preceding example could be written as

```
BOUNDS
*   Free variables
 FR MINSURF   'DEFAULT'
*   Boundaries
 DO i         1                         P+1
 XX MINSURF   X(1,i)     1.0
 XX MINSURF   X(P+1,i)   1.0
 XX MINSURF   X(i,1)     1.0
 XX MINSURF   X(i,P+1)   1.0
 OD i
```

which is both shorter and more elegant. The idea is that we say, in the line

```
 FR MINSURF   'DEFAULT'
```

that all variables are free. We may then modify the status of some variables (fix the boundary ones), thereby overriding the default. (Of course, the default must be set *before* any other bound specification.) Similarly, the CONSTANTS section can be written as

```
CONSTANTS
    MINSURF   'DEFAULT' -1.0
```

instead of explicitly specifying all constants individually.

Defaults can be set for a number of things within the SDIF problem specification file. Possible uses of this technique are given in Table 2.7.

The syntax for default setting is quite simple. The idea is that *the line defining a default value for a class of objects is identical to the line that would be used to assign a value to a single member of this class, except that the name of the class member is replaced by the string* "'DEFAULT'". In the line

```
 FR MINSURF   'DEFAULT'
```

the string 'DEFAULT' replaces the name of the variable, X say, that would be declared free by the line

```
 FR MINSURF   X
```

Similarly, one may declare that all groups are by default of the least-squares type (SQUARE) by the line

```
 T  'DEFAULT' SQUARE
```

Section	Uses	Field
CONSTANTS	Set default value for group constants	3
	Set default value for group constants	3
BOUNDS	Set all variables to be bounded below	3
	Set all variables to be bounded above	3
	Set all variables to fixed to a value	3
	Set all variables to be free	3
	Set all variables to be unbounded above	3
	Set all variables to be unbounded below	3
START POINT	Set default starting value for all variables	3
ELEMENT USES	Set default type for elements	2
GROUP USES	Set default type for groups	2

Table 2.7: Possible default settings

because declaring the group AGROUP to be of the same type would be written as

```
 T  AGROUP    SQUARE
```

The lines

```
 XL PROBLEM   'DEFAULT' 2.0
 ZU PROBLEM   'DEFAULT'                      UBOUND
```

are valid default definitions within the BOUNDS section, while

```
 X  PROBLEM   'DEFAULT'  5.0
```

or

```
 Z  PROBLEM   'DEFAULT'                      DEFVALUE
```

are appropriate in the CONSTANTS section.

The syntax thus implies that the string "'DEFAULT'" may appear in fields 2 or 3, depending on the section. The exact field where it must appear is detailed, section by section, in the third column of Table 2.7.

2.7.3 Summary

This section has introduced the use of indexed variables, loops on indices and default setting.

Indexed variables permit the problem description to follow its mathematical expression much more closely, especially if the problem at hand arises from the discretization of some infinite dimensional computation. Variational calculations, processes governed by partial or ordinary differential equations, regular network and boundary value determination are just a few of the application areas where this facility is most useful. Indexed variables contain up to three (integer) indices within `()`, and these indices are separated by commas, as in `X(i,j,k)`. Any SDIF line referring to an indexed variable must start with a code whose first letter is either `X` or `Z`.

The appeal of indexed variables is clearly dependent on the facility that allows loops on indices to be written within a SDIF file. The syntax for these loops is similar to that used in Fortran: they start with a `DO` line and end with an `OD` line. It is also possible to nest loops up to three levels, and also to define the increment by which the loop index is increased each time the loop is executed.

The third important novelty in this section is the setting of default values for possibly large sets of objects, such as constants, bounds, starting point components or element/group types. The syntax used for this purpose is identical that that used for defining a single object, except that the object name is replaced by the string `'DEFAULT'`: all relevant objects then take this default unless explicitly declared to have a different value at a later stage in the specification.

The tools introduced in this section are crucial for the specification of many large scale nonlinear programming problems.

2.8 More Advanced Techniques

You now possess the basic skills required to specify a large number of problems, from small simple ones to large structured cases. However, there is still more to learn. The additional concepts and constructs that we will introduce in this section may not be absolutely necessary for the statement of your favourite problem, but they may also be quite helpful.

2.8.1 Elemental and Group Parameters

The first additional construct that we will examine is the use of *parameters* in element and group types. Consider the simple problem of fitting, in the least-squares sense, an exponential to a set of points on the bisector of the positive orthant of the plane. Mathematically, the problem is expressed as

$$\min_{\alpha,\beta} \sum_{i=0}^{p} \left(\alpha e^{ih\beta} - ih\right)^2, \tag{2.8.46}$$

given some positive integer p and some stepsize $h > 0$. This could be stated using the techniques developed in the previous sections, but would then require the definition of $p+1$ distinct element types of the form

$$ve^{ihw} \quad (i = 0, \ldots, p). \tag{2.8.47}$$

This is most definitely a waste of energy, especially if p exceeds two or three. An alternative is to consider i as a *parameter* associated with an element of the form

$$ve^{ihw}, \tag{2.8.48}$$

and then to assign to this parameter the required values when the problem variables α and β are assigned to the elemental variable v and w respectively. Such parameters are called *elemental parameters*. Practically speaking, we first have to declare that our (unique) element type has a parameter (called RI because it is the real value of the otherwise integer i). This is done using an EP line, as in

```
ELEMENT TYPE
 EV EXPIH      V                        W
 EP EXPIH      RI
```

where the name of the considered element type is repeated in field 2[29]. The next stage is to assign the correct values to this parameter in the ELEMENT USES section. This is written as

```
ELEMENT USES
 DO i          1                        P
 RI Reali      i
 XT E(i)       EXPIH
 ZV E(i)       V                        ALPHA
 ZV E(i)       W                        BETA
 ZP E(i)       RI                       Reali
 OD i
```

where the ZP code indicates that the value of the elemental Parameter is to be found in field 5 (hence the Z). We would use a similar XP code if this value was to be read in the numerical field 4.

```
NAME           EXPFIT

*   A simple exponential fit in 2 variables

*   Number of points
 IE P                     10
*   Stepsize
 RE H                     0.25
*   Other parameters
 IE 1                     1
VARIABLES
    ALPHA
    BETA
GROUPS
 DO i          1                        P
 XN R(i)
```

[29]just as for the declaration of internal variables on the IV- lines.

```
 OD i
CONSTANTS
 DO i         1                         P
 RI Reali     i
 R* Reali*H   Reali                     H
 Z  EXPFIT    R(i)                      Reali*H
 OD i
BOUNDS
*   Free variables inside the unit square
 FR EXPFIT    'DEFAULT'
ELEMENT TYPE
 EV EXPIH     V                         W
 EP EXPIH     RI
ELEMENT USES
 DO i         1                         P
 RI Reali     i
 XT E(i)      EXPIH
 ZV E(i)      V                         ALPHA
 ZV E(i)      W                         BETA
 ZP E(i)      RI                        Reali
 OD i
GROUP TYPE
 GV L2        GVAR
GROUP USES
 XT 'DEFAULT' L2
 DO i         1                         P
 XE R(i)      E(i)
 OD i
ENDATA

ELEMENTS      EXPFIT
TEMPORARIES
 R  IH
 R  EXPWIH
 M  EXP
INDIVIDUALS
 T  EXPIH
 A  IH                  0.25 * RI
 A  EXPWIH              EXP( W * IH )
 F                      V * EXPWIH
 G  V                   EXPWIH
 G  W                   V * IH * EXPWIH
 H  V         W         IH * EXPWIH
 H  W         W         V * IH * IH * EXPWIH
ENDATA

GROUPS        EXPFIT
INDIVIDUALS
 T  L2
```

```
 F                          GVAR**2
 G                          2.0 * GVAR
 H                          2.0
ENDATA
```

A similar situation could occur, when parameters are wanted within the group functions. An example might be the robust regression problem already considered in Section 2.5.3, where the parameter k associated with the Huber function $h_k(\cdot)$ could be dependent on the particular observation[30] and therefore vary from group to group.

Such parameters, called *group parameters*, are introduced in a way entirely similar to that of the elemental parameters, the codes EP (in the ELEMENT TYPE section), P, XP and ZP (in the ELEMENT USES section) being replaced by GP (in the GROUP TYPE section), P, XP and ZP (in the GROUP USES section) respectively. For instance, the GROUP TYPE section appropriate for declaring k to be a parameter for our robust regression problem is

```
GROUP TYPE
 GV HUBER     ALPHA
 GP HUBER     HUBERK
```

2.8.2 Scaling the Problem Variables

We already discussed in Section 2.5.2 how to scale the constraints and the objective function of a problem so that their respective size are comparable[31] by using the 'SCALE' reserved name. MPS, and therefore SDIF, also provide a way to scale the *problem variables* much in the same way. If the variables of a problem are expected to take values that differ by orders of magnitude in the region of interest (that is "between" the starting point and the solution), it is indeed often useful to *scale* them. This is to say that each variable is *divided* by a suitable fixed scalar. If we consider, for instance, the problem of minimizing the simple two dimensional quadratic function

$$q(x_1, x_2) = 10^{-10}x_1^2 + 10^{20}x_2^2 \tag{2.8.49}$$

the actual value of x_1 will not, in typical double precision say, contribute to that of the objective function, except when x_2 is itself extremely tiny. The simple change of variables

$$y_1 = \frac{x_1}{10^5}, \quad y_2 = \frac{x_2}{10^{-10}} \tag{2.8.50}$$

[30]This parameter indeed represents a level above which "outliers" are less penalized in the fit. It is therefore linked to the quality of the observations, which may itself vary from observation to observation.

[31]Nearly always a good idea, but not always easy to do.

produces the function

$$q(x_1, x_2) = y_1^2 + y_2^2 \tag{2.8.51}$$

which may be substantially easier to minimize on a computer. The required values of x_1 and x_2 may then be safely recovered from the relations (2.8.50). A typical case where a scaling of this type is often necessary is in calculations for electronic circuits. Indeed, if resistances are measured in ohms and capacitances in farads, typical values for resistance and capacitance values may be 10^5 and 10^{-12} respectively! It is[32] much safer from the numerical point of view to use scaled units such as megohms or picofarads...

The change of variable (2.8.50) is easy to specify in MPS and in the case of SDIF, if the entries appear in the linear elements. The reserved word `'SCALE'` is again used for this purpose, but now appears in the `VARIABLES` section. The `VARIABLES` section for the simple problem of minimizing (2.8.49) (assuming we had used a group of type `SQUARE` to ensure that `X1` and `X2` appear as linear elements) may then look like

```
VARIABLES
 X  X1          'SCALE'   1.0D5
 X  X2          'SCALE'   1.0D-10
```

An efficient optimization algorithm such as LANCELOT will of course use the transformed variables in its inner calculations and return the optimal values to the user after the necessary back transformation: the change of variable is entirely implicit and hidden from the user's sight (and mind) within the numerical algorithm itself.

2.8.3 Two-sided Inequality Constraints

It sometimes happens that inequality constraints are specified both with an upper and a lower limiting value. For instance, the water level in a reservoir must always be nonnegative, but it must also not exceed the height of the dam that closes the valley (unless the safety of the valley inhabitants is of no consequence). Mathematically, if $h(r, f)$ represents the water height in the reservoir as a function of rainfall and outflow at the dam, we have a constraint of the form

$$0 \le h(r, f) \le h_{\text{disaster}} \tag{2.8.52}$$

where h_{disaster} is the height beyond which water starts flooding the valley downstream. Of course, one could state (2.8.52) as *two* constraints of the form

$$h(r, f) \ge 0 \text{ and } h(r, f) \le h_{\text{disaster}}, \tag{2.8.53}$$

[32] Quite obviously!

but this is slightly inelegant, and also inefficient because one has to introduce $h(r, f)$ twice in the formulation. It is much nicer to specify one of the constraints (2.8.53), the first, say, as an ordinary inequality constraint, and then to add the constraint that the village stays dry after the fact, by specifying *another value for the "other side" of the constraint.* In practice, the first step would require the GROUPS and CONSTANTS sections to contain lines of the form

```
GROUPS
 XG H2OLEVEL
CONSTANTS
 X  RESERVOIR H2OLEVEL   0.0
```

(where the complicated nonlinear element function giving $h(r, f)$ is later assigned to the group H2OLEVEL)[33]. It is now possible to specify the upper bound on the water height by introducing yet another new section, the RANGES section, where "other sides" of inequality constraints[34] may be given. The syntax of the RANGES section is entirely similar to that of the CONSTANTS, so that the RANGES section for our example reads

```
RANGES
 Z  RESERVOIR H2OLEVEL                  HDISASTER
```

The value HDISASTER is called the *range value* associated with the H2OLEVEL constraint. As the constraint H2OLEVEL has already been declared in the GROUPS section to be a "greater-than-or-equal-to" constraint (with associated value zero, or more generally, HMUDDY, say, given in the CONSTANTS section), the value given in the RANGES section to the *same constraint group* is automatically associated with a "less-than-or-equal-to" constraint involving the same group. *The value specified in the* RANGES *section is always associated with an inequality constraint of the type* opposed *to that specified for the relevant constraint within the* GROUPS *section.*

We have said that the syntax of the RANGES section is identical to that of the CONSTANTS section. The codes X and Z are thus available (depending on the field in which the actual value is to be found), but defaults can also be set for the ranges values, and loops can be used. As for constants, a name is associated with a vector of range value (RESERVOIR in our example). The attentive reader may also guess that the initial default values have to be chosen as to have no impact at all on the original inequality constraints. Hence, the default range value is set to $+\infty$ for "greater than" groups, and to $-\infty$ for "less than" groups[35].

[33] The CONSTANTS section is also optional in our example, because it merely specifies a zero value, which is identical to the default for this section.

[34] This type of constraint is called a "range constraint" in the MPS jargon.

[35] If you wish to modify defaults for range values, be sure that you understand what you are doing...

This may seem simple, but this appearance is quite deceptive. In fact, the RANGES section only allows the specification of constraints of the type

$$0 \left\{ \begin{array}{c} \leq \\ = \\ \geq \end{array} \right\} \frac{1}{s_i} g_i \left(\sum_{j \in J_i} w_{ij} f_j(x) + a_i^T x - b_i \right) \left\{ \begin{array}{c} \leq \\ = \\ \geq \end{array} \right\} r_i \; (i \in I_I), \tag{2.8.54}$$

for possibly infinite values of r_i. Our previous example was chosen so that the lower bound of zero coincides with the left-hand-side of (2.8.54), making the specification easier. We now illustrate the potential complications by assuming that the water level must not only be nonnegative, but also above the level of the mud at the bottom of the reservoir. It may then be impossible to specify the new constraint

$$h_{\text{muddy}} \leq h(r, f) \leq h_{\text{disaster}} \tag{2.8.55}$$

as a single constraint with two sides, except if the group type associated with this constraint is linear! This is because the only freedom we have to specify this new lower bound is via the constant b_i which appears *inside* the group value. More precisely, assume that the group is trivial; it is then enough to replace (2.8.55) by

$$0 \leq h(r, f) - h_{\text{muddy}} \leq h_{\text{disaster}} - h_{\text{muddy}}, \tag{2.8.56}$$

and we therefore find the form of (2.8.54) again. We then may write the following lines

```
GROUPS
 XG H2OLEVEL
CONSTANTS
 Z  RESERVOIR H2OLEVEL                        -HMUDDY
RANGES
 R- HUSEFUL   HDISASTER                       HMUDDY
 Z  RESERVOIR H2OLEVEL                        HUSEFUL
```

On the other hand, if the constraint group type is nonlinear, it may not be possible to incorporate the constant $-h_{\text{muddy}}$ in the group value... and one therefore has to specify two separate constraints in this unlucky (and, fortunately, infrequent) case.

2.8.4 Bounds on the Objective Function Value

There is one feature of the SDIF syntax that is not used by the current version of LANCELOT. Its purpose is to allow the knowledgeable user to specify a priori known limits on the possible values of the objective function. For instance, the objective function of a least squares fitting problem, being a sum of squared terms, will never be negative. This is *not a constraint* that one *imposes* on the objective function, but merely a way to specify a priori *knowledge* of the problem. Because the S(D)IF is not only intended to work only in tandem with

LANCELOT, and because some algorithms could well exploit this knowledge, a syntax is provided to state these "objective function bounds". They appear in a specialized section, called the OBJECT BOUND section. A typical section (for a least-squares problem) is given by

```
OBJECT BOUND
 LO PROBNAME             0.0
```

where PROBNAME is the name of a *set* of objective bounds, just as they are sets of bounds, constants or ranges. Again, we use the convention of specifying the problem name as set name.

The code LO allows to specify a LOwer bound on the objective function, but other codes can be used to specify lower and upper bounds, just as in the BOUNDS section. The possible codes and their meaning are given in Table 2.8. The codes XL and XU are provided for the (somewhat farfetched) case where the objective function name is an array name; they are otherwise equivalent to LO and UP respectively.

Code	Meaning
LO XL	specifies a numerical lower bound in field 4
UP XU	specifies a numerical upper bound in field 4
ZL	specifies a parametric lower bound in field 5
ZU	specifies a parametric upper bound in field 5

Table 2.8: Codes for specifying bounds on the objective function

2.8.5 Column Oriented Formulation

Let us now consider a type of problem that frequently arises in large scale nonlinear optimization: a nonlinear network problem[36]. Consider the network shown at Figure 2.2 and assume that we are interested in computing the best (in a measure described below) flows on each of its arcs so as to ensure flow conservation at all internal nodes and so as to satisfy a demand of 10 units at node 1 and a supply of 10 units at node 4. In addition, we also impose some bounds on the flows.

More precisely, we consider the mathematical programming problem given by

$$\min \quad x_2 e^{x_1+x_3} + (x_3 x_4)^2 + (x_3 - x_5)^2 \tag{2.8.57}$$

[36] We will however only consider a small one to avoid unnecessary tree cutting in the rain forests.

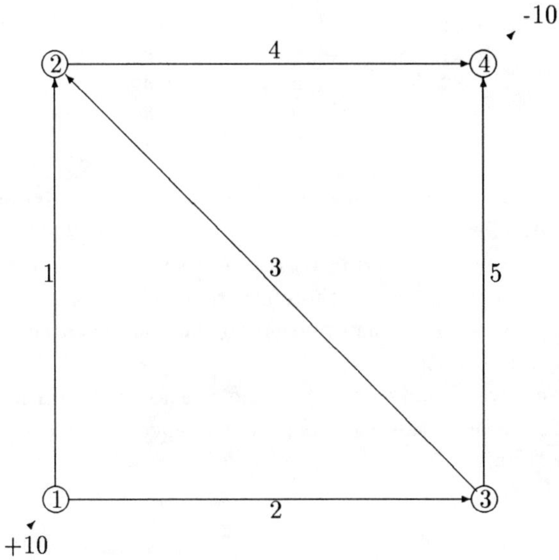

Figure 2.2: A small network with associated supply and demand

subject to the constraints

$$\begin{array}{rrrrrcl} x_1 & +x_2 & & & & = & 10, \\ -x_1 & & -x_3 & +x_4 & & = & 0, \\ & -x_2 & +x_3 & & +x_5 & = & 0, \\ & & & -x_4 & -x_5 & = & -10 \end{array} \qquad (2.8.58)$$

and the bounds

$$2 \leq x_1 \leq 4, \quad 6 \leq x_2 \leq 8 \text{ and } 0 \leq x_3 \leq 5. \qquad (2.8.59)$$

We have used the "node egoistic" convention that a flow leaving a node is given a negative value while a flow arriving at a node is given a positive value.

In (2.8.58), each equation corresponds to the conservation of flow at a node, which dictates that the total flow arriving at a node is equal to the flow that leaves it. In the form (2.8.57)–(2.8.59), the problem is well suited to an input of the type we have used up to now. However, it is often more natural to forget about the equations (2.8.58) and to consider Figure 2.2 instead. In this figure, the arcs are more pre-eminent, and we may be tempted to describe the network arc by arc, especially since there is a problem variable associated with each one of them. It is then natural to specify the arcs (the variables) by indicating their origin and destinations nodes. Assume that we wish to call the flows `X1` to `X5` and the nodes `N1` to `N4`. we may then say, in a new form of the `VARIABLES` section,

```
VARIABLES
    X1          N1        -1.0          N2        1.0
    X2          N1        -1.0          N3        1.0
    X3          N3        -1.0          N2        1.0
    X4          N2        -1.0          N4        1.0
    X5          N3        -1.0          N4        1.0
```

In this form, there is one line per arc in the network, and its origin and destination are easily identifiable. A brief comparison with (2.8.58) quickly reveals that each line of this new section correspond to a *column* of this set of linear equalities. This formulation is therefore referred to as the *column oriented formulation* of the problem, as opposed to the *row oriented formulation* that we have always used until now.

Because we have already described the relation between nodes and arcs in the VARIABLES section, there is no need to repeat it in the GROUPS section, which is then simply

```
GROUPS
 DO I           1                            NNODES
 XE  N(I)
 OD I
```

where we have assumed that the integer parameters 1 and NNODES have already been assigned the values of the constant 1 and of the number of nodes in the network respectively. We use XE codes because each of the flow conservation laws is an equality. This seems simple enough, but there is one additional catch: we have used the name of the constraints in the VARIABLES section, which requires that these names are known by then! This explains why the GROUPS section must precede the VARIABLES section in a column oriented formulation, the result thus being

```
NAME           SMALLNET
 IE 1                     1
 IE NNODES                4
GROUPS
 DO I           1                            NNODES
 XE  N(I)
 OD I
VARIABLES
    X1          N1        -1.0          N2        1.0
    X2          N1        -1.0          N3        1.0
    X3          N3        -1.0          N2        1.0
    X4          N2        -1.0          N4        1.0
    X5          N3        -1.0          N4        1.0
CONSTANTS
    SMALLNET    N1        10.0
    SMALLNET    N4        -10.0
```

where we have added the CONSTANTS section, which indicates the demand and supply at nodes 1 and 4.

Of course, column oriented formulation is not restricted to network problems: it can be used in general as a substitute for the row oriented description. They cannot be mixed, however, if only because one has to choose an order for the GROUPS and VARIABLES sections. Just as in the row oriented formulation, it is only possible to specify the linear part of the constraints in the fashion described above, the nonlinear elements always being assigned to the relevant groups in the GROUP USES section. In fact, except for the GROUPS and VARIABLES sections, the syntax of a column oriented problem description is identical to that of the corresponding row oriented one.

Which of the two formulation is better depends very much on the problem itself.

- A column oriented description might be more appropriate for problems with a large number of linear constraints.[37] Network problems are a special case of this class.
- A row orientation is probably more appropriate for more nonlinear problems, where constraints are usually expressed row by row, after one knows what the problem variables are.

Obviously, the SIF and LANCELOT do not care which orientation is actually chosen.

2.8.6 External Functions

When evaluating the function and derivative values of a nonlinear element or group, it may occasionally happen that you wish to use a function for which there is no Fortran supplied intrinsic version available. For instance, one might require the value of the Error function,[38]

$$e(x) = \frac{2}{\pi} \int_0^x e^{-t^2}\, dt \qquad (2.8.60)$$

for some argument x. You needn't necessarily despair. There are provisions within the SIF to handle such difficulties.

Firstly, you will need to know, or know a friend who knows, or know a friend who knows a book which gives, a computable approximation to your required function *in Fortran*. Let us suppose we have found a Fortran program which does the job for us

[37] Only the column oriented formulation is allowed by MPS, the standard for linear programming input.

[38] Some useful, but nonstandard, dialects of Fortran 77 supply an intrinsic error function!

```
      DOUBLE PRECISION FUNCTION E( X )
      DOUBLE PRECISION X
C
C  This function returns the value of the error function e( x )
C
      ...
      ...
      E = ...
      RETURN
      END
```

Now all we have to do is to define the function E as an external Real Function in the TEMPORARIES section of the relevant SEIF/SGIF file by writing

```
TEMPORARIES
 ...
 R  E
 F  E
 ...
```

and we can then compute the value of $e(\frac{1}{2})$

```
 ...
 A  EHALF                  E( 5.0D-1 )
 ...
```

in either the GLOBALS or INDIVIDUALS section, remembering of course that we must have defined the temporary variable EHALF to be real. Notice that E has to be declared to be both real using the code R and an external function with the code F.

Needless-to-say, you must remember to include the Fortran definition of $e(x)$ along with your SDIF, SEIF and SGIF files otherwise it will be difficult to guess what function you had in mind!

It is, of course, permissible to define integer or logical external functions, remebering that the R code would then be either I or L as appropriate. It is also possible to assign values to parameters via the argument list of a function. For instance, suppose we have an external Fortran function given by

```
      DOUBLE PRECISION FUNCTION F( X, INFORM )
      DOUBLE PRECISION X
      INTEGER INFORM
C
C  This function returns the value of f( x ) and a
C  suitable return code INFORM.
C
      ...
      F = ...
      INFORM = 1
      RETURN
      END
```

Then we can evaluate $f(1)$ and obtain its return code by writing

```
TEMPORARIES
 I  INFORM
 R  FONE
 R  F
 F  F
GLOBALS
 A  FONE                  F( 1.0D+0, INFORM )
```

The return code INFORM will have been assigned the integer value 1by the call to F.

2.8.7 The Order of the SIF Sections

The introduction of the column oriented formulation and of the resulting permutation between the VARIABLES and GROUPS section raises the long eluded question of the order in which the SDIF sections are deemed to appear in the problem description file.

This order and the order of the associated keywords are not left to your imagination but comes in only two types: one for row oriented input and the other for column oriented. They are both given in complete detail in Table 2.9, where optional sections are identified with keywords within brackets[39].

In this Table, [SEIF] means that the complete SEIF file is optional, but the fact that the ELEMENTS keyword is not within brackets indicates that it is mandatory *if the SEIF file is present at all.* The same comment applies to the SGIF keywords and sections.

Note the last entry in the table, "external functions". This is where we would place Fortran code to evaluate any or all of those naughty functions which are not supplied as intrinsics with Fortran 77, see Section 2.8.6.

2.8.8 Ending all Loops at Once

We mentioned in Section 2.7.1 that nested loops on integer parameters were allowed up to a maximum of three levels. Of course, nested loops must use different loop parameters, and we might have a VARIABLES section like

```
VARIABLES
 DO I         1                        NX
 DO J         1                        NY
 DO K         1                        NZ
 X  X(I,J,K)
 OD K
```

[39] Which means, of course, that the sections having keywords without brackets are mandatory.

file	row oriented	column oriented
SDIF	NAME	
	VARIABLES	GROUPS
	GROUPS	VARIABLES
	[CONSTANTS]	
	[RANGES]	
	[BOUNDS]	
	[ELEMENT TYPE]	
	[ELEMENT USES]	
	[GROUP TYPE]	
	[GROUP USES]	
	[OBJECT BOUND]	
	ENDATA	
[SEIF]	ELEMENTS	
	[TEMPORARIES]	
	[GLOBALS]	
	[INDIVIDUALS]	
	ENDATA	
[SGIF]	GROUPS	
	[TEMPORARIES]	
	[GLOBALS]	
	[INDIVIDUALS]	
	ENDATA	
[external functions]	[one or more external functions]	

Table 2.9: Possible orders for the SDIF/SEIF/SGIF keywords and sections

```
 OD J
 OD I
```

to define variables corresponding to points on a three dimensional grid, say. The SDIF syntax provides one additional facility for such cases: it allows one to *terminate all currently open loops at once*, in a single statement. This statement only contains the code ND, so that the above specification of variables is entirely equivalent to

```
VARIABLES
 DO I         1                         NX
 DO J         1                         NY
 DO K         1                         NZ
 X  X(I,J,K)
 ND
```

The use of the ND statement of course results in more compact problem specification, but is also less explicit than individual loop termination. The choice between both techniques therefore mostly depends on the readability of a particular problem file: the ND loop terminator is quite acceptable when the context is clear...

2.8.9 Specifying Starting Values for Lagrange Multipliers

We now present a feature for mathematical programming experts. They indeed know that most algorithms for solving constrained problems make some use of the so-called *Lagrange multipliers*[40], which are the components of the vector λ in the expression of the Lagrangian

$$L(x, \lambda) = f(x) + \lambda^T c(x) \tag{2.8.61}$$

associated with the problem

$$\min f(x) \tag{2.8.62}$$

such that

$$c(x) \left\{ \begin{array}{c} \leq \\ = \\ \geq \end{array} \right\} 0 \tag{2.8.63}$$

It is not our intention to start here a discussion on the use, importance and meaning of these multipliers[41]. We just emphasize that it is very often helpful (from the algorithmic point of view) to know an approximation of their value at the problem's solution, especially if the starting point is close to to this solution. It then make sense to specify these "starting values" for the Lagrange multipliers as well as the starting values for the problem variables. This can be done in the START POINT section, using the codes XM or ZM, where the M stands for Multiplier and the X or Z as usual depends on the actual field in which the associated value is to be found. (The codes X and M are also available for the case where none of the names used is an array name.) Of course, multiplier starting values can be defaulted, just as for problem variables. A rather complete START POINT section for a constrained problem may then look like

```
START POINT
 ZV STARTP    'DEFAULT'                X0-VALUE
 XM STARTP    'DEFAULT' 0.0
 XV STARTP    X1        1.0
 ZM STARTP    CONSTR1                  LAMBDA1
```

where the real parameters LAMBDA1 and X0-VALUE contain suitable starting values for the Lagrange multiplier associated with the constraint CONSTR1 and default values for the problem variables respectively.

[40] Sometimes also called *shadow prices* in the econometric jargon associated with linear programming.

[41] Are you a mathematical programming expert?

If you are really short on file space, there is even the possibility of mixing problem variable and Lagrange multiplier starting values on the same line, using a code consisting only of blank characters. One may then have a START POINT section containing a line of the form

```
        CONPROB   X1          3.14                LAMBDA1   2.71
```

but this kind of mix is prohibited on lines that indicate default values.

It is important to note at this point that the initial default value for starting Lagrange multipliers is zero, just as for the problem variables themselves.

2.8.10 Using Free Format in SIF Files

So far, we have been quite specific on the format that we allow for writing the lines of a problem description in SIF. We used fields spanning well-defined ranges of columns within the line, each of these fields being used for specific purposes. The rigidity of this *fixed format* can sometimes be considered an intolerable burden[42]. We discuss in this paragraph a second form of input, which we will refer to as the *free format*, although the freedom here is not quite complete[43].

The idea of the free format is that users should be able to arrange the problem data in the way they prefer, writing several SIF statements on a single line or spacing the fields in an entirely different fashion. The SIF free format however imposes a crucial restriction: *every line of a free format problem specification must be such that it can be translated unambiguously to one (or more) fixed format line.*

The first difficulty, if we are to allow data to appear anywhere on a line, is to define separators between fields. In fixed format, these are unnecessary, because fields are identified by the range of columns in which they lie, but free format requires *explicit separators*, whose purpose is to separate *strings* on a given line. A line may contain at most 160 characters. The natural separator in a problem description file is probably the blank character, but, unfortunately, the MPS standard allows significant blanks within valid names! In fact, we have made all ASCII characters valid in SDIF names, which leaves none for the separators... The only way out is to exclude special characters from the free format files, except when they are meant as explicit separators.

“ ” (blank) indicates that a string is finished and that the next string is about to begin. Several consecutive blanks are interpreted as a single blank, allowing users to space the strings as they please on the line.

---

[42] Despite our experience that fixed format often helps in keeping the problem files reasonably organized and readable...

[43] But what in life is?

";" indicates that a string is finished (and another is to follow), but also indicates that the next string should start a new line in the fixed format translation of the line. The semi-colon is thus used to separate "fixed format statements", much in the same way that it separates C statements, for instance.

"_" indicates an empty string (with its preceding and following separators). For instance, ___ indicates three empty fields.

"$" indicates that what is left of the current line is to be interpreted as comment (i.e. neglected).

How does one decide in which format a particular problem description is written? The rule is that every SIF file is in fixed format in the beginning. This fixed format can be converted to free format anywhere in the file by inserting the line

```
FREE FORMAT
```

Any line following this declaration is assumed to be in free format, until one encounters the line

```
FIXED FORMAT
```

(after which fixed format is again assumed) or the end of the current SIF file.

We now illustrate our new technique by re-writing a free format variant of the specification for the LSQFIT problem of page 39 as follows:

```
NAME          LSQFIT
FREE FORMAT
$   An elementary constrained linear least-squares fit
RE X1_0.1;RE X2_0.3;RE X3_0.5;RE X4_0.7;RE X5_0.9          $Data
RE Y1_0.25;RE Y2_0.3;RE Y3_0.625;RE Y4_0.701;RE Y5_1.0 $Observed
IE 1_1;IE 5_5;RE C_0.85                                 $Constants
VARIABLES;_a;_b
GROUPS;XL Cons a 1.0 b 1.0                             $Constraint
DO I 1_5;ZN Obj(I) a_X(I);XN Obj(I) b 1.0;XN Obj(I) 'SCALE' 2.0;ND
CONSTANTS;DO I 1 5;Z LSQFIT Obj(I)_Y(I);ND;Z LSQFIT Cons_C
BOUNDS;XR LSQFIT b
GROUP TYPE;GV SQUARE ALPHA
GROUP USES;DO I 1 5;XT Obj(I) SQUARE;ND
ENDATA

GROUPS        LSQFIT
FREE FORMAT
INDIVIDUALS;T SQUARE;F__ALPHA * ALPHA;G__ALPHA + ALPHA;H__2.0
ENDATA
```

This is of course shorter (even discounting the use of loops in this version), but also shows that free format, if not carefully laid out, can make a problem file very much harder to read...

Again, it is hard in general to recommend one of the free or the fixed format in favour of the other. Both have their advantages, the free format being mostly useful for its conciseness. Much may depend on how constrained you feel in the columns oriented fixed format. Also notice that, unfortunately, free format does not allow for extra features like very long names or numbers with 24 digits, which is probably its most limiting feature.

2.8.11 MPS Features Kept for Compatibility

The last paragraph of this section is devoted to the description of certain features of the SDIF that we do not particularly recommend, but that have been kept in the standard for compatibility with MPS.

The first of these features is the existence of alternative equivalent definitions for certain keywords. They are listed in Table 2.10

Keyword used above	Equivalent keywords
GROUPS	CONSTRAINTS
	ROWS
VARIABLES	COLUMNS
CONSTANTS	RHS
	RHS'

Table 2.10: Equivalent keywords

For instance, ROWS can be used anywhere GROUPS appears: they are entirely equivalent.

The second feature is a further facility present in the MPS standard: the ability to define new constraints by linear combination of other constraints. We admit to not liking this technique very much, but it is nevertheless available in SDIF for the sake of compatibility. The idea is rather simple. Assume we have already written the beginning of the GROUPS section as

```
GROUPS
 XL CONS1
 XG CONS2      X         1.0            Y          2.0
 XE CONS3
```

where we have two purely nonlinear groups, the first, CONS1, being of the "less than" type and the second, CONS3, being of the "equals" type. We also have another "greater than" group, named CONS2, in which the variables X and Y appear linearly, and which may also contain nonlinear elements. We then *create a new group by taking a linear combination of two already existing groups*, CONS1 and CONS2 in our case. This is done by writing

```
 DL MESS       CONS1     5.0            CONS2      -1.0
```

in which case the new "less than" constraint `MESS` is defined by

$$\texttt{MESS} = 5 \times \texttt{CONS1} - \texttt{CONS2}$$

The constant associated with this additional constraint is then defined in the `CONSTANTS` section, as for any other group.

The codes `DN`, `DL`, `DG` and `DE` are available to create objective groups, "less than", "greater than" and equality constraints respectively. Of course, all groups whose names appear in fields 3 and 5 of these `D` lines must have been already defined in the `GROUPS` section.

2.8.12 Summary

In this section, we have introduced a number of advanced (?) features present in the SIF. As the features are relatively uncorrelated, it is rather difficult to summarize the whole section. We have been considering

- the definition and use of parameters in group and element functions,
- a remedy for the possible bad scaling of the problem's variables,
- the introduction of two-sided inequality constraints,
- the specification of a priori known bounds on the objective function values,
- the column oriented problem formulation, as opposed to the row oriented formulation,
- the definition and use of external functions,
- a special code (`ND`) for ending all open loops at once,
- ways to initialize the Lagrange multipliers associated with constraints of the problem,
- the free format that allows greater freedom in the layout of the problem file, and, finally,
- some MPS features that have been kept in the SDIF to ensure compatibility between the two standards.

2.9 Some Typical SIF Examples

Although it has been necessary to introduce an exact specification of the allowed input formats, we feel that the best way to learn how to input problems is by studying specific examples. Consequently, we present here three further examples which use many of the salient features of the standard. We hope that you will be encouraged to trying converting your favourite test examples into the standard form.

2.9.1 A Simple Constrained Problem

Firstly, we give an SIF file for the 65th constrained example collected by Hock and Schittkowski [38, p. 87]. The objective function is

$$(x_1 - x_2)^2 + (x_1 + x_2 - 10)^2/9 + (x_3 - 5)^2.$$

There is a single inequality constraint

$$x_1^2 + x_2^2 + x_3^2 \leq 48$$

and additional simple bounds

$$-4.5 \leq x_1 \leq 4.5, \quad -4.5 \leq x_2 \leq 4.5, \quad -5 \leq x_3 \leq 5$$

on the variables. The values $x_1 = -5$, $x_2 = 5$, $x_3 = 0$ give the starting point for the minimization. Noting the use of range transformations, an appropriate SIF file might be:

```
NAME          HS65

*   Number of variables

 IE N                     3

*   Useful parameters

 IE 1                     1
 IE 3                     3
 RE NINE                  9.0
 RD 1/9          NINE     1.0

VARIABLES

    X1
    X2
    X3

GROUPS

*   Objective function

 N  OBJ

*    Constraints functions

 L  CON

CONSTANTS

    HS65         CON      48.0

BOUNDS

 LO HS65         X1       -4.5
 UP HS65         X1       4.5
```

```
 LO HS65      X2        -4.5
 UP HS65      X2        4.5

 LO HS65      X3        -5.0
 UP HS65      X3        5.0

START POINT

    HS65      X1        -5.0            X2        5.0
    HS65      X3        0.0

ELEMENT TYPE

 EV DIFSQR    V1                        V2
 IV DIFSQR    U

 EV SUMSQR    V1                        V2
 IV SUMSQR    U

 EV SQ-5      V
 EV SQ        V

ELEMENT USES

*   Nonlinear elements for the objective function

 T  O1        DIFSQR
 V  O1        V1                        X1
 V  O1        V2                        X2

 T  O2        SUMSQR
 V  O2        V1                        X1
 V  O2        V2                        X2

 T  O3        SQ-5
 V  O3        V                         X3

*   Nonlinear elements for the constraint function

 DO I         1                         3

 XT C(I)      SQ
 ZV C(I)      V                         X(I)

 ND

GROUP USES

 E  OBJ       O1                        O3
 ZE OBJ       O2                        1/9
 E  CON       C1                        C2
 E  CON       C3

ENDATA

*   Specify the form of the different element types

ELEMENTS      HS65

TEMPORARIES

 R  DIF
```

```
INDIVIDUALS

*  square of V, where V = U1 - U2

 T  DIFSQR

 R  U          V1         1.0            V2         -1.0
 F                        U * U
 G  U                     U + U
 H  U          U          2.0

*  square of ( U - 10 ), where U = V1 + V2

 T  SUMSQR

 R  U          V1         1.0            V2         1.0
 A  DIF                   U - 10
 F                        DIF * DIF
 G  U                     DIF + DIF
 H  U          U          2.0

*  square of ( V - 5 )

 T  SQ-5

 A  DIF                   V - 5
 F                        DIF * DIF
 G  V                     DIF + DIF
 H  V          V          2.0

*  square of V

 T  SQ

 F                        V * V
 G  V                     V + V
 H  V          V          2.0

ENDATA

*  Specify the form of the different group types

GROUPS         HS65

*  All groups are trivial

ENDATA
```

2.9.2 A System of Nonlinear Equations with Single-indexed Variables

Here we give an SIF file for the discrete boundary value problem in n variables given by Moré, Garbow and Hillstrom [45, p.27, problem 28]. We set the problem up as a system of nonlinear equations (constraints)

$$2x_i - x_{i-1} - x_{i+1} + \frac{h^2}{2}(x_i + ih + 1)^3 = 0, \quad (i = 1, \ldots, n),$$

where $h = 1/(n+1)$ and $x_0 = x_{n+1} = 0$. We specify $n = 8$ and use the starting point $x_i = ih(ih-1)$ for $i = 1, \ldots, n$. Note, there is no explicit objective function. An appropriate SIF file might be:

```
NAME          BDVLE

*   N is the number of internal discretization points

 IE N                    8

*   Define useful parameters

 IE 0                    0
 IE 1                    1
 IA N+1         N        1

 RI RN+1        N+1
 RD H           RN+1     1.0
 R* H2          H                        H
 RM HALFH2      H2       0.5

VARIABLES

 DO I           0                        N+1

 X  X(I)

 ND

GROUPS

 DO I           1                        N
 IA I-1         I        -1
 IA I+1         I         1

 XE G(I)        X(I-1)   -1.0            X(I)      2.0
 XE G(I)        X(I+1)   -1.0

 ND

BOUNDS

 FR BDVLE       'DEFAULT'
 XX BDVLE       X(0)     0.0
 XX BDVLE       X(N+1)   0.0

START POINT

 X  BDVLE       X(0)     0.0             X(N+1)    0.0

 DO I           1                        N

 RI RI          I
 R* IH          RI                       H
 RA IH-1        IH       -1.0
 R* TI          IH                       IH-1

 Z  BDVLE       X(I)                     TI

 ND
```

```
ELEMENT TYPE

 EV WCUBE     V
 EP WCUBE     B

ELEMENT USES

 DO I         1                          N

 RI REALI     I
 R* IH        REALI                      H
 RA IH+1      IH         1.0

 XT E(I)      WCUBE
 ZV E(I)      V                          X(I)
 ZP E(I)      B                          IH+1

 ND

GROUP USES

 DO I         1                          N

 ZE G(I)      E(I)                       HALFH2

 ND

ENDATA

*   Specify the form of the different element types

ELEMENTS      BDVLE

*   the cube of (V + B)

TEMPORARIES

 R  VPLUSB

INDIVIDUALS

 T  WCUBE
 A  VPLUSB               V + B
 F                       VPLUSB**3
 G  V                    3.0 * VPLUSB**2
 H  V         V          6.0 * VPLUSB

ENDATA

*  Specify the form of the different group types

GROUPS        BDVLE

*  all groups are trivial

ENDATA
```

2.9.3 A Constrained Problem with Triple Indexed Variables

Finally, we give an SIF file for the problem of determining the maximum growth possible when performing Gaussian elimination with complete pivoting (see [33]). Starting with an n by n real matrix $X^{(1)}$, we let $X^{(k)}$ be the matrix which remains to be eliminated after $k-1$ steps of Gaussian elimination without pivoting. Let $x_{i,j,k}$ be the (i,j)-th entry of $X^{(k)}$. We thus wish to maximize $x_{n,n,n}$ subject to the restrictions that the matrices $X^{(k)}$ and $X^{(k+1)}$ are related to each other by elimination restrictions, that the largest element in the bottom $n-k-1$ block of $X^{(k)}$ occurs in position (k,k,k) and that the initial matrix $X^{(1)}$ is scaled so that the largest entry in magnitude is 1. This leads to the problem

$$\text{minimize} - x_{n,n,n}$$

subject to the elimination constraints

$$x_{i,j,k+1} - x_{i,j,k} + \frac{x_{i,k,k} x_{k,j,k}}{x_{k,k,k}} = 0 \text{ for } k \le i \le n, k \le j \le n \text{ and } k = 1, \ldots, n-1,$$

constraints which make the signs of the pivots unique,

$$x_{k,k,k} \ge 0 \text{ for } k = 1, \ldots, n,$$

a normalizing constraint,

$$x_{1,1,1} = 1,$$

and complete pivoting constraints

$$-1 \le x_{i,j,1} \le 1 \text{ for } 1 \le i \le n, 1 \le j \le n$$

and

$$-x_{k,k,k} \le x_{i,j,k} \le x_{k,k,k} \text{ for } k \le i \le n, k \le j \le n \text{ and } k = 2, \ldots, n-1.$$

These details, along with a suitable starting point, are given for the case $n = 6$ in the following SIF file:

```
NAME          GAUSS

*   size of the matrices = n

 IE N                     6

*   other parameter definitions

 IE 1                     1
 IE 2                     2
 IA N-1       N           -1

VARIABLES

 DO K         1                         N
 DO J         K                         N
```

```
 DO I         K                            N

 X  X(I,J,K)

 ND

GROUPS

*  objective function

 DO K         N                            N

 XN OBJ       X(K,K,K)  -1.0

 ND

*   elimination constraints

 DO K         1                            N-1

 IA K+        K          1

 DO I         K+                           N
 DO J         K+                           N

 XE E(I,J,K)  X(I,J,K+) 1.0                X(I,J,K)  -1.0

 ND

*   complete pivoting constraints (submatrices 2 to n-1)

 DO K         2                            N-1
 DO I         K                            N
 DO J         K                            N

 XL M(I,J,K)  X(I,J,K)  1.0                X(K,K,K)  -1.0
 XG P(I,J,K)  X(I,J,K)  1.0                X(K,K,K)  1.0

 ND

BOUNDS

*   default = free variables

 FR GAUSS     'DEFAULT'

*   complete pivoting constraints (submatrix 1)

 DO I         1                            N
 DO J         1                            N

 XL GAUSS     X(I,J,1)  -1.0
 XU GAUSS     X(I,J,1)  1.0

 ND

*   ensure pivotal elements are nonnegative

 DO K         1                            N

 XL GAUSS     X(K,K,K)  0.0
```

```
 ND

*   normalize first pivot

 FX GAUSS     X1,1,1    1.0

START POINT

*   default value for starting point component

 V  GAUSS     'DEFAULT' 0.01

*   Set initial matrices to perturbed identities

 DO K         1                        N
 DO I         K                        N

 X  GAUSS     X(I,I,K)  1.0

 ND

ELEMENT TYPE

 EV ELIM      V1                       V2
 EV ELIM      V3

ELEMENT USES

 DO K         1                        N-1

 IA K+        K         1

 DO I         K+                       N
 DO J         K+                       N

 XT A(I,J,K)  ELIM
 ZV A(I,J,K)  V1                       X(I,K,K)
 ZV A(I,J,K)  V2                       X(K,J,K)
 ZV A(I,J,K)  V3                       X(K,K,K)

 ND

GROUP USES

 DO K         1                        N-1

 IA K+        K         1

 DO I         K+                       N
 DO J         K+                       N

 XE E(I,J,K)  A(I,J,K)

 ND

ENDATA

* Specify the form of the different element types

ELEMENTS      GAUSS

TEMPORARIES
```

```
 R  VALUE
 R  V3SQ

INDIVIDUALS

 T  ELIM

 A  VALUE                  V1 * V2 / V3
 A  V3SQ                   V3 * V3

 F                         VALUE

 G  V1                     V2 / V3
 G  V2                     V1 / V3
 G  V3                     - VALUE / V3

 H  V1        V2           1.0 / V3
 H  V1        V3           - V2 / V3SQ
 H  V2        V3           - V1 / V3SQ
 H  V3        V3           2.0 * VALUE / V3SQ * V3

ENDATA

*  Specify the form of the different group types

GROUPS        GAUSS

* all groups are trivial

ENDATA
```

2.9.4 A Test Problem Collection

A collection of SIF files for many sets of standard test problems is currently being compiled. To date, we have translated, amongst others, the problems given by Moré, Garbow and Hillstrom [45], Toint [52], Moré [44] and Buckley [3]. Work is proceeding on the problems given in the books by Hock and Schittkowski [38] and Schittkowski [50]. It is the authors intention to put the resulting files in the public domain under the name CUTE, for Constrained and Unconstrained Test Examples.

2.10 A Complete Template for SIF Syntax

In this last section of the introduction to the SIF, we finally provide a complete template containing all possible types of valid SIF lines (in fixed format). This template is intended as an aid to allow you to verify that the lines that you are undoubtly starting to write[44] are syntactically correct. This template is included in machine readable form with the LANCELOT distribution tape.

In this template, we use the conventions defined in Table 2.11.

[44] We can hear you!

Name	Object named	Max length (in chars)	Fortran restr.
prb-name	the problem	8	
b-name	a vector of bounds	10	
c-name	a vector of constants	10	
el-name	an element	10	
ep-nam	an elemental parameter	6	y
et-name	an element type	10	
ev-nam	an elemental variable	6	y
f-name	a function (see Table 2.3)	6	y
gt-name	a group type	10	
gp-nam	a group parameter	6	y
gv-nam	the group variable	6	y
gr-name	a group	10	
ip-name	an integer parameter	10	
it-nam	a temporary integer parameter	6	y
iv-nam	an internal variable	6	y
mf-nam	a machine supplied Fortran function	6	y
numerical-vl	a number	12	
lt-name	a temporary logical parameter	6	y
o-name	a vector of bounds on the objective	10	
pv-name	a problem variable	10	
r-name	a vector of ranges	10	
rt-name	a temporary real parameter	6	y
rpa-name	a real array of parameters	10	
rp-name	a real parameter	10	
s-name	a vector of starting point	10	
uf-nam	a user supplied Fortran function	6	

Table 2.11: Naming conventions for the SIF template

```
NAME          prb-name
 IE ip-name             numerical-vl
 IR ip-name   rp-name
 IA ip-name   ip-name   numerical-vl
 IS ip-name   ip-name   numerical-vl
 IM ip-name   ip-name   numerical-vl
 ID ip-name   ip-name   numerical-vl
 I= ip-name   ip-name
 I+ ip-name   ip-name                  ip-name
 I- ip-name   ip-name                  ip-name
 I* ip-name   ip-name                  ip-name
 I/ ip-name   ip-name                  ip-name
 RE rp-name             numerical-vl
 RI rp-name   ip-name
 RA rp-name   rp-name   numerical-vl
 RS rp-name   rp-name   numerical-vl
 RM rp-name   rp-name   numerical-vl
 RD rp-name   rp-name   numerical-vl
 RF rp-name   f-name    numerical-vl
 R= rp-name   rp-name
 R+ rp-name   rp-name                  rp-name
 R- rp-name   rp-name                  rp-name
 R* rp-name   rp-name                  rp-name
 R/ rp-name   rp-name                  rp-name
 R( rp-name   f-name                   rp-name
 AE rpa-name            numerical-vl
 AI rpa-name  ip-name
 AA rpa-name  rpa-name  numerical-vl
 AS rpa-name  rpa-name  numerical-vl
 AM rpa-name  rpa-name  numerical-vl
 AD rpa-name  rpa-name  numerical-vl
 AF rpa-name  f-name    numerical-vl
 A= rpa-name  rpa-name
 A+ rpa-name  rpa-name                 rpa-name
 A- rpa-name  rpa-name                 rpa-name
 A* rpa-name  rpa-name                 rpa-name
 A/ rpa-name  rpa-name                 rpa-name
 A( rpa-name  f-name                   rpa-name
 DO ip-name   ip-name                  ip-name
 DI ip-name   ip-name
 OD ip-name
 ND
GROUPS
ROWS
CONSTRAINTS
 N  gr-name
 N  gr-name   pv-name   numerical-vl
 N  gr-name   pv-name   numerical-vl   pv-name    numerical-vl
 N  gr-name   'SCALE'   numerical-vl
 G  gr-name
 G  gr-name   pv-name   numerical-vl
 G  gr-name   pv-name   numerical-vl   pv-name    numerical-vl
 G  gr-name   'SCALE'   numerical-vl
 L  gr-name
 L  gr-name   pv-name   numerical-vl
 L  gr-name   pv-name   numerical-vl   pv-name    numerical-vl
 L  gr-name   'SCALE'   numerical-vl
 E  gr-name
 E  gr-name   pv-name   numerical-vl
 E  gr-name   pv-name   numerical-vl   pv-name    numerical-vl
 E  gr-name   'SCALE'   numerical-vl
 XN gr-name
```

```
 XN gr-name   pv-name   numerical-vl
 XN gr-name   pv-name   numerical-vl   pv-name     numerical-vl
 XN gr-name   'SCALE'   numerical-vl
 XG gr-name
 XG gr-name   pv-name   numerical-vl
 XG gr-name   pv-name   numerical-vl   pv-name     numerical-vl
 XG gr-name   'SCALE'   numerical-vl
 XL gr-name
 XL gr-name   pv-name   numerical-vl
 XL gr-name   pv-name   numerical-vl   pv-name     numerical-vl
 XL gr-name   'SCALE'   numerical-vl
 XE gr-name
 XE gr-name   pv-name   numerical-vl
 XE gr-name   pv-name   numerical-vl   pv-name     numerical-vl
 XE gr-name   'SCALE'   numerical-vl
 ZN gr-name
 ZN gr-name   pv-name                  rpa-name
 ZN gr-name   'SCALE'                  rpa-name
 ZG gr-name
 ZG gr-name   pv-name                  rpa-name
 ZG gr-name   'SCALE'                  rpa-name
 ZL gr-name
 ZL gr-name   pv-name                  rpa-name
 ZL gr-name   'SCALE'                  rpa-name
 ZE gr-name
 ZE gr-name   pv-name                  rpa-name
 ZE gr-name   'SCALE'                  rpa-name
 DN gr-name   gr-name   numerical-vl
 DN gr-name   gr-name   numerical-vl   gr-name     numerical-vl
 DG gr-name   gr-name   numerical-vl
 DG gr-name   gr-name   numerical-vl   gr-name     numerical-vl
 DL gr-name   gr-name   numerical-vl
 DL gr-name   gr-name   numerical-vl   gr-name     numerical-vl
 DE gr-name   gr-name   numerical-vl
 DE gr-name   gr-name   numerical-vl   gr-name     numerical-vl
VARIABLES
COLUMNS
    pv-name
    pv-name   gr-name   numerical-vl
    pv-name   gr-name   numerical-vl   gr-name     numerical-vl
    pv-name   'SCALE'   numerical-vl
 X  pv-name
 X  pv-name   gr-name   numerical-vl
 X  pv-name   gr-name   numerical-vl   gr-name     numerical-vl
 X  pv-name   'SCALE'   numerical-vl
 Z  pv-name   gr-name                  rpa-name
 Z  pv-name   'SCALE'                  rpa-name
CONSTANTS
RHS
RHS'
    c-name    'DEFAULT' numerical-vl
 X  c-name    'DEFAULT' numerical-vl
 Z  c-name    'DEFAULT'                rpa-name
    c-name    gr-name   numerical-vl
    c-name    gr-name   numerical-vl   gr-name     numerical-vl
 X  c-name    gr-name   numerical-vl
 X  c-name    gr-name   numerical-vl   gr-name     numerical-vl
 Z  c-name    gr-name                  rpa-name
RANGES
    r-name    'DEFAULT' numerical-vl
 X  r-name    'DEFAULT' numerical-vl
 Z  r-name    'DEFAULT'                rpa-name
```

```
   r-name    gr-name   numerical-vl
   r-name    gr-name   numerical-vl   gr-name    numerical-vl
 X r-name    gr-name   numerical-vl
 X r-name    gr-name   numerical-vl   gr-name    numerical-vl
 Z r-name    gr-name                  rpa-name
BOUNDS
 LO b-name   'DEFAULT' numerical-vl
 UP b-name   'DEFAULT' numerical-vl
 FX b-name   'DEFAULT' numerical-vl
 FR b-name   'DEFAULT'
 MI b-name   'DEFAULT'
 PL b-name   'DEFAULT'
 XL b-name   'DEFAULT' numerical-vl
 XU b-name   'DEFAULT' numerical-vl
 XX b-name   'DEFAULT' numerical-vl
 XM b-name   'DEFAULT'
 XP b-name   'DEFAULT'
 ZL b-name   'DEFAULT'                rpa-name
 ZU b-name   'DEFAULT'                rpa-name
 ZX b-name   'DEFAULT'                rpa-name
 LO b-name   pv-name   numerical-vl
 UP b-name   pv-name   numerical-vl
 FX b-name   pv-name   numerical-vl
 FR b-name   pv-name
 MI b-name   pv-name
 PL b-name   pv-name
 XL b-name   pv-name   numerical-vl
 XU b-name   pv-name   numerical-vl
 XX b-name   pv-name   numerical-vl
 XR b-name   pv-name
 XM b-name   pv-name
 XP b-name   pv-name
 ZL b-name   pv-name                  rpa-name
 ZU b-name   pv-name                  rpa-name
 ZX b-name   pv-name                  rpa-name
START POINT
 V  s-name   'DEFAULT' numerical-vl
 XV s-name   'DEFAULT' numerical-vl
 ZV s-name   'DEFAULT'                rpa-name
 M  s-name   'DEFAULT' numerical-vl
 XM s-name   'DEFAULT' numerical-vl
 ZM s-name   'DEFAULT'                rpa-name
    s-name   'DEFAULT' numerical-vl
 X  s-name   'DEFAULT' numerical-vl
 Z  s-name   'DEFAULT'                rpa-name
 V  s-name   pv-name   numerical-vl
 V  s-name   pv-name   numerical-vl   pv-name   numerical-vl
 XV s-name   pv-name   numerical-vl
 XV s-name   pv-name   numerical-vl   pv-name   numerical-vl
 ZV s-name   pv-name                  rpa-name
 M  s-name   pv-name   numerical-vl
 M  s-name   pv-name   numerical-vl   pv-name   numerical-vl
 XM s-name   pv-name   numerical-vl
 XM s-name   pv-name   numerical-vl   pv-name   numerical-vl
 ZM s-name   pv-name                  rpa-name
    s-name   pv-name   numerical-vl
    s-name   pv-name   numerical-vl   pv-name   numerical-vl
    s-name   gr-name   numerical-vl
    s-name   gr-name   numerical-vl   gr-name   numerical-vl
    s-name   pv-name   numerical-vl   gr-name   numerical-vl
    s-name   gr-name   numerical-vl   pv-name   numerical-vl
 X  s-name   pv-name   numerical-vl
```

```
 X  s-name    pv-name   numerical-vl   pv-name   numerical-vl
 X  s-name    gr-name   numerical-vl
 X  s-name    gr-name   numerical-vl   gr-name   numerical-vl
 X  s-name    pv-name   numerical-vl   gr-name   numerical-vl
 X  s-name    gr-name   numerical-vl   pv-name   numerical-vl
 Z  s-name    pv-name                  rpa-name
 Z  s-name    gr-name                  rpa-name
ELEMENT TYPE
 EV et-name   ev-nam
 EV et-name   ev-nam                   ev-nam
 IV et-name   iv-nam
 IV et-name   iv-nam                   iv-nam
 EP et-name   ep-nam
 EP et-name   ep-nam                   ep-nam
ELEMENT USES
 T  'DEFAULT' et-name
 XT 'DEFAULT' et-name
 T  el-name   et-name
 XT el-name   et-name
 V  el-name   ev-nam                   pv-name
 ZV el-name   ev-nam                   pv-name
 P  el-name   ep-nam    numerical-vl
 P  el-name   ep-nam    numerical-vl   ep-name   numerical-vl
 XP el-name   ep-nam    numerical-vl
 XP el-name   ep-nam    numerical-vl   ep-name   numerical-vl
 ZP el-name   ep-nam                   rpa-name
GROUP TYPE
 GV gt-name   gv-nam
 GP gt-name   gp-nam
 GP gt-name   gp-nam                   gp-nam
GROUP USES
 T  'DEFAULT' gt-name
 XT 'DEFAULT' gt-name
 T  gr-name   gt-name
 XT gr-name   gt-name
 E  gr-name   el-name
 E  gr-name   el-name   numerical-vl
 E  gr-name   el-name                  el-name
 E  gr-name   el-name   numerical-vl   el-name
 E  gr-name   el-name   numerical-vl   el-name   numerical-vl
 XE gr-name   el-name                  el-name   numerical-vl
 XE gr-name   el-name
 XE gr-name   el-name   numerical-vl
 XE gr-name   el-name                  el-name
 XE gr-name   el-name   numerical-vl   el-name
 XE gr-name   el-name   numerical-vl   el-name   numerical-vl
 XE gr-name   el-name                  el-name   numerical-vl
 ZE gr-name   el-name                  rpa-name
 P  gr-name   gp-nam    numerical-vl
 P  gr-name   gp-nam    numerical-vl   gp-nam    numerical-vl
 XP gr-name   gp-nam    numerical-vl
 XP gr-name   gp-nam    numerical-vl   gp-nam    numerical-vl
 ZP gr-name   gp-name                  rpa-name
OBJECT BOUND
 LO o-name              numerical-vl
 UP o-name              numerical-vl
 XL o-name              numerical-vl
 XU o-name              numerical-vl
 ZL o-name                             rpa-name
 ZU o-name                             rpa-name
ENDATA
```

```
ELEMENTS        prb-name
TEMPORARIES
 I  it-nam
 R  rt-nam
 L  lt-nam
 M  mf-nam
 F  uf-nam
GLOBALS
 A  it-nam                fortran-expression
 A  rt-nam                fortran-expression
 A  lt-nam                fortran-expression
 A+                       fortran-expression
 I  lt-nam      it-nam    fortran-expression
 I  lt-nam      rt-nam    fortran-expression
 I  lt-nam      lt-nam    fortran-expression
 I+                       fortran-expression
 E  lt-nam      it-nam    fortran-expression
 E  lt-nam      rt-nam    fortran-expression
 E  lt-nam      lt-nam    fortran-expression
 E+                       fortran-expression
INDIVIDUALS
 T  et-name
 R  iv-nam      ev-nam    numerical-vl   ev-nam     numerical-vl
 R  iv-nam      ev-nam    numerical-vl
 A  it-nam                fortran-expression
 A  rt-nam                fortran-expression
 A  lt-nam                fortran-expression
 A+                       fortran-expression
 I  lt-nam      it-nam    fortran-expression
 I  lt-nam      rt-nam    fortran-expression
 I  lt-nam      lt-nam    fortran-expression
 I+                       fortran-expression
 E  lt-nam      it-nam    fortran-expression
 E  lt-nam      rt-nam    fortran-expression
 E  lt-nam      lt-nam    fortran-expression
 E+                       fortran-expression
 F                        fortran-expression
 F+                       fortran-expression
 G  iv-nam                fortran-expression
 G  ev-nam                fortran-expression
 G+                       fortran-expression
 H  iv-nam      iv-nam    fortran-expression
 H  ev-nam      ev-nam    fortran-expression
 H+                       fortran-expression
ENDATA

GROUPS          prb-name
TEMPORARIES
 I  it-nam
 R  rt-nam
 L  lt-nam
 M  mf-nam
 F  uf-nam
GLOBALS
 A  it-nam                fortran-expression
 A  rt-nam                fortran-expression
 A  lt-nam                fortran-expression
 A+                       fortran-expression
 I  lt-nam      it-nam    fortran-expression
 I  lt-nam      rt-nam    fortran-expression
 I  lt-nam      lt-nam    fortran-expression
 I+                       fortran-expression
```

```
 E  lt-nam    it-nam    fortran-expression
 E  lt-nam    rt-nam    fortran-expression
 E  lt-nam    lt-nam    fortran-expression
 E+                     fortran-expression
INDIVIDUALS
 T  et-name
 R  iv-nam    ev-nam    numerical-vl   ev-nam    numerical-vl
 R  iv-nam    ev-nam    numerical-vl
 A  it-nam              fortran-expression
 A  rt-nam              fortran-expression
 A  lt-nam              fortran-expression
 A+                     fortran-expression
 I  lt-nam    it-nam    fortran-expression
 I  lt-nam    rt-nam    fortran-expression
 I  lt-nam    lt-nam    fortran-expression
 I+                     fortran-expression
 E  lt-nam    it-nam    fortran-expression
 E  lt-nam    rt-nam    fortran-expression
 E  lt-nam    lt-nam    fortran-expression
 E+                     fortran-expression
 F                      fortran-expression
 F+                     fortran-expression
 G                      fortran-expression
 G+                     fortran-expression
 H                      fortran-expression
 H+                     fortran-expression
ENDATA
```

Chapter 3. A Comprehensive Description of the Mathematical Algorithms Used in LANCELOT

3.1 Introduction

In this chapter, we describe the numerical methods which lie at the heart of the first release of LANCELOT. At this time, LANCELOT comprises two different optimization algorithms, one for solving the simple-bound constrained minimization problem and the other for the more general nonlinearly constrained problem. We shall give a description of the bound constrained technique, LANCELOT/SBMIN, and of the nonlinearly constrained method, LANCELOT/-AUGLG. We shall refer to these two methods as SBMIN and AUGLG for short.

The convergence of the algorithm at the heart of SBMIN has been analysed in a previous paper [8] and numerical experience with a small-scale prototype of the code was reported by [9]. Further details of a large-scale prototype were given by [7].

The algorithm which forms the basis of AUGLG has been analysed in [12] and [13]. This latter algorithm makes repeated use of the the simple-bound constrained algorithm at the heart of SBMIN.

The convergence properties of both methods are good. The algorithms were developed with the solution of large-scale problems in mind. Arguments in favour of our, as opposed to other, approaches are given by [10] and [12].

All of the algorithms described in this chapter have been implemented in standard Fortran 77 and are designed to be portable. There are both single and double precision versions available. Wherever feasible, the codes are self-contained. However, optional use is made of a few Harwell Subroutine Library codes. If users wish to make use of these options, they will need to obtain them directly from Harwell (see [36]). It is our hope to replace the Harwell codes in future releases.

In Section 3.2 we describe SBMIN in general terms. More details, in particular aspects related to large-scale computation, are given in Section 3.3. A general description of AUGLG follows in Section 3.4. Problem scaling issues are considered in Section 3.5.

3.2 A General Description of SBMIN

LANCELOT/SBMIN is a method for solving the bound-constrained minimization problem,

$$\underset{x \in \Re^n}{\text{minimize}} \quad f(x) \tag{3.2.1}$$

subject to the simple bound constraints

$$l_i \leq x_i \leq u_i, \quad 1 \leq i \leq n. \tag{3.2.2}$$

Here, f is assumed to be twice-continuously differentiable and any of the bounds in (3.2.2) may be infinite. We will denote the vector of first partial derivatives, $\nabla_x f(x)$, by $g(x)$ and the Hessian matrix, $\nabla_{xx} f(x)$, will be denoted by $H(x)$. We shall refer to the set of points which satisfy (3.2.2) as the *feasible box* and any point lying in the feasible box is said to be *feasible.*

SBMIN is an iterative method. At the end of the k-th iteration, an estimate of the solution, $x^{(k)}$, satisfying the simple bounds (3.2.2), is given. The purpose of the $(k+1)$-st iteration is to find a feasible iterate $x^{(k+1)}$ which is a significant improvement on $x^{(k)}$. The algorithm underpinning SBMIN is so designed that, under reasonable assumptions, any limit point of the sequence $\{x^{(k)}\}$ converges to a constrained stationary point of the problem (3.2.1, 3.2.2), see [8].

In the $(k+1)$-st iteration, we build a quadratic model of our (possibly) nonlinear objective function, $f(x)$. This model takes the form

$$m^{(k)}(x) = f(x^{(k)}) + g(x^{(k)})^T(x - x^{(k)}) + \tfrac{1}{2}(x - x^{(k)})^T B^{(k)}(x - x^{(k)}), \tag{3.2.3}$$

where $B^{(k)}$ is a symmetric approximation to the Hessian matrix $H(x^{(k)})$. We also define a scalar $\Delta^{(k)}$, the *trust-region radius*, which defines the *trust region,*

$$\|x - x^{(k)}\| \leq \Delta^{(k)}, \tag{3.2.4}$$

within which we trust that the values of $m^{(k)}(x)$ and $f(x)$ will generally agree. An appropriate range of values for the trust-region radius is accumulated as the minimization proceeds.

The $(k+1)$-st iteration proceeds in a number of stages. These may be summarized, in order, as:

1. Test for convergence, see Section 3.2.1.

2. Find an approximation to the generalized Cauchy point of the quadratic model, within the intersection of the feasible box and the trust region, see Section 3.2.2.

3. Obtain a new point which further reduces the quadratic model within the intersection of the feasible box and the trust region, see Section 3.2.3.

4. Test whether there is a general agreement between the values of the model and true objective function at the new point. If so, accept the new point as the next iterate. Otherwise, retain the existing iterate as the next iterate. Adjust the trust region radius as appropriate. See Section 3.2.4.

Clearly, this is a very sketchy description — indeed, some of the terminology has not even been defined — and we need to elaborate further.

3.2.1 The Test for Convergence

The first-order necessary conditions for a feasible point x^* to solve the problem (3.2.1)–(3.2.2) require that the projected gradient at x^* be zero. The *projected gradient* of $f(x)$ into the feasible box (3.2.2) is defined to be

$$x - P(x - g(x), l, u), \tag{3.2.5}$$

where the projection operator, $P(x, l, u)$ is defined componentwise by

$$P(x, l, u)_i = \begin{cases} l_i & \text{if } x_i < l_i \\ u_i & \text{if } x_i > u_i \\ x_i & \text{otherwise.} \end{cases} \tag{3.2.6}$$

Notice that, in the unconstrained case where all the bounds (3.2.2) are infinite, this is merely the well-known condition that the gradient must vanish at a minimizer.

One may then gauge the convergence of the method by the size of the projected gradient at $x^{(k)}$. For instance, one might stop if the condition

$$\|x^{(k)} - P(x^{(k)} - g(x^{(k)}), l, u)\| \leq \epsilon_g, \tag{3.2.7}$$

holds for some appropriate small convergence tolerance ϵ_g. Alternatively, following [18, p. 160], one might consider the relative projected gradient with i-th component

$$r_i(x) = \frac{(x_i - P(x - g(x), l, u)_i) \max(x_i, x_i^{\text{typical}})}{\max(f(x), f^{\text{typical}})} \tag{3.2.8}$$

for "typical" values x^{typical} and f^{typical} (which may, of course, be sometimes difficult to choose). In this case, it might be appropriate to stop if the condition

$$\|r(x^{(k)})\| \leq \epsilon_r, \tag{3.2.9}$$

is satisfied for some appropriate small convergence tolerance ϵ_r.

3.2.2 The Generalized Cauchy Point

The approximate minimization of the quadratic model (3.2.3) within the intersection of the feasible box and the trust region at the $(k+1)$-st iteration is accomplished in two stages. In the first, we obtain an approximation to the so called generalized *Cauchy point*. This point is important for two reasons. Firstly, convergence of the algorithm to a point at which the projected gradient is zero can be guaranteed provided the value of the quadratic model at the end of the iteration is no larger than that at the generalized Cauchy point (see [8]). Secondly, the set of variables which lie on their bounds at the generalized Cauchy point as the algorithm proceeds often provide an excellent prediction of those which will ultimately be fixed at the solution to the problem. It is thus important that an adequate approximation to such a point be identified. Let $D^{(k)}$ be a positive definite diagonal scaling matrix and let $\bar{g}^{(k)} = D^{(k)} g(x^{(k)})$. Now define the *Cauchy arc*, $x^{(k)}(t)$, by

$$x^{(k)}(t) = P(x^{(k)} - t\bar{g}^{(k)}, l, u) \tag{3.2.10}$$

for all values of the parameter $t \geq 0$. Considering the arc as t increases from zero, the generalized Cauchy point is defined to be $x(t_e)$, where t_e is the first local minimizer of $m^{(k)}(x^{(k)}(t))$, the quadratic model along the arc, subject to the trust region constraint (3.2.4) being satisfied at $x^{(k)}(t)$.

In [8], the convergence of an algorithm, in which the exact value of the generalized Cauchy point is determined, is analysed. An efficient algorithm for this calculation, when $\|.\|$ is the infinity-norm, is given in [9]. We give further details in Section 3.3.6.1.

It is not necessary that the generalized Cauchy point be calculated exactly. Indeed, a number of authors have considered approximations which are sufficient to guarantee convergence (see [4], [5], [6], [43], [53]). We consider the approximation suggested by [43]. Let $\gamma > 0$, $0 < \beta < 1$ and $0 < \sigma < 1$. Then we choose the approximation $x(t_i)$, where t_i is of the form $\gamma\beta^{m_s}$ and where m_s is the smallest nonnegative integer for which

$$m^{(k)}(x^{(k)}(t)) \leq m^{(k)}(x^{(k)}) + \sigma g(x^{(k)})^T (x^{(k)}(t) - x^{(k)}) \tag{3.2.11}$$

and

$$\|x^{(k)}(t) - x^{(k)}\| \leq \beta \Delta^{(k)}. \tag{3.2.12}$$

We shall refer to such a point as an *approximate* generalized Cauchy point. The above algorithm is known as a *backtracking linesearch*. Notice, however, that the generalized Cauchy point may not satisfy equation (3.2.11). Thus there is a possibility of different behaviour from algorithms which use the true and approximate generalized Cauchy points. We consider the approximate generalized Cauchy point further in Section 3.3.6.2.

3.2.3 Beyond the Generalized Cauchy Point

We have ensured that the algorithm will converge by determining a suitable approximation to the generalized Cauchy point. We are now concerned that the algorithm should converge at a reasonable rate. This is achieved by, if necessary, further reducing the quadratic model. Those variables which lie on their bounds at the approximation to the generalized Cauchy point are fixed. Attempts are then made to reduce the quadratic model by changing the values of the remaining free variables. This may be achieved in a number of ways. We concentrate here on one. Let $x^{(k,1)}$ be the obtained approximation to the generalized Cauchy point obtained and let $x^{(k,j)}, j = 2, 3, \ldots$ be distinct points such that:

- $x^{(k,j)}$ lies within the intersection of the feasible box and the trust region;
- those variables which lie on a bound at $x^{(k,1)}$ lie on the same bound at $x^{(k,j)}$;
- $x^{(k,j+1)}$ is constructed from $x^{(k,j)}$ by

  1. determining a nonzero search direction $p^{(k,j)}$ for which
  $$\nabla_x m^{(k)}(x^{(k,j)})^T p^{(k,j)} < 0; \qquad (3.2.13)$$
  2. finding a steplength $\alpha^{(k,j)} > 0$ which minimizes $m^{(k)}(x^{(k,j)} + \alpha p^{(k,j)})$ within the intersection of the feasible box and the trust region; and
  3. setting
  $$x^{(k,j+1)} = x^{(k,j)} + \alpha^{(k,j)} p^{(k,j)}. \qquad (3.2.14)$$

Notice that the one dimensional minimizer of $m^{(k)}(x^{(k,j)} + \alpha p^{(k,j)})$ within the intersection of the feasible box and the trust region is easy to compute. For, let the gradient of the model at $x^{(k,j)}$ be

$$g^{(k,j)} \equiv g(x^{(k)}) + B^{(k)}(x^{(k,j)} - x^{(k)}). \qquad (3.2.15)$$

Then the unconstrained line minimizer of the quadratic model is given by

$$\alpha_0^{(k,j)} = \begin{cases} -g^{(k,j)T} p^{(k,j)} / p^{(k,j)T} B^{(k)} p^{(k,j)} & \text{if } p^{(k,j)T} B^{(k)} p^{(k,j)} > 0, \\ \infty & \text{otherwise.} \end{cases} \qquad (3.2.16)$$

Furthermore, the maximum feasible step allowed by the bound (3.2.2) on the i-th variable is

$$\alpha_i^{(k,j)} = \begin{cases} (l_i - x_i^{(k,j)})/p_i^{(k,j)} & \text{if } p_i^{(k,j)} < 0, \\ (u_i - x_i^{(k,j)})/p_i^{(k,j)} & \text{if } p_i^{(k,j)} > 0, \\ \infty & \text{otherwise.} \end{cases} \qquad (3.2.17)$$

for $1 \le i \le n$. Finally, the trust region imposes a bound $\alpha_{n+1}^{(k,j)} > 0$, where

$$\|x^{(k,j)} + \alpha_{n+1}^{(k,j)} p^{(k,j)}\| = \Delta^{(k)}, \tag{3.2.18}$$

on the step. Therefore the required steplength is

$$\alpha^{(k,j)} = \min_{0 \le i \le n+1} \alpha_i^{(k,j)}. \tag{3.2.19}$$

This process is stopped when the norm of the free gradient of the model at $x^{(k,j)}$ is sufficiently small. The *free gradient of the model* is

$$Q(\nabla_x m^{(k)}(x^{(k,j)}), x^{(k,j)}, l, u), \tag{3.2.20}$$

where the operator

$$Q(y,x,l,u)_i = \begin{cases} y_i & \text{if } l_i < x_i < u_i, \\ 0 & \text{otherwise}, \end{cases} \tag{3.2.21}$$

zeros components of the gradient corresponding to variables which lie on their bounds. It is known (see [41]) that if

$$\|Q(\nabla_x m^{(k)}(x^{(k,j)}), x^{(k,j)}, l, u)\| = o(\|Q(\nabla_x m^{(k)}(x^{(k)}), x^{(k,1)}, l, u)\|), \tag{3.2.22}$$

the convergence rate of the method is asymptotically superlinear. Notice that the free gradient at the final point is measured against the components of the gradient of the model at the original point which were free at the approximation to the generalized Cauchy point. In SBMIN, we aim to stop when

$$\|Q(\nabla_x m^{(k)}(x^{(k,j)}), x^{(k,j)}, l, u)\| \le \|Q(\nabla_x m^{(k)}(x^{(k)}), x^{(k,1)}, l, u)\|^{1.5}. \tag{3.2.23}$$

There is a lot of flexibility in obtaining a search direction which satisfies (3.2.13). One would normally think of determining such a direction by finding an approximation to the minimizer of the quadratic subproblem (3.2.3) where certain of the variables are fixed on their bounds but the constraints on the remaining variables are ignored. Specifically, let $\mathcal{I}^{(k,j)}$ be a set of indices of the variables which are to be fixed, let e_i be the i-th column of the n by n identity matrix I and let $\bar{I}^{(k,j)}$ be the matrix made up of columns e_i, $i \notin \mathcal{I}^{(k,j)}$. Now define

$$\bar{g}^{(k,j)} \equiv \bar{I}^{(k,j)T} g^{(k,j)} \text{ and } \bar{B}^{(k,j)} \equiv \bar{I}^{(k,j)T} B^{(k,j)} \bar{I}^{(k,j)}. \tag{3.2.24}$$

Then the quadratic model (3.2.3) at $x^{(k,j)} + p$, considered as a function of the free variables $\bar{p} \equiv \bar{I}^{(k,j)T} p$ is,

$$\bar{m}^{(k,j)}(\bar{p}) = m^{(k)}(x^{(k,j)}) + \bar{g}^{(k,j)T}\bar{p} + \tfrac{1}{2}\bar{p}^T \bar{B}^{(k,j)}\bar{p}. \tag{3.2.25}$$

We may attempt to minimize (3.2.25) using either a direct or iterative method.

In a direct minimization of (3.2.25), one factorizes the coefficient matrix $\bar{B}^{(k,j)}$. If the factors indicate that the matrix is positive definite, the Newton equations

$$\bar{B}^{(k,j)}\bar{p}^{(k,j)} = -\bar{g}^{(k,j)} \tag{3.2.26}$$

may be solved and the required search direction $p^{(k,j)} = \bar{I}^{(k,j)}\bar{p}^{(k,j)}$ recovered. If, on the other hand, the matrix is merely positive semi-definite, a direction of linear infinite descent (dolid) or a weak solution to the Newton equations can be determined. Finally, if the matrix is truly indefinite, a direction of negative curvature (donc) may be obtained. We consider this further in Section 3.3.8.

In an iterative minimization of (3.2.25), the index set $\mathcal{I}^{(k,j)}$ may stay constant over a number of iterations, while at each iteration the search direction may be calculated from the current model gradient and Hessian $\bar{B}^{(k,j)}$ and previous search directions. A typical iterative method is the method of conjugate gradients. This method is considered in Section 3.3.10. The convergence of such a method may be accelerated by preconditioning, see Sections 3.3.10.1, 3.3.10.4 and 3.3.10.2. In fact the boundary between a good preconditioned iterative method and a direct method is quite blurred.

3.2.4 Accepting the New Point and Other Bookkeeping

A point $x^{(k,j)}$, which gives a sufficient reduction in the quadratic model, has thus been found. Of course, we are really interested in reducing the true objective function $f(x)$, not the model. The success or failure of the iteration may be measured by computing the ratio of the actual reduction in the objective function to that predicted by the model

$$\rho^{(k)} = \frac{f(x^{(k)}) - f(x^{(k,j)})}{m^{(k)}(x^{(k)}) - m^{(k)}(x^{(k,j)})}. \tag{3.2.27}$$

If this ratio is negative or small relative to one, the iteration should be viewed as a failure. We call such an iteration*unsuccessful.* Conversely if the ratio is large, the model predicted a reasonable decrease in the objective function and the iteration has been *successful* (even though, if the ratio is significantly larger than one, the model was not an accurate approximation to f).

Let $0 < \mu < 1$. We choose to update $x^{(k+1)}$ according to the recipe

$$x^{(k+1)} = \begin{cases} x^{(k,j)} & \text{if } \rho^{(k)} > \mu, \\ x^{(k)} & \text{otherwise.} \end{cases} \tag{3.2.28}$$

We note that, as $\mu > 0$, this recipe may prevent the algorithm from accepting the lowest point encountered so far as the new iterate. One could circumvent this drawback by explicitly keeping the overall lowest point found and returning it as the required solution on exit from SBMIN. However, this requires additional storage and has not proved beneficial in our experience.

We also need to update the trust-region radius. Again, if ρ is small or negative, the model has not predicted the behaviour of the true objective function well, while if the ratio is close to one, a good prediction has been obtained. In the former case, the region in which the quadratic approximation is trusted must be reduced and the radius consequently decreased. In the latter case, their may be reason to believe that the region in which we may trust our model is larger than we previously thought and the radius is increased as a consequence.

Again let $\mu < \eta < 1$ and $0 < \gamma_0 < 1 \le \gamma_2$. We chose to update the trust region radius according to the formula

$$\Delta^{(k+1)} = \begin{cases} \gamma_0^{(k)}\Delta^{(k)} & \text{if } \rho^{(k)} \le \mu, \\ \Delta^{(k)} & \text{if } \mu < \rho^{(k)} < \eta, \\ \gamma_2^{(k)}\Delta^{(k)} & \text{otherwise,} \end{cases} \tag{3.2.29}$$

where $\gamma_0^{(k)} \in [\gamma_0, 1)$ and $\gamma_2^{(k)} \in [1, \gamma_2]$. Further details are given in Section 3.3.4.

If the iterate has changed, we need to recompute the gradient at the new point. We may also wish to form a new second derivative approximation.

3.3 A Further Description of SBMIN

The method outlined in Section 3.2 is appropriate regardless of the size of the problem. In order to derive an efficient algorithm for large-scale calculations, we first need to know how to handle the structure typically inherent in functions of many variables.

3.3.1 Group Partial Separability

A function $f(x)$ is said to be *group partially separable* if:

1. the function can be expressed in the form

$$f(x) = \sum_{i=1}^{n_g} g_i(\alpha_i(x)); \tag{3.3.30}$$

2. each of the *group functions* $g_i(\alpha)$ is a twice continuously differentiable function of the single variable α;

3. the function

$$\alpha_i(x) = \sum_{j \in \mathcal{J}_i} w_{i,j} f_j(x^{[j]}) + a_i^T x - b_i \tag{3.3.31}$$

is known as the i-th *group*;

4. each of the index sets $\mathcal{J}_i$ is a subset of $\{1, \ldots, n_e\}$;

5. each of the *nonlinear element functions* f_j is a twice continuously differentiable function of a subset $x^{[j]}$ of the variables x. Each function is assumed to have a large invariant subspace. Usually, this is manifested by $x^{[j]}$ comprising a small fraction of the variables x;

6. the gradient a_i of each of the *linear element functions* $a_i^T x - b_i$ is, in general, sparse; and

7. the $w_{i,j}$ are known as element *weights*.

This structure is extremely general. Indeed, any function with a continuous, sparse Hessian matrix may be written in this form (see [34]). A more thorough introduction to group partial separability is given by [10]. SBMIN assumes that the objective function $f(x)$ is of this form.

3.3.2 Derivatives and their Approximations

One of the main advantages of group partial separable structure is that it considerably simplifies the calculation of derivatives of $f(x)$. If we consider (3.3.30) and (3.3.31), we see that we merely need to supply derivatives of the nonlinear element and group functions. SBMIN then assembles the required gradient and, possibly, Hessian matrix of f from this information.

3.3.2.1 Derivatives of $f(x)$

The gradient of (3.3.30) is given by

$$\nabla_x f(x) = \sum_{i=1}^{n_g} g_i'(\alpha_i(x)) \nabla_x \alpha_i(x), \tag{3.3.32}$$

where the gradient of the i-th group is

$$\nabla_x \alpha_i(x) = \sum_{j \in \mathcal{J}_i} w_{i,j} \nabla_x f_j(x^{[j]}) + a_i. \tag{3.3.33}$$

Similarly, the Hessian matrix of the same function is given by

$$\nabla_{xx} f(x) = \sum_{i=1}^{n_g} g_i''(\alpha_i(x)) \nabla_x \alpha_i(x) (\nabla_x \alpha_i(x))^T + \sum_{i=1}^{n_g} g_i'(\alpha_i(x)) \nabla_{xx} \alpha_i(x), \tag{3.3.34}$$

where the Hessian matrix of the i-th group is

$$\nabla_{xx} \alpha_i(x) = \sum_{j \in \mathcal{J}_i} w_{i,j} \nabla_{xx} f_j(x^{[j]}). \tag{3.3.35}$$

Notice that the Hessian matrix is the sum of two different types of terms. The first is a sum of rank-one terms only involving first derivatives of the nonlinear element functions. The second involves second derivatives of the nonlinear elements.

We shall assume that the first and second derivatives of the group functions are available. This is frequently the case in practice and is assumed by SBMIN.

The quadratic model (3.2.3) uses the gradient of f and a approximation to its Hessian matrix. In SBMIN, we have a couple of options.

- We can calculate the true first and second derivatives of each nonlinear element and group function and use the exact Hessian $B^{(k)} = \nabla_{xx} f(x^{(k)})$.
- We can calculate the true first and second derivatives of each group function, calculate the first derivatives of the nonlinear elements but use approximations, $B_i^{[j](k)}$, to their second derivatives. We then use the approximation

$$B^{(k)} = \sum_{i=1}^{n_g} g_i''(\alpha_i(x^{(k)})) \nabla_x \alpha_i(x^{(k)}) (\nabla_x \alpha_i(x^{(k)}))^T + \sum_{i=1}^{n_g} g_i'(\alpha_i(x^{(k)})) B_i^{(k)}, \tag{3.3.36}$$

where $B_i^{(k)}$ satisfies

$$B_i^{(k)} = \sum_{j \in \mathcal{J}_i} w_{i,j} B^{[j](k)} \tag{3.3.37}$$

for some suitable matrices $B^{[j](k)}$, see Section 3.3.2.3.

We strongly recommend the use of exact second derivatives whenever they are available. SBMIN fully exploits this information. In our experience, exact second derivatives are often available, either by direct calculation or by using automatic differentiation tools.

3.3.2.2 Derivatives of the Group and Nonlinear Element Functions

The derivatives of f_j need only be found with respect to the variables $x^{[j]}$, the remaining derivatives being zero. There are frequently, moreover, two further savings to be made.

Firstly, although the variables used by individual f_j may differ, the structure of many of the nonlinear elements may be the same. For instance, $f_1(x^{[1]})$ might be $x_1 x_2$ and $f_2(x^{[2]})$ might be $x_3 x_4$; both functions are of the generic *type* $e(v_1, v_2) = v_1 v_2$. If we need the derivatives of these two nonlinear elements, all we have to compute are the derivatives of the generic function $e(v_1, v_2)$ and to associate the variables x_1 and x_2 and x_3 and x_4 with v_1

and v_2, respectively (A more realistic example is described by [10]). In general, we associate each nonlinear element function f_j, $1 \leq j \leq n_e$ with a generic type of nonlinear function from the set $\{e_m(v^{[m]})\}$, $1 \leq m \leq m_e$, via a suitable mapping $m(j)$. We then only need to know the derivatives of the different types of nonlinear element functions and remember which variables $x^{[j]}$ are associated with the relevant *elemental variables* $v^{[m]}$. Secondly, computing the derivatives of a given type of nonlinear element function may be further simplified. For example, suppose that a nonlinear element is of type $e(v_1, v_2, v_3) = ((v_1 + v_2)(v_2 - v_3))^2$. Although e is a function of three elemental variables, it only really depends on two internal variables $u_1 = v_1 + v_2$ and $u_2 = v_2 - v_3$; in this case, $e(v_1, v_2, v_3) = \hat{e}(u_1, u_2) = (u_1 u_2)^2$. The nonlinear information is still conveyed in $\hat{e}$. The derivatives with respect to the variables v may be calculated by computing the derivatives with respect to the variables u and then using the chain rule of partial differentiation to recover the required ones.

In general, we might obtain *internal* variables $u^{[m]}$ from the elemental variables $v^{[m]}$ by the linear *range* transformation

$$u^{[m]} = W^{[m]} v^{[m]}. \tag{3.3.38}$$

The transformation is only *useful* if the transformation matrix $W^{[m]}$ has fewer rows than columns. Note that the transformation is far from unique and, in many cases, there is no useful transformation (in which case the trivial transformation, with $W^{[m]} = I$, suffices). The gradient and Hessian matrix of a nonlinear element type $e_m(v^{[m]}) \equiv \hat{e}_m(u^{[m]})$, with respect to the elemental variables, may be calculated from those with respect to the internal variables as

$$\nabla_v e_m(v^{[m]}) = W^{[m]T} \nabla_u \hat{e}_m(u^{[m]}) \tag{3.3.39}$$

and

$$\nabla_{vv} e_m(v^{[m]}) = W^{[m]T} \nabla_{uu} \hat{e}_m(u^{[m]}) W^{[m]}. \tag{3.3.40}$$

Thus, in order to calculate these derivatives, we have merely to calculate the derivatives with respect to the internal variables and record the range transformations made. The remaining part of the calculation is automated within SBMIN. Now consider derivatives of the group functions. The first point that we made above, that many functions with different arguments are of the same generic type, applies equally to the group functions $g_i(\alpha)$. Thus we associate each group function g_i, $1 \leq i \leq n_g$ with a generic type of group function from the set $\{h_\ell(\alpha)\}$, $1 \leq \ell \leq m_g$, via a suitable mapping $\ell(i)$. We then only need to provide the derivatives of the generic group types.

3.3.2.3 Approximating the Derivatives of Nonlinear Element Functions

We see in (3.3.39) that the nonlinear information in the first derivatives of the nonlinear element function $e_m(v^{[m]}) \equiv \hat{e}_m(u^{[m]})$ is conveyed by the vector $\nabla_u \hat{e}_m(u^{[m]})$. If we seek an approximation $g^{[j](k)}$ to $\nabla_x f_j(x^{[j](k)})$ in (3.3.33), we should first determine the type of nonlinear element j, $m \equiv m(j)$ say. Now we approximate $\nabla_u \hat{e}_m(x^{[j](k)})$ by a suitable vector $\hat{g}^{[j](k)}$ and form

$$g^{[j](k)} = W^{[m]T} \hat{g}^{[j](k)}. \tag{3.3.41}$$

The approximations $\hat{g}^{[j](k)}$ would normally be formed by finite differences. As the dimension of $u^{[m]}$ is assumed to be small (f_j has a large invariant subspace), this is realistic even for large problems.

A finite difference approximation to the i-th component of the required gradient is usually calculated by forward differences

$$\frac{\hat{e}_m(x^{[j](k)} + \delta_i e_i) - \hat{e}_m(x^{[j](k)})}{\delta_i} \tag{3.3.42}$$

or central differences

$$\frac{\hat{e}_m(x^{[j](k)} + \delta_i e_i) - \hat{e}_m(x^{[j](k)} - \delta_i e_i)}{2\delta_i} \tag{3.3.43}$$

for some suitable small positive number δ_i. In SBMIN, the less expensive forward differences are prefered until the projected gradient of the objective function is small when a switch is made to central differences. Further details of such procedures are given by [18] and [31]. Turning to second derivatives, we see in (3.3.40) that the nonlinear information in the second derivatives of the nonlinear element function $e_m(v^{[m]}) \equiv \hat{e}_m(u^{[m]})$ is contained in the matrix $\nabla_{uu} \hat{e}_m(u^{[m]})$. As before, if we seek an approximation $B^{[j](k)}$ to $\nabla_{xx} f_j(x^{[j](k)})$ in (3.3.37), we should first determine the type of nonlinear element j, $m \equiv m(j)$ say. Now we approximate its Hessian $\nabla_{uu} \hat{e}_m(x^{[j](k)})$ by a suitable matrix $\hat{B}^{[j](k)}$ and form

$$B^{[j](k)} = W^{[m]T} \hat{B}^{[j](k)} W^{[m]}. \tag{3.3.44}$$

The approximations $\hat{B}^{[j](k)}$ might be formed by

1. finite differences; or

2. secant updating formulae, using information from previous iterations,

As the dimension of $u^{[m]}$ is assumed to be small (f_j has a large invariant subspace), both of these options are realistic even for large problems. This technique is known as partitioned updating [35].

A finite difference approximation is usually calculated as $\frac{1}{2}(B + B^T)$, where the i-th column of B is

$$\frac{\nabla_u \hat{e}_m(x^{[j](k)} + \delta_i e_i) - \nabla_u \hat{e}_m(x^{[j](k)})}{\delta_i} \tag{3.3.45}$$

for some suitable small positive number δ_i. Further details of this procedure are given by [18] and [31].

A secant approximation is any matrix chosen to satisfy the secant equation

$$\hat{B}^{[j](k)} \hat{s}^{[j](k)} = \hat{y}^{[j](k)}, \tag{3.3.46}$$

where

$$\hat{s}^{[j](k)} = W^{[m]}(x^{[j](k)} - x^{[j](k-1)}) \tag{3.3.47}$$

and

$$\hat{y}^{[j](k)} = \nabla_u \hat{e}_m(x^{[j](k)}) - \nabla_u \hat{e}_m(x^{[j](k-1)}). \tag{3.3.48}$$

The matrix is normally chosen to be symmetric. Sometimes, positive definiteness is also imposed, although there is little justification for this in our circumstances. Examples of general secant updates are the Powell-symmetric-Broyden (PSB) and symmetric-rank-one (SR1) updates. Positive definite updates include the Broyden-Fletcher-Goldfarb-Shanno (BFGS) and Davidon-Fletcher-Powell (DFP) updates. See [18], [23] and [31] for further details.

One could, of course, think of approximating $B^{[j](k)}$ directly. However, the reader is cautioned that both finite difference and secant methods may be adversely affected by such an approximation. With finite differences, there are two distinct drawbacks. Firstly, more gradient evaluations will be required if we approximate $B^{[j](k)}$ rather than the smaller $\hat{B}^{[j](k)}$. Secondly, ill-conditioning in $B^{[j](k)}$ may be entirely due to poor-conditioning of $W^{[m]}$. In this case, approximating $\hat{B}^{[j](k)}$ by differences is likely to be more accurate. On the other hand, secant updating formulae may be inappropriate when approximating singular matrices. Note that (3.3.44) implies that $B^{[j](k)}$ is itself singular whenever the range transformation is useful. Consequently the convergence of SBMIN may be severely impaired when this transformation is neglected.

3.3.3 Required Data Structures

In Section 3.3.1, we saw that group partial separability is a useful structure to consider when dealing with large problems. In this section, we consider a data structure which is sufficient to pass this structure to a minimization program, such as LANCELOT. In the following discussion, we indicate the relevant arguments of the subroutine SBMIN (see Chapter 8) in parantheses.

We need the following:

- The number of variables, n (`N`), the number of group functions, n_g (`NG`), and the number of nonlinear element functions, n_e (`NEL`).

- The number of distinct types of group functions, m_g, and of nonlinear elements, m_e.

- An integer array giving the type, $\ell(i)$, $1 \leq i \leq n_g$, for each group function.

- An integer array (`IELING`) holding the elements of the sets $\mathcal{J}_i$, $1 \leq i \leq n_g$, each set appearing as a contiguous block, with the elements in $\mathcal{J}_i$ appearing directly before those for in $\mathcal{J}_{i+1}$. A second integer array (`ISTADG`) then gives pointers to the start of the elements in each $\mathcal{J}_i$ in the first array.

- A real array holding the weights, $w_{i,j}$, associated with the nonlinear elements in each group (`ESCALE`), the weights from the i-th group appearing as a contiguous block in the order they are indexed in $\mathcal{J}_i$, $1 \leq i \leq n_g$, with those from group i appearing directly before those from group $i+1$.

- An integer array giving the type, $m(j)$, $1 \leq j \leq n_e$, for each nonlinear element function.

- An integer array (`IELVAR`) holding the indices of the variables $x^{[j]}$, $1 \leq j \leq n_e$, each set of indices appearing as a contiguous block, with those from $x^{[j]}$ appearing directly before those from $x^{[j+1]}$. A second integer (`ISTAEV`) array then gives pointers to the start of the indices from each $x^{[j]}$ in the first array.

- An integer array holding (`ICNA`) the indices of the variables which appear (have nonzero coefficients) in the i-th linear element, $1 \leq i \leq n_e$, together with a chained real array (`A`) holding the associated coefficients of a_i. The indices and coefficients from a single element appear in a contiguous block, with those from the i-th linear element appearing directly before those from the $i+1$-st linear element. A second integer array (`ISTADA`) then gives pointers to the start of the indices of the variables (and the associated coefficients) from each linear element in the real and first integer arrays.

- A real array (`B`) containing the constant terms b_i from each group.

- A logical array (`INTREP`) indicating which of the different types of nonlinear element uses a nontrivial range transformation between elemental and internal variables.

Furthermore, subroutines which calculate the function and derivative values of the different types of group and nonlinear element functions, along with the nontrivial range transformations, are required.

3.3.4 The Trust Region

There is a lot of flexibility in the trust-region updating rule (3.2.29). Many trust region methods make the simple choices $\gamma_0^{(k)} = \gamma_0$ and $\gamma_2^{(k)} = \gamma_2$ at every iteration. In SBMIN, there are a few refinements which we have found quite useful. Firstly, although we may increase the radius by a factor of as much as γ_2 when $\rho^{(k)} \geq \eta$, there is little point in doing so if the step $\|x^{(k,j)} - x^{(k)}\|$ is small relative to the current radius. We really need to give the step an opportunity to grow, not the radius. With this in mind, we choose

$$\gamma_2^{(k)} = \max\left[1, \gamma_2 \frac{\|x^{(k,j)} - x^{(k)}\|}{\Delta^{(k)}}\right] \tag{3.3.49}$$

whenever $\rho^{(k)} \geq \eta$.

Secondly, if there is little agreement between the model and the real functions at the trial point $x^{(k,j)}$, and the current step is small relative to the trust-region radius, there is little point in reducing the radius by less than the amount needed to exclude $x^{(k,j)}$ from the new trust region. We thus pick

$$\gamma_0^{(k)} = \max\left[\gamma_0, \gamma_1 \frac{\|x^{(k,j)} - x^{(k)}\|}{\Delta^{(k)}}\right] \tag{3.3.50}$$

for some $\gamma_0 \leq \gamma_1 < 1$, whenever $0 \leq \rho^{(k)} \leq \mu$. If $\rho^{(k)} < 0$, the agreement between the model and real functions is poor and more drastic action is needed. In this case, we consider how large a radius would allow us to make a very successful step — a step for which $\rho^{(k)} \geq \eta$ — if the step were in the direction we have just attempted to move and if f were a quadratic. A simple calculation reveals that we should choose $\gamma_0^{(k)}$ as

$$\max\left[\gamma_0, \frac{(1-\eta) g(x^{(k)})^T (x^{(k,j)} - x^{(k)})}{(1-\eta)(f(x^{(k)}) + g(x^{(k)})^T (x^{(k,j)} - x^{(k)})) + \eta m^{(k)}(x^{(k,j)}) - f(x^{(k,j)})}\right] \tag{3.3.51}$$

whenever $\rho^{(k)} < 0$. It is straightforward to show that such a choice satisfies the inequality $\gamma_0^{(k)} \in [\gamma_0, 1)$.

We choose $\gamma_0 = 0.0625$, $\gamma_1 = 0.5$, $\gamma_2 = 2$, $\mu = 0.25$ and, $\eta = 0.75$ in SBMIN.

A good choice of the initial trust-region radius is sometimes difficult and may be left to the user. As an alternative, choosing the initial radius to be proportional to $\|g(x^{(0)})\|$ often proves satisfactory. Experiments with more sophisticated tests, in the case where a lower bound on f within the feasible region is known *ab initio*, proved less reliable. In particular trying to estimate the maximum step allowable if f were linear often did not help matters.

We thus allow users to specify an initial trust-region radius and fall back on the value

$$\Delta^{(0)} = 0.1\|g(x^{(0)})\| \tag{3.3.52}$$

within SBMIN if they decline.

We realize that the mechanism for contracting the trust-region radius is more sophisticated than that for expanding it. Consequently, we feel that it is appropriate to choose the initial radius on the large side rather than conservatively small.

Another consequence of this sophistication is that the generalized Cauchy point in an unsuccessful iteration is frequently excluded from the trust region at the next iteration. Thus a new Cauchy point must be calculated and it does not pay to retain information about the existing one. We had hoped to avoid computing the new Cauchy point whenever this was possible but our experience has been that this saving is rarely achieved.

3.3.5 Matrix-Vector Products

In the course of an iteration of SBMIN, we will need to calculate the value and derivatives of the model (3.2.3) at a number of points. This requires that we form the product of the matrix $B^{(k)}$ with a specified vector, in this case $x - x^{(k)}$. It turns out that, in most cases, we never need to form the matrix $B^{(k)}$, merely to be able to form matrix-vector products using it (the exception to this is when we use a direct method, or a preconditioned iterative method based on either a factorization or an incomplete factorization of $B^{(k)}$, to improve upon the Cauchy point, see Sections 3.3.8 and 3.3.10.4).

If we examine (3.3.33), (3.3.34), (3.3.35), (3.3.36), (3.3.39)–(3.3.40) and (3.3.44) we see that rather than requiring the explicit matrices $W^{[m]}$, we only need a method for forming the matrix-vector product between $W^{[m]}$, and its transpose, and a given vector. In SBMIN, the user is asked to write a subroutine to achieve this. We shall call this subroutine **RANGES**.

As we remarked earlier, $B^{(k)}$ is represented as the sum of two terms, one being a sum of rank-one matrices,

$$\sum_{i=1}^{n_g} g_i''(\alpha_i(x^{(k)}))\nabla_x \alpha_i(x^{(k)})(\nabla_x \alpha_i(x^{(k)}))^T, \tag{3.3.53}$$

and the other being a weighted sum of low rank element Hessian approximations,

$$\sum_{i=1}^{n_g} g_i'(\alpha_i(x^{(k)})) \sum_{j \in \mathcal{J}_i} w_{i,j} B^{[j](k)}. \tag{3.3.54}$$

When forming the matrix vector product $B^{(k)}v$, it is important to differentiate between the case when v is "super-sparse", that is v has a very small number of nonzeros, and when it is not worthwhile to exploit the structure of v. We shall see, in Section 3.3.6.1, that products involving super-sparse vectors frequently occur. The main benefit of super-sparsity is that many of the terms in the products of (3.3.53) and (3.3.54) with v vanish. Specifically, if a group does not involve any variable whose index corresponds to a nonzero of v, there is no contribution to the product from that group. In SBMIN, a list of the groups which use each variable in turn is maintained. This list is then consulted for each nonzero in v, in turn, to decide which terms from (3.3.53) and (3.3.54) to include.

Let $J(x)$ denote the Jacobian matrix of the vector of group functions. That is, $J(x)^T$ is the matrix whose i-th column is $\nabla_x \alpha_i(x)$ for $1 \leq i \leq n_g$. We shall refer to $J(x)$ as the *group Jacobian* for short. The matrix-vector product between (3.3.53) and the vector v can then be written as

$$J(x^{(k)})^T G''(x^{(k)}) J(x^{(k)}) v \tag{3.3.55}$$

where $G''(x)$ is the diagonal matrix whose i-th diagonal entry is $g_i''(\alpha_i(x))$. As matrix-vector products of the form (3.3.55) may be required many times in a single iteration of SBMIN, it is normally worth assembling the group Jacobian from its constituent parts and storing it within a suitable sparse data structure. The product (3.3.55) may be decomposed into three parts; one involving a matrix-vector product with J, one with its transpose and a third with the diagonal matrix G''.

If v is super-sparse, we only need access to the columns of J corresponding to the nonzeros in v when forming the product $J(x^{(k)})v$. We thus store J by columns, that is, we store an array of integer-real pairs, where the integer records the row index and the real holds the value of each nonzero of J, the entries in column j immediately preceding those from column $j+1$. A second integer array is used to hold pointers to the starting position in the array of integer-real pairs of the entries for each column in turn. Note that, although the numerical values of each may change at every iteration, the row indices are calculated just once at the start of the minimization.

Although the product $J(x^{(k)})v$ is likely to be denser than v, it is frequently not much so. Thus the vector $G''(x^{(k)})J(x^{(k)})v$ is often also super-sparse. As we form $J(x^{(k)})v$, we make a list of its nonzeros. Using this list, we now form the nonzeros of $G''(x^{(k)})J(x^{(k)})v$. To complete the product (3.3.55) as efficiently as possible, we now need access to the columns of $J(x)^T$ (i.e., the rows of $J(x)$) corresponding to the nonzeros in $G''(x^{(k)})J(x^{(k)})v$. To this end, we use a partial row-wise storage scheme where we store the column indices of J by row (but not the corresponding values). We also keep an integer array which gives the position in the column-wise storage scheme of J of each entry in the partial row-wise scheme. Thus we can obtain the numerical value of each entry in the row-wise scheme by looking up its address in the column-wise scheme. Notice, once again, that these integer arrays need only be set once at the start of the minimization. Using the partial row-wise scheme, we may now complete the product (3.3.55).

The product between (3.3.54) and a super-sparse vector is performed in a more straightforward fashion. The nonzeros of v are considered one at a time. For the first, the nonlinear elements which use the variable are marked. The product between the element Hessian matrices $B^{[j](k)}$ and their elemental variables for the marked variables are formed, multiplied by their appropriate weights and the first derivatives of the relevant group function and added to the required vector. The same procedure is followed for the second and subsequent nonzeros in v except that now, nonlinear elements that have already been used are excluded. When there is a nontrivial range transformation, $W^{[m]}$, between elemental and internal variables, the product between (3.3.44) and the components of $v^{[m]}$ needs to be performed in three stages. The subroutine RANGES will have to be called twice and a single product between the small dense matrix $\hat{B}^{[j](k)}$ and a small vector obtained.

One drawback of this approach is that the RANGE subroutine may be called many times in a single matrix-vector product. As there is often a significant overhead with subroutine calls, it may sometimes pay not to transform to internal variables even though a suitable transformation is available. This is especially true when exact second derivatives are used. This overhead may be removed with many modern compilers where subroutines may be put "in-line".

3.3.6 The Cauchy Point

As every iteration of SBMIN requires that we find some approximation to the generalized Cauchy point, we try to ensure that this calculation is as efficient as possible.

3.3.6.1 The Exact Cauchy Point

An algorithm for finding the generalized Cauchy point is given in [9, Section 3]. In essence, the Cauchy arc (3.2.10) is followed from $x^{(k)}$ until the model value starts to rise. The arc is continuous and piecewise linear. The breakpoints which divide the linear segments occur when further increasing the parameter t would violate a bound on one or more variables. These variables are thereafter fixed at their bounds. On each linear segment, the model is quadratic and, consequently, its behaviour is easy to predict. The slope and curvature along the arc are usually discontinuous at the breakpoints but these values may be efficiently calculated. Rather than calculating the slope and curvature afresh at each breakpoint, the existing values are updated from their previous values using a simple recurrence relationship. The recurrence requires the product of $B^{(k)}$ with a super-sparse vector, whose nonzeros occur in positions corresponding to variables which become fixed to bounds at the breakpoint.

To summarize, we must calculate the breakpoints and investigate the behaviour of the model between successive pairs of breakpoints until the model starts to increase. In SBMIN, the breakpoints are calculated and then sorted in increasing order using the Heapsort algorithm of [54]. Heapsort is appropriate here as we do not know *a priori* how many segments we will need to search before we find the generalized Cauchy point. The product of $B^{(k)}$ with a super-sparse vector v, required when calculating the slope and curvature of the model in each segment, is computed as described in Section 3.3.5.

3.3.6.2 An Approximate Cauchy Point

The calculation of a suitable approximation to the generalized Cauchy point, described in Section 3.2.2, is very different from the calculation of the exact point. Rather than starting at $x^{(k)}$ and moving along the arc, we now start at some point on the arc and retrace our steps back towards $x^{(k)}$. The scheme described in Section 3.2.2 may jump over many breakpoints between successive evaluations of the model. There is, therefore, not the same scope for updating the model as there is when the exact generalized Cauchy point is calculated. In particular, the calculation of the model usually requires a matrix vector product between $B^{(k)}$ and a vector which is not super-sparse. There is thus a trade off between a more expensive matrix-vector product and the ability to leap many breakpoints at once. We now refine the algorithm described in Section 3.2.2 a little. In the latter stages of a run of SBMIN, we would often expect that the variables which are fixed on bounds at the solution would have been identified and thus that the generalized Cauchy point lies between $x^{(k)}$ and the first breakpoint along the Cauchy arc. Therefore, we always look in this interval for the generalized Cauchy point and only start the backtracking linesearch if the interval does not contain the required point.

In SBMIN, we select $\sigma = 0.1$ for the test (3.2.11). The backtracking linesearch is started from the smaller of the last breakpoint on the arc and the step to the line minimizer of the quadratic model if no breakpoints had interfered. A similar starting value has been proposed by [46]. We realize that this is somewhat arbitrary but this strategy has performed reasonably in practice.

3.3.7 Beyond the Cauchy Point

In Section 3.2.3, we gave a general framework for obtaining a new iterate that is "better" than the generalized Cauchy point. At each stage, an approximation to the minimizer of the model is sought while some of the variables are held fixed at bounds. This set of fixed variables, $\mathcal{I}^{(k,j)}$, always includes those which were fixed at the approximation to the generalized Cauchy point. In SBMIN, we also include all variables which encounter bounds at $x^{(k,j)}$, for $j > 0$ up until the test (3.2.23) is satisfied. Then, optionally, we may free all variables except those which were fixed at the approximation to the generalized Cauchy point and perform one or more further cycles. This optional process is terminated when releasing variables does nothing to improve the model value and is detected when (3.2.23) and

$$Q(\nabla_x m^{(k)}(x^{(k,j)}), x^{(k,j)}, l, u) = Q(\nabla_x m^{(k)}(x^{(k,j)}), x^{(k,1)}, l, u) \qquad (3.3.56)$$

are satisfied. We might also put "special steps" into each cycle with the view of improving convergence. For instance, at the start of each cycle, we could compute a new generalized Cauchy point for the model fixing the variables which were on a bound at the original Cauchy point. This recursive use of SBMIN is guaranteed to satisfy (3.2.23) if a sufficient number of cycles are performed.

Notice that a single cycle is sufficient to guarantee asymptotic superlinear convergence of the method. However, on degenerate problems, we have found that additional cycles sometimes help in reducing the overall number of outer iterations, k, of the method. We allow the use of both a single cycle (known as an inexact solution to the model problem) and multiple cycles interleaved with extra Cauchy steps (an exact solution) within SBMIN.

So far, we have ignored the influence of the trust region. When the trust region boundary is encountered, we might stop. However, if we are using an infinity-norm trust region and encounter the boundary of this region, we might equally fix the variables on the boundary and continue.

3.3.8 Beyond the Cauchy Point - Direct Methods

Having determined the set $\mathcal{I}^{(k,j)}$, we remarked in Section 3.2.3 that we should now find an approximation to the minimizer of (3.2.25). The nature of the quadratic model restricted to the subset of free variables is determined by the eigenvalue distribution (or inertia) of the matrix $\bar{B}^{(k,j)}$. To summarize:

- If all the eigenvalues of $\bar{B}^{(k,j)}$ are strictly positive, the unique minimizer of (3.2.25) is given as the solution to the Newton equations (3.2.26).

- If all the eigenvalues of $\bar{B}^{(k,j)}$ are nonnegative, but some are zero, and $\bar{g}^{(k,j)}$ lies in the range of $\bar{B}^{(k,j)}$, there are an infinite number of solutions to the Newton equations (3.2.26). Each solution is a weak minimizer of the model.

- If all the eigenvalues of $\bar{B}^{(k,j)}$ are nonnegative, but some are zero, while $\bar{g}^{(k,j)}$ does not lie in the range of $\bar{B}^{(k,j)}$, the quadratic model is unbounded from below. There is then a vector $\bar{p}^{(k,j)}$ for which $\bar{g}^{(k,j)T}\bar{p}^{(k,j)} < 0$ and $\bar{B}^{(k,j)}\bar{p}^{(k,j)} = 0$. This vector is known as a *direction of linear infinite descent* (dolid) as the model decreases as a linear function of α for steps $\alpha\bar{p}^{(k,j)}$ as α increases.

- If $\bar{B}^{(k,j)}$ has negative eigenvalues, the model is unbounded from below. There is a vector $\bar{p}^{(k,j)}$ for which $\bar{g}^{(k,j)T}\bar{p}^{(k,j)} \leq 0$ and $\bar{p}^{(k,j)T}\bar{B}^{(k,j)}\bar{p}^{(k,j)} < 0$. This vector is known as a *direction of negative curvature* (donc). The model decreases as a quadratic function of α for steps $\alpha\bar{p}^{(k,j)}$ as α increases.

The use of a sparse multifrontal direct method to solve large-scale optimization problems has been advocated by [7]. Briefly, the matrix $\bar{B}^{(k,j)}$ is factorized using the Harwell Subroutine Library code MA27 [21], [22] as

$$\bar{B}^{(k,j)} = \bar{\Pi}^{(k,j)} \bar{L}^{(k,j)} \bar{D}^{(k,j)} \bar{L}^{(k,j)T} \bar{\Pi}^{(k,j)T}, \tag{3.3.57}$$

where $\bar{\Pi}^{(k,j)}$ is a permutation matrix, $\bar{L}^{(k,j)}$ is unit lower triangular and $\bar{D}^{(k,j)}$ is block diagonal with 1 by 1 and 2 by 2 diagonal blocks. The inertia of $\bar{B}^{(k,j)}$ and $\bar{D}^{(k,j)}$ are identical. The factorization is used to compute a search direction $\bar{p}^{(k,j)}$ appropriate for any of the four cases mentioned above.

A key factor is that the Newton direction is always chosen if $\bar{B}^{(k,j)}$ is positive definite. An alternative direct method which shares this property is based on the modified Cholesky methods of [28] and [51]. Here, we form a factorization

$$\bar{B}^{(k,j)} + \bar{E}^{(k,j)} = \bar{L}^{(k,j)} \bar{D}^{(k,j)} \bar{L}^{(k,j)T}, \tag{3.3.58}$$

where $\bar{L}^{(k,j)}$ is unit lower triangular, $\bar{D}^{(k,j)}$ is positive definite and diagonal, and $\bar{E}^{(k,j)}$ is positive semi-definite, diagonal and nonzero only when $\bar{B}^{(k,j)}$ is not (sufficiently) positive definite. It is straightforward to modify MA27 to achieve this factorization. Now, the modified Newton equations,

$$(\bar{B}^{(k,j)} + \bar{E}^{(k,j)})\bar{p}^{(k,j)} = -\bar{g}^{(k,j)} \tag{3.3.59}$$

are solved to obtain a suitable search direction. Notice, furthermore, that

$$\begin{aligned} \bar{m}^{(k,j)}(\bar{p}^{(k,j)}) &\leq m^{(k)}(x^{(k,j)}) + \bar{g}^{(k,j)T}\bar{p}^{(k,j)} + \tfrac{1}{2}\bar{p}^{(k,j)T}(\bar{B}^{(k,j)} + \bar{E}^{(k,j)})\bar{p}^{(k,j)} \\ &\leq \bar{m}^{(k,j)}(\bar{p}^{(k,1)}). \end{aligned} \tag{3.3.60}$$

We stress that an advantage of these techniques is that $B^{(k)}$ will typically not be modified as we approach the solution to the problem. Moreover, provided the trust-region radius is sufficiently large that the Newton step (3.2.26) may be taken, we would also expect to take very few (indeed, in the nondegenerate case, one) inner-iterations before (3.2.23) is satisfied.

3.3.9 Sequences of Closely Related Problems

We have seen that consecutive index sets $\mathcal{I}^{(k,j)}$ and $\mathcal{I}^{(k,j+1)}$ are often closely related. Typically, $\mathcal{I}^{(k,j+1)}$ will contain the elements of $\mathcal{I}^{(k,j)}$ and the indices of variables which hit bounds in the $j+1$-st inner-iteration.

As it may be relatively expensive to calculate a fresh factorization of $\bar{B}^{(k,j)}$ at each stage, we consider an alternative. This alternative is based on the work of [1], [29] and [30].

Suppose that inner iterations j_1 and $j > j_1$ belong to the same cycle (see Section 3.2.3). Suppose furthermore that $\bar{B}^{(k,j_1)}$ is positive definite, that we have factorized it, and we now wish to solve (3.2.26) at the j-th inner iteration. The set $\mathcal{I}^{(k,j)}$ differs from $\mathcal{I}^{(k,j_1)}$ by the indices of variables that have hit bounds since moving from $x^{(k,j_1)}$. Let the $n^{(k,j_1)}$ free variables at $x^{(k,j_1)}$ be $\bar{x} \equiv \bar{I}^{(k,j_1)T}x$. Now define $\mathcal{F}^{(k,j)}$ to be the indices of those free variables which have been fixed by the start of j-th inner iteration. Furthermore, let e_i be the i-th column of the $n^{(k,j_1)}$ by $n^{(k,j_1)}$ identity matrix and define $\hat{I}^{(k,j)}$ and $\tilde{I}^{(k,j)}$ as the matrices made up of columns e_i, $i \in \mathcal{F}^{(k,j)}$ and $i \notin \mathcal{F}^{(k,j)}$, respectively. Then the Newton equations (3.2.26) may be rewritten as the augmented system

$$\begin{pmatrix} \bar{B}^{(k,j_1)} & \hat{I}^{(k,j)} \\ \hat{I}^{(k,j)T} & 0 \end{pmatrix} \begin{pmatrix} \tilde{p}^{(k,j)} \\ \tilde{q}^{(k,j)} \end{pmatrix} = - \begin{pmatrix} \tilde{g}^{(k,j)} \\ 0 \end{pmatrix} \tag{3.3.61}$$

where $\tilde{g}^{(k,j)} = \tilde{I}^{(k,j)} \bar{g}^{(k,j)}$, $\tilde{q}^{(k,j)}$ is an auxiliary vector and the required solution is recovered as $\bar{p}^{(k,j)} = \tilde{I}^{(k,j)T} \tilde{p}^{(k,j)}$. The significant advantage of (3.3.61) is that it may be solved just knowing factorizations of $\bar{B}^{(k,j_1)}$ and the symmetric Schur complement matrix

$$\bar{S}^{(k,j)} \equiv \hat{I}^{(k,j)T} \bar{B}^{(k,j_1)-1} \hat{I}^{(k,j)}. \tag{3.3.62}$$

For (3.3.61) may be rearranged so that

$$\bar{S}^{(k,j)} \tilde{q}^{(k,j)} = -\hat{I}^{(k,j)T} \bar{B}^{(k,j_1)-1} \tilde{g}^{(k,j)} \tag{3.3.63}$$

and

$$\bar{B}^{(k,j_1)} \tilde{p}^{(k,j)} = -\tilde{g}^{(k,j)} - \hat{I}^{(k,j)} \tilde{q}^{(k,j)}. \tag{3.3.64}$$

Provided $\mathcal{F}^{(k,j)}$ is small, the Schur complement will also be small and may be stored and factorized as a dense matrix. Notice furthermore that $\bar{S}^{(k,j)}$ expands as j increases but that the expansion at each inner iteration involves merely the addition of a few extra rows and columns. The factorization of the Schur complement may therefore be updated rather than recomputed at inner iteration j giving considerable computational savings. As $\bar{S}^{(k,j)}$ is symmetric and positive definite, a Cholesky factorization is appropriate. See [31, Section 2.2.5.2], for further details. (A subroutine for performing the Schur complement update, in these and in more general circumstances, may also be found as MA39 in the Harwell Subroutine Library.)

If an inner iteration cycle starts at iteration j_s, $\bar{B}^{(k,j_s)}$ is factorized. As we have seen above, in the case $j_1 = j_s$, we may solve the related Newton equations (3.2.26) at all inner iterations $j_s \leq j \leq j_e$ for which there is room to store the Schur complement matrix. More importantly, it is worth our while doing so provided it takes less time to update the Schur complement and solve the Newton equations via (3.3.63) and (3.3.64) than it would if we were to factorize $\bar{B}^{(k,j)}$ and solve (3.2.26) at inner iteration j_e. In SBMIN, estimates of these times are obtained and the Schur complement update is always preferred when it is more efficient and there is sufficient storage. When this ceases to be the case, $\bar{B}^{(k,j)}$ is refactorized and j_1 set to the current inner iteration counter j.

We have assumed that $\bar{B}^{(k,1)}$, and thus $\bar{B}^{(k,j)}$ for $j > 1$, is positive definite. This stops us from using the Schur complement method when $\bar{B}^{(k,1)}$

is indefinite. We might, however, choose to modify the first $\bar{B}^{(k,j)}$ of each cycle according to the formula (3.3.58) and to use this in place of $\bar{B}^{(k,j_1)}$ in the previous discussion. It is important to observe, however, that if we do this, the search directions we generate will not necessarily be the same that were generated if the methods of Section 3.3.8 are used directly. In particular, if $\bar{B}^{(k,j_s)}$ is not positive definite, and consequently modified, while $\bar{B}^{(k,j)}$ is positive definite for some $j_s \leq j \leq j_e$, there is certainly no guarantee that the modification to $\bar{B}^{(k,j_s)}$ does not filter through into $\bar{S}^{(k,j)}$.

3.3.10 Beyond the Cauchy Point - Iterative Methods

There is often a very fine distinction between iterative and direct methods. In fact, in finite precision arithmetic, one is often advised to perform iterative refinement with direct methods to obtain accurate solutions. However, iterative methods are in general more flexible in the sense that it may be impossible to use direct methods as there is insufficient storage on a user's machine, while iterative methods can be adapted to use whatever space is available. This flexibility has its drawbacks as well. In particular, the convergence of iterative methods on difficult problems can be severely impaired unless considerable care is taken.

In SBMIN, the iterative method of choice is the method of conjugate gradients (see, for example, [31, Section 4.8.3], or [32, Sections 10.2 and 10.3]. Such a method attempts to find a stationary point of a quadratic function, in our case (3.2.25), by generating a sequence of (conjugate) search directions, $\bar{p}^{(k,j)}$. If $\bar{B}^{(k,j)}$ is not positive definite, the conjugate gradient may terminate with a donc or dolid.

The convergence of the conjugate gradient method may be enhanced by preconditioning the coefficient matrix $\bar{B}^{(k,j)}$. A preconditioner is a symmetric, positive definite matrix $\bar{P}^{(k,j)}$ which is chosen to make the eigenvalues of the product $\bar{P}^{(k,j)-1}\bar{B}^{(k,j)}$ cluster around as few distinct values as possible. If $\bar{B}^{(k,j)}$ were positive definite, the ideal preconditioner would be $\bar{B}^{(k,j)}$ itself. However, we have to bear in mind that at each step of the preconditioned conjugate gradient method we have to solve sets of linear equations of the form

$$\bar{P}^{(k,j)}\bar{z}^{(k,j)} = -\bar{r}^{(k,j)}, \tag{3.3.65}$$

for given vectors $\bar{r}^{(k,j)}$ and required solutions $\bar{z}^{(k,j)}$. Thus there is normally a compromise between using a good approximation of $\bar{B}^{(k,j)}$, with the associated difficulties of finding and storing its factorization, and a poor approximation, where many conjugate gradient iterations may be required. Choosing a good preconditioner for a given problem is considered to be an art. Certain classes of problems, in particular those associated with fluid flows, have been much analysed and reasonable preconditioners designed.

In SBMIN, we have tried to supply a reasonable cross-section of widely used preconditioners. We recognize that users may have a better idea of a good preconditioner for their problem by allowing them to solve (3.3.65) outside SBMIN.

3.3.10.1 Diagonal Preconditioners

Certainly the simplest preconditioner of all is the choice $\bar{P}^{(k,j)} = I$. A far more useful preconditioner is often provided by the choice $\bar{P}^{(k,j)} = \bar{D}^{(k,j)}$, where $\bar{D}^{(k,j)}$ is a diagonal matrix whose i-th diagonal is

$$\bar{D}_{ii}^{(k,j)} = \max(\bar{B}_{ii}^{(k,j)}, \epsilon), \tag{3.3.66}$$

for some suitable small number ϵ. In SBMIN, ϵ is set to the cube root of the machine precision.

3.3.10.2 Band Preconditioners

Many application areas give rise to problems whose Hessian matrices are banded. A band matrix is a matrix B for which $b_{ij} = 0$ for all $|i - j| \leq m_b$. The smallest integer m_b for which this is so is known as the semi-bandwidth of the matrix. The significant property as far as we are concerned is that, if B is positive definite, the Cholesky factors fit within the band. Moreover, clever storage schemes have been constructed to make the factorization and subsequent solutions extremely efficient (see, for example, [27, Chapter 4], and [20, Section 10.2]).

We offer a band preconditioner within SBMIN. This works in two stages. The desired semi-bandwidth, m_b, is assumed to have been specified. Firstly, the band matrix $\bar{M}^{(k,j)}$, with semi-bandwidth m_b, is chosen so that

$$\bar{M}_{il}^{(k,j)} = \bar{B}_{il}^{(k,j)} \text{ for all } |i - l| \leq m_b. \tag{3.3.67}$$

Secondly, we obtain a modified Cholesky factorization of $\bar{M}_{il}^{(k,j)}$, just as described in Section 3.3.8. That is we form a factorization

$$\bar{P}^{(k,j)} \equiv \bar{M}^{(k,j)} + \bar{E}^{(k,j)} = \bar{L}^{(k,j)}\bar{D}^{(k,j)}\bar{L}^{(k,j)T}, \tag{3.3.68}$$

where $\bar{L}^{(k,j)}$ is unit lower triangular with semi-bandwidth m_b, $\bar{D}^{(k,j)}$ is positive definite and diagonal, and $\bar{E}^{(k,j)}$ is positive semi-definite, diagonal and nonzero only when $\bar{M}^{(k,j)}$ is not (sufficiently) positive definite.

When $\bar{B}^{(k,j)}$ is positive definite and m_b is chosen large enough, $\bar{P}^{(k,j)}$ is an exact preconditioner, that is, the preconditioned conjugate gradient method will converge in a single iteration. The effect of the preconditioner in other cases is unknown. However, one would suspect that if the essential part of $\bar{B}^{(k,j)}$ is contained within the band, the preconditioner will be successful.

3.3.10.3 Incomplete Factorization Preconditioners

In some applications, although it is possible to store the matrix $\bar{B}^{(k,j)}$, there is so much fill-in during the factorization (3.3.57) that it proves impossible to store the resulting factor $\bar{L}^{(k,j)}$. Such cases arise frequently if the underlying problem has connections with two and, particularly, three dimensional partial differential equations.

It is sometimes possible to construct good preconditioners for such problems by either rejecting any fill-in in the factors at all or by tolerating a modest amount. Such incomplete factorization preconditioners are very popular with researchers in partial differential equations and it is possible to get off-the-shelf software to form them. We include the example MA31, due to [47], from the Harwell Subroutine Library in SBMIN.

3.3.10.4 Full-Matrix Preconditioners

Finally, as we alluded to in Section 3.3.10, if space permits, one can always use a complete factorization of $\bar{B}^{(k,j)}$ as a preconditioner. This is only possible if $\bar{B}^{(k,j)}$ is positive definite, but it is of course also possible to use one of the modifications suggested in Section 3.3.8 as a preconditioner. In fact, we allow the modification (3.3.58) within SBMIN.

3.3.10.5 Expanding Band Preconditioners

One further possibility is to use an expanding band preconditioner. Consider the band matrix $\bar{M}^{(k,j)}$ given by (3.3.67), where the semi-bandwidth m_b satisfies the inequalities $0 \leq m_{bmin} \leq m_b \leq m_{bmax} \leq n$. Here m_{bmax} is some upper bound on the allowable bandwidth chosen so that a sparse factorization of $\bar{M}^{(k,j)}$ is practicable. The lower bound m_{bmin} is selected so that the band matrix with exactly that semi-bandwidth provides a useful preconditioner. (Of course, both of these bounds are difficult, if not impossible, to pick *a priori* and one might just choose $m_{bmin} = 0$ and $m_{bmax} = n$.) The idea is now to select the semi-bandwidth $m_b^{(k)}$ at each iteration to reflect the speed and accuracy which one wants from the preconditioned conjugate gradient method. In particular, if low accuracy is required, a preconditioner with a small semi-bandwidth (such as a diagonal preconditioner) is often very effective. But if high accuracy is desired, it may be better to pick a preconditioner which is a good approximation to $\bar{B}^{(k,j)}$.

Having obtained the preconditioner, we obtain a modified Cholesky factorization of $\bar{M}_{il}^{(k,j)}$, just as described in Section 3.3.8. However, unlike the band method described in Section 3.3.10.2, the matrix and its factorization are stored as a general sparse, rather than band, matrix.

In SBMIN, we use the following very simple recipe to select the semi-bandwidth. Pick

$$m_b^{(k)} = \begin{cases} m_{bmax} & \text{if } \|x^{(k)} - P(x^{(k)} - g(x^{(k)}), l, u)\| \leq 10^{-2}, \\ m_{bmax}/2 & \text{if } 10^{-2} < \|x^{(k)} - P(x^{(k)} - g(x^{(k)}), l, u)\| \leq 10^{-1}, \\ m_{bmax}/5 & \text{otherwise.} \end{cases} \tag{3.3.69}$$

We realize that further sophistication may be necessary but have found that this simple scheme is effective in practice. We normally choose $m_{bmax} = n$.

3.3.10.6 Sequences of Closely Related Preconditioners

In Section 3.3.9, we saw that economies may be made when sequences of closely related problems are solved. If a sequence of inner iterations are performed within any of these iterative methods, related equations of the form (3.3.65) may be solved many times. We may therefore use the Schur complement method described in Section 3.3.9 as an alternative to forming and factorizing a new preconditioner at each inner iteration at which the free variables change. Such a scheme is used extensively within SBMIN.

3.3.11 Assembling Matrices

Whether we use a direct method or a preconditioned conjugate gradient method to move beyond the generalized Cauchy point, we need to assemble part or all of the matrix $B^{(k)}$ before we can factorize it. (Strictly, this need not be the case. The Harwell Subroutine Library code MA27 is a multifrontal method. Such methods were originally designed for finite element problems and are capable of working with unassembled element matrices including those of the form (3.3.36). However, MA27 does not have this feature. It is intended that its successor, MA47, will have (J. K. Reid, private communication, 1990). Assembly is performed in the obvious way using the relationship (3.3.36).

3.3.12 Reverse Communication

SBMIN makes extensive use of reverse communication to obtain function and derivative information for the nonlinear elements and nontrivial groups at trial points. This frees the user from having to specify function evaluation routines with fixed parameter lists and makes it easier to embed SBMIN within other packages. In particular, it makes it easy to call SBMIN from within AUGLG and ultimately from within the whole LANCELOT package.

3.4 A General Description of AUGLG

LANCELOT/AUGLG is a method for solving the generally-constrained minimization problem,

$$\underset{x \in \Re^n}{\text{minimize}} \quad f(x) \tag{3.4.70}$$

subject to the general (possibly nonlinear) constraints

$$c_j(x) = 0, \quad 1 \le j \le m, \tag{3.4.71}$$

and the simple bounds

$$l_i \le x_i \le u_i, \quad 1 \le i \le n. \tag{3.4.72}$$

Here, f and the c_j are all assumed to be twice-continuously differentiable and any of the bounds in (3.4.72) may be infinite.

There is no loss in assuming that all the general constraints are equations, as inequality constraints may easily be transformed to equations by the addition of extra slack or surplus variables (see, for example, [31, Section 5.6]). Indeed, LANCELOT automatically transforms inequality constraints to equations. This technique is extensively used in simplex-like methods for large-scale linear and nonlinear programs. We note that barrier-type methods avoid introducing additional variables by treating inequality constraints explicitly (see, for example [31, Section 6.2.1.2]).

AUGLG makes repeated use of SBMIN. The objective function and general constraints are combined into a composite function, the *augmented Lagrangian function*,

$$\Phi(x, \lambda, S, \mu) = f(x) + \sum_{i=1}^{m} \lambda_i c_i(x) + \frac{1}{2\mu} \sum_{i=1}^{m} s_{ii} c_i(x)^2, \tag{3.4.73}$$

where the components λ_i of the vector λ are known as *Lagrange multiplier estimates*, the entries s_{ii} of the diagonal matrix S are positive scaling factors, and μ is known as the *penalty parameter*.

The constrained minimization problem (3.4.70)–(3.4.72) is now solved by finding approximate minimizers of Φ for a carefully constructed sequence of Lagrange multiplier estimates, constraint scaling factors and penalty parameters.

The $k+1$-st major iteration of AUGLG is made up of three steps. At the start of the iteration, Lagrange multiplier estimates, $\lambda^{(k)}$, constraint scaling factors, $S^{(k)}$, and a penalty parameter $\mu^{(k)}$ are given. The steps performed may be summarized, in order, as follows:

1. Test for convergence, see Section 3.4.1.

2. Use SBMIN to find an approximate minimizer, $x^{(k+1)}$, of the augmented Lagrangian function $\Phi(x, \lambda^{(k)}, S^{(k)}, \mu^{(k)})$ in the feasible box, (3.4.72), see Section 3.4.2.

3. Update the Lagrange multiplier estimates, constraint scaling factors, penalty parameter and convergence tolerances, see Section 3.4.3.

We now consider these steps in more detail.

3.4.1 Convergence of the Augmented Lagrangian Method

The first-order necessary conditions for a feasible point x^* to solve the problem (3.4.70)–(3.4.72) require that there are *Lagrange multipliers*, λ^*, for which the projected gradient of the Lagrangian function at x^* and λ^* and the general constraints (3.4.71) at x^* vanish. The *Lagrangian function* is the function

$$L(x, \lambda) = f(x) + \sum_{i=1}^{m} \lambda_i c_i(x). \tag{3.4.74}$$

The projected gradient is given by (3.2.5) but now with the gradient of the Lagrangian function, $\nabla_x L(x, \lambda)$, replacing the gradient of f.

One may then assess the convergence of the augmented Lagrangian method by the size of the projected gradient and constraints at $x^{(k)}$ and $\lambda^{(k)}$. For instance, one might stop if the conditions

$$\|x^{(k)} - P(x^{(k)} - \nabla_x L(x^{(k)}, \lambda^{(k)}), l, u)\| \leq \epsilon_l, \tag{3.4.75}$$

and

$$\|c(x^{(k)})\| \leq \epsilon_c, \tag{3.4.76}$$

hold for some appropriate small convergence tolerances ϵ_l and ϵ_c.

Alternatively, one might consider the values of the relative constraint functions, $c^{\text{relative}}(x)$, where

$$c_i^{\text{relative}}(x) = c_i(x)/c_i^{\text{typical}} \tag{3.4.77}$$

for "typical" values c^{typical}, and the relative projected gradient of the Lagrangian and stop when both of these quantities are small in norm.

3.4.2 Minimizing the Augmented Lagrangian Function

The convergence of augmented Lagrangian methods is guaranteed, under very weak assumptions, if the penalty parameter is gradually reduced to zero, almost regardless of the values of the Lagrange multiplier estimates (see [12, Lemma 4.3 and Theorem 4.4]). The Lagrange multiplier estimates may even diverge provided that $\mu^{(k)}\|\lambda^{(k)}\|$ converges to zero). However, it becomes more difficult to minimize (3.4.73) when $\mu^{(k)}$ is small. Fortunately, a judicious choice of the Lagrange multiplier estimates also ensures convergence for *fixed* μ provided $x^{(k)}$ is close to x^*. Thus $\mu^{(k)}$ is allowed to decrease until we are sure that we are in a neighbourhood of x^* from which point $\mu^{(k)}$ is left unchanged but the Lagrange multipliers adjusted to ensure ultimate convergence. We can gauge whether we are in such a neighbourhood by monitoring the expected decrease in $\|c(x^{(k)})\|$.

At each iteration, we exit from SBMIN when the condition

$$\|x^{(k+1)} - P(x^{(k+1)} - \nabla_x \Phi(x^{(k+1)}, \lambda^{(k)}, S^{(k)}, \mu^{(k)}), l, u)\| \leq \omega^{(k)} \quad (3.4.78)$$

is satisfied for some tolerance $\omega^{(k)}$. We then test whether

$$\|c(x^{(k+1)})\| \leq \eta^{(k)} \quad (3.4.79)$$

holds, for some other tolerance $\eta^{(k)}$. If (3.4.79) is satisfied, we leave the penalty parameter unchanged but update the Lagrange multiplier estimates. Otherwise, we reduce the penalty parameter while leaving the Lagrange multiplier estimates as they are.

3.4.3 Updates

It is straightforward to pick $\mu^{(k)}$, $\lambda^{(k)}$, $\omega^{(k)}$ and $\eta^{(k)}$ to ensure convergence of the above scheme. Moreover this can normally be done so as to guarantee that the penalty parameter remains bounded away from zero (see [12, Theorem 5.5]).

We start by selecting positive tolerances μ_0, ω_0, η_0, τ, α_ω, β_ω, α_η and β_η satisfying

$$\mu_0 < 1, \ \alpha_\eta < \min(1, \alpha_\omega) \text{ and } \beta_\eta < \min(1, \beta_\omega). \quad (3.4.80)$$

In AUGLG, we choose $\tau = 0.1$, $\alpha_\omega = 1$, $\beta_\omega = 1$, $\alpha_\eta = 0.1$ and $\beta_\eta = 0.9$. Now, we set

$$\begin{aligned} \mu^{(0)} &= \mu_0 \\ \omega^{(0)} &= \omega_0 \left(\mu^{(0)}\right)^{\alpha_\omega} \\ \eta^{(0)} &= \eta_0 \left(\mu^{(0)}\right)^{\alpha_\eta}. \end{aligned} \quad (3.4.81)$$

We also pick initial Lagrange multiplier estimates $\lambda^{(0)}$ and scaling matrix $S^{(0)}$. In the absence of a better choice, AUGLG selects $\lambda^{(0)} = 0$ and $S^{(0)} = I$.

The parameters are updated in different ways depending whether or not (3.4.79) is satisfied. When (3.4.79) holds, the next vector of Lagrange multiplier estimates is chosen as

$$\lambda^{(k+1)} = \lambda^{(k)} + S^{(k)} c(x^{(k)}) / \mu^{(k)}; \quad (3.4.82)$$

these are often known as first order multiplier estimates. The remaining parameters are chosen as

$$\begin{aligned} \mu^{(k+1)} &= \mu^{(k)} \\ \omega^{(k+1)} &= \omega^{(k)} \left(\mu^{(k+1)}\right)^{\beta_\omega} \\ \eta^{(k+1)} &= \eta^{(k)} \left(\mu^{(k+1)}\right)^{\beta_\eta}. \end{aligned} \tag{3.4.83}$$

When (3.4.79) does not hold, the penalty parameter is reduced. We set

$$\begin{aligned} \mu^{(k+1)} &= \tau\mu^{(k)} \\ \omega^{(k+1)} &= \omega_0 \left(\mu^{(k+1)}\right)^{\alpha_\omega} \\ \eta^{(k+1)} &= \eta_0 \left(\mu^{(k+1)}\right)^{\alpha_\eta}. \end{aligned} \tag{3.4.84}$$

The Lagrange multiplier estimates are left unchanged so that $\lambda^{(k+1)} = \lambda^{(k)}$.

Notice that, if (3.4.82) holds,

$$\nabla_x \Phi(x^{(k+1)}, \lambda^{(k)}, S^{(k)}, \mu^{(k)}) \equiv \nabla_x L(x^{(k+1)}, \lambda^{(k+1)}). \tag{3.4.85}$$

In this case, the overall convergence tests (3.4.75) and (3.4.76) are satisfied at the start of iteration $k+1$ provided that $\omega^{(k)} \leq \epsilon_l$ and $\eta^{(k)} \leq \epsilon_c$. Automatic choices of the initial penalty parameter in penalty function methods are notoriously hard to justify. On the other hand, the choice is less critical if the constraints are well scaled. By default, AUGLG selects $\mu_0 = 0.1$; this may be overruled by the user. We also choose $\omega_0 = 1$ and $\eta_0 = 0.1258925$.

3.4.4 Structural Consideration for AUGLG

The method outlined in Section 3.4 is appropriate regardless of the size of the problem. In order to derive an efficient algorithm for large-scale calculations, we again need to recognize the structure inherent in functions of many variables. However, because the overall problem (3.4.70)–(3.4.72) is more complicated than (3.2.1)–(3.2.2), and because we wish to use SBMIN as a "black box" solver at each iteration of AUGLG, we have to reduce the allowed complexity of each of the functions c_j. With this in mind, we only allow each of the problem functions in (3.4.71) to have a single group.

3.5 Constraint and Variable Scaling

In almost all that we have said up until now, there has been an implicit assumption that the values of the problem variables are all typically the same size. Similarly, it has been assumed that all the constraint functions are of similar magnitude. It may happen in practice that this is not so. The convergence of the method may be adversely affected by poor scaling and it should be avoided if at all possible. There are two alternatives. A user may manually rescale the variables and constraints so that the scaled quantities are of roughly the same size. Alternatively, the user might rely on an automatic rescaling algorithm.

There has been a fair amount written on scaling (see, for example, [31, Section 8.7] and [18, Section 7.1], and there is some consensus that it is extremely difficult to design a general purpose automatic scheme, especially for highly nonlinear problems. Notwithstanding, we still feel that such a scheme should be provided as an option.

Both AUGLG and SBMIN allow the user to specify variable and constraint scalings as input parameters and the scalings are then used implicitly by the codes. We may thus construct automatic scalings independent of the minimization routines. We proceed as follows.

Consider the vector valued function

$$v(x) = \begin{pmatrix} c_1(x) \\ \vdots \\ c_m(x) \\ f(x) \end{pmatrix}. \tag{3.5.86}$$

Let $F(x)$ denote the Jacobian matrix, $F_{ij}(x) = \delta v_i(x)/\delta x_j$. This matrix reflects the changes in the elements of v which are likely for small identical changes in x. If we were to change to variables $\hat{x} = D_x x$ and rescale v as $\hat{v} = D_v v$, where D_x and D_v are positive definite and diagonal, the Jacobian matrix of $\hat{v}$ with respect to the variables $\hat{x}$ is

$$D_v F(x) D_x^{-1}. \tag{3.5.87}$$

Ideally, we would like to choose the scalings so that the rows and columns of (3.5.87) are of roughly equal norm for all x in the feasible box (3.2.2). However, this is in general impossible for nonlinear functions and we must accept a compromise.

Let $x^{typical}$ be a typical value of x within the feasible box. We now apply the matrix equilibration algorithm of [15] to $F(x^{typical})$ to derive suitable scaling matrices D_x and D_v to equilibrate (3.5.87). The first m diagonals of D_v are rescaled by the constant $1/\max_{1\leq i\leq m}(D_v)_{ii}$. The resulting matrices D_x and the first m components of the rescaled D_v are now passed as input parameters to the minimizer. The Curtis-Reid algorithm is implemented as MC19 in the Harwell Subroutine Library. This automatic scaling procedure is available as an option within LANCELOT.

Chapter 4. The LANCELOT Specification File

Up until now we have given details of the algorithms used within the package but have not yet indicated how the user is able to select specific options. This is the purpose of the present chapter.

4.1 Keywords

LANCELOT operates under the control of a method-specification, or spec, file, usually called SPEC.SPC[1] This file is read when LANCELOT is first called and controls a number of attributes of the optimizer used.

A method specification file comprises a collection of keywords. Each keyword must occur on a separate line. The position on the line is unimportant. Blank lines are ignored. Only the first 8 characters of the keyword are significant and the remainder may be ignored. Any keyword may appear as a mixture of lower and upper case characters. Some keywords have associated integer or real data; this data must follow the keyword on the same line and the two items must be separated by one or more blanks. Each keyword specified overrides or reinforces a default definition. Contradictions are resolved by the last mentioned keyword taking precedence.

Many of the concepts used here are explained in the methods section (Section 3), where detailed references are given.

4.1.1 The Start and End of the File

keywords:

`BEGIN` or `BEGIN-SPECIFICATION`
`END` or `END-SPECIFICATION`

A method specification file must start with the word `BEGIN` or `BEGIN-SPECIFICATION`. Likewise, a spec file must be terminated by the word `END` or `END-SPECIFICATION`. The remaining keywords must lie between these two words.

[1] On VM/CMS systems, the file will be SPEC SPC.

4.1.2 Minimizer or Maximizer?

keyword:

```
MAXIMIZER-SOUGHT
```

It is assumed that the user wishes to find the minimum value of the given objective function. If, instead, the maximum value is desired, the keyword `MAXIMIZER-SOUGHT` should be specified.

4.1.3 Output Required

keywords:

```
PRINT-LEVEL (integer)
START-PRINTING-AT-ITERATION (integer)
STOP-PRINTING-AT-ITERATION (integer)
ITERATIONS-BETWEEN-PRINTING (integer)
```

These words control the amount of output that is required at each iteration of the optimization procedure. The `PRINT-LEVEL` keyword indicates the level of printing required. If no printing is required at all, the integer value following the keyword should be set non-positive. A single line summary of a specific iteration results if the integer is set to 1; this summary is expanded slightly when the integer is set to 2. Details of the current estimate of the optimum value are printed if the integer is larger than 2 and an increasing level of output is available by selecting larger integer values. A full debugging print occurs when the integer is set larger than 100.

The printing starts when the iteration specified on the `START-PRINTING--AT-ITERATION` iteration is reached and stops after the `STOP-PRINTING-AT--ITERATION` iteration is passed. Printing occurs at regular intervals. The exact interval may be specified using the `ITERATIONS-BETWEEN-PRINTING` option. The default settings are

```
PRINT-LEVEL 1
START-PRINTING-AT-ITERATION 0
STOP-PRINTING-AT-ITERATION 100
ITERATIONS-BETWEEN-PRINTING 1
```

and the printing is sent to the standard output channel.

4.1.4 The Number of Iterations Allowed

keyword:

```
MAXIMUM-NUMBER-OF-ITERATIONS (integer)
```

The optimization will terminate after a specified number of iterations unless convergence has already taken place. This uppper limit is specified using the `MAXIMUM-NUMBER-OF-ITERATIONS` keyword, the limit following the keyword. A default setting of

```
MAXIMUM-NUMBER-OF-ITERATIONS 100
```

is used.

4.1.5 Saving Intermediate Data

keyword:

```
SAVE-DATA-EVERY (integer)
```

The values of the problem's variables, the values of the Lagrange multipliers associated with the problem's constraints (if any) and the value of the penalty parameter (if meaningful) are always saved in the `SAVEDATA.d`[2] file (see page 149) at the end of optimization. This data can also be saved at regular intervals (expressed as SBMIN iterations) at the user's request. This is specified using the `SAVE-DATA-EVERY` keyword, with the desired interval following the keyword. A zero value means that the data is only saved at the end of the optimization. A default setting of

```
SAVE-DATA-EVERY 0
```

is used.

4.1.6 Checking Derivatives

keywords:

```
CHECK-ALL-DERIVATIVES
CHECK-DERIVATIVES
CHECK-ELEMENT-DERIVATIVES
CHECK-GROUP-DERIVATIVES
IGNORE-DERIVATIVE-WARNINGS
IGNORE-ELEMENT-DERIVATIVE-WARNINGS
IGNORE-GROUP-DERIVATIVE-WARNINGS
```

The derivatives of the nonlinear element and group functions specified in the SEIF and SGIF files may be checked to reduce the likelihood that programming errors have been made. Numerical estimates of the derivatives are compared

---

[2] `SAVEDATA D` on VM/CMS systems.

with their exact representations and warning messages issued if any of the derivatives appear to be inaccurate. The derivatives of each of the nonlinear elements and groups used are checked if the `CHECK-ALL-DERIVATIVES` keyword is specified. To save effort when many of the elements or groups are of the same type, the derivatives of a single example of each different type of nonlinear element and group supplied are checked if the `CHECK-DERIVATIVES` keyword is specified. The nonlinear element values are checked if `CHECK-ELEMENT-DERIVATIVES` is specified while the `CHECK-GROUP-DERIVATIVES` keyword ensures that the group function derivatives are checked. The default is that no derivative checking is performed but it is recommended that newly specified problems are always checked for accuracy.

It is sometimes possible that a derivative is reported to be inaccurate when in fact it has been coded correctly - this is a consequence of truncation errors in the numerical tests performed. LANCELOT will not allow the optimization to proceed if a derivative error is detected unless the user specifically requests it using one of the `IGNORE-DERIVATIVE-WARNINGS`, `IGNORE-ELEMENT--DERIVATIVE-WARNINGS` or `IGNORE-GROUP-DERIVATIVE-WARNINGS` keywords. All warnings are ignored when the first of these keywords is present while only element and group warnings, respectively, are overridden with the latter two keywords.

4.1.7 Finite Difference Gradients

keyword:

```
FINITE-DIFFERENCE-GRADIENTS
```

The user is assumed to have provided analytic expressions for the values of the nonlinear element functions in the SEIF file. The specification of analytic first derivatives is optional, but recommended whenever possible. If the user is unable to provide analytic first derivatives of the nonlinear element functions, the derivatives will be approximated by finite differences whenever the `FINITE-DIFFERENCE-GRADIENTS` keyword is specified. The user should note, however, that the accuracy of the solution obtained by LANCELOT may be compromised if finite-difference approximations are used.

4.1.8 The Second Derivative Approximations

keywords:

```
EXACT-SECOND-DERIVATIVES-USED
BFGS-APPROXIMATION-USED
DFP-APPROXIMATION-USED
PSB-APPROXIMATION-USED
SR1-APPROXIMATION-USED
```

The user is frequently able to provide analytic expressions for the first derivatives of the nonlinear element functions in the SEIF file. The specification of exact second derivatives is optional, but recommended whenever possible. If exact second derivatives have been supplied, the keyword EXACT-SECOND--DERIVATIVES-USED should be included. In their absence four different second derivative approximations may be used. These are the Broyden-Fletcher-Goldfarb-Shanno (BFGS), Davidon-Fletcher-Powell (DFP), Powell-symmetric-Broyden (PSB) and symmetric rank-one (SR1) approximations which may be invoked using the BFGS-APPROXIMATION-USED, DFP-APPROXIMATION-USED, PSB-APPROXIMATION-USED and SR1-APPROXIMATION-USED keywords, respectively. A default of

```
SR1-APPROXIMATION-USED
```

is supplied.

4.1.9 Problem Scaling

keywords:

```
USE-SCALING-FACTORS
USE-CONSTRAINT-SCALING-FACTORS
USE-VARIABLE-SCALING-FACTORS
GET-SCALING-FACTORS
PRINT-SCALING-FACTORS
```

It may happen that a user's problem uses variables or general constraints whose normal numerical values are wildly different in magnitude. This poor scaling can cause difficulties for LANCELOT. As a precaution, the user may ask that the problem be (implicitly) rescaled before the optimization commences. The scaling algorithm used is not foolproof and should only really be used if the user encounters difficulties when solving the problem unscaled. It is frequently far better for the user to rescale his or her problem him or herself, so that all the variables have roughly the same value at a "typical point"; the same advice hold for constraint functions.

LANCELOT's scaling algorithm is invoked by the GET-SCALING-FACTORS keyword. This produces appropriate scale factors. These values may be printed by specifying the PRINT-SCALING-FACTORS keyword and are applied to the problem during the optimization when USE-SCALING-FACTORS is specified. The scale factors are applied to the general constraints and variables separately when the USE-CONSTRAINT-SCALING-FACTORS and USE-VARIABLE-SCALING--FACTORS keywords are, respectively, included. The default is that no scaling is performed.

4.1.10 Constraint Accuracy

keyword:

`CONSTRAINT-ACCURACY-REQUIRED (real)`

The optimization will be terminated only when the norm of the constraint functions, if any, is smaller than a pre-specified amount (see method section). This accuracy may be supplied as a real number that follows the keyword `CONSTRAINT-ACCURACY-REQUIRED`. A default

```
CONSTRAINT-ACCURACY-REQUIRED 0.00001
```

is supplied.

4.1.11 Gradient Accuracy

keyword:

`GRADIENT-ACCURACY-REQUIRED (real)`

The optimization will be terminated only when the norm of the projected gradient of the merit function - the objective function in the bound-constrained case and the augmented Lagrangian function when there are more general constraints present - is smaller than a pre-specified amount (see method section). This accuracy may be supplied as a real number that follows the keyword `GRADIENT-ACCURACY-REQUIRED`. A default

```
GRADIENT-ACCURACY-REQUIRED 0.00001
```

is supplied.

4.1.12 The Penalty Parameter

keyword:

`INITIIAL-PENALTY-PARAMETER (real)`

If there are general constraints present, the constraints are combined with the objective function in an augmented Lagrangian function (see method section). The weighting factor by which the squares of the constraints are initially divided, the penalty parameter, may be supplied as a real number that follows the keyword `INITIAL-PENALTY-PARAMETER`. A default

```
INITIAL-PENALTY-PARAMETER 0.1
```

is used.

4.1.13 Controlling the Penalty Parameter

keywords:

```
DECREASE-PENALTY-PARAMETER-UNTIL< (real)
FIRST-CONSTRAINT-ACCURACY-REQUIRED (real)
FIRST-GRADIENT-ACCURACY-REQUIRED (real)
```

When general constraints are present, a sequence of augmented Lagrangian functions are (approximately) optimized until the solution to the original problem is determined. At the end of each (approximate) optimization, either the penalty parameter is reduced or improved Lagrange multiplier estimates are accepted. The method is designed so that ultimately only the latter possibility may occur. When the algorithm is far from the solution to the original problem, there may not be any point in accepting Lagrange multiplier estimates as they may be very inaccurate and it may be preferable to reduce the penalty parameter instead.

If the user includes the `DECREASE-PENALTY-PARAMETER-UNTIL<` keyword, no multiplier estimates will be accepted until the penalty parameter has been reduced to below the real value following the keyword. Likewise, no multiplier estimates will be accepted until the norms of the projected gradients and constraints are smaller than the values following the keywords `FIRST-CONSTRAINT-ACCURACY-REQUIRED` and `FIRST-GRADIENT-ACCURACY-REQUIRED` respectively. Defaults of

```
DECREASE-PENALTY-PARAMETER-UNTIL< 0.1
FIRST-CONSTRAINT-ACCURACY-REQUIRED 0.1
FIRST-GRADIENT-ACCURACY-REQUIRED 0.1
```

are set.

4.1.14 The Trust Region

keywords:

```
INFINITY-NORM-TRUST-REGION-USED
TWO-NORM-TRUST-REGION-USED
```

Each iteration of LANCELOT involves finding an approximation to the optimium value of a suitable quadratic model of the merit function within a suitably shaped region surrounding the current iterate. This region is known as the trust region. The user may choose between a "box" shaped, infinity norm region, which is specified by the keyword `INFINITY-NORM-TRUST-REGION-USED` and a "spherical" shaped region, which arises when the `TWO-NORM-TRUST-REGION-USED` keyword is used. The default is for

INFINITY-NORM-TRUST-REGION-USED

to be specified.

4.1.15 The Trust Region Radius

keyword:

TRUST-REGION-RADIUS (real)

The initial trust region radius is normally set internally by LANCELOT. However, the user may have good reason for overriding this value. If the user wishes to specify an initial radius, the value of the radius should follow the keyword TRUST-REGION-RADIUS. A nonpositive value will be replaced by a value chosen by LANCELOT. The default is for LANCELOT to select the initial radius.

4.1.16 Solving the Inner-Iteration Subproblem

keyword:

SOLVE-BQP-ACCURATELY

The user may also chose whether or not to solve the subproblem which arises at each iteration accurately or whether to accept an approximation (the approximation improves as the solution to the true problem is approached). The default is that an approximation suffices, but an accurate subproblem solution will be found if the SOLVE-BQP-ACCURATELY keyword is present.

4.1.17 The Cauchy Point

keywords:

EXACT-CAUCHY-POINT-REQUIRED
INEXACT-CAUCHY-POINT-REQUIRED

The (approximate) solution of the subproblem at each iteration is divided into two stages (see method section). In the first, a one dimensional search is made along a piecewise linear arc, the Cauchy arc. The first local optimizer of the model function along the arc is known as the Cauchy point. The user may choose whether this point should be found accurately, by using the EXACT-CAUCHY-POINT-REQUIRED keyword, or whether a suitable approximation suffices, by including the keyword INEXACT-CAUCHY-POINT-REQUIRED. A default

EXACT-CAUCHY-POINT-REQUIRED

is used.

4.1.18 The Linear Equation Solver

keywords:

`CG-METHOD-USED`
`DIAGONAL-PRECONDITIONED-CG-SOLVER-USED` or
`MUNKSGAARDS-PRECONDITIONED-CG-SOLVER-USED` or
`EXPANDING-BAND-PRECONDITIONED-CG-SOLVER-USED` or
`FULL-MATRIX-PRECONDITIONED-CG-SOLVER-USED` or
`GILL-MURRAY-PONCELEON-SAUNDERS-PRECONDITIONED-CG-SOLVER-USED` or
`MODIFIED-MA27-PRECONDITIONED-CG-SOLVER-USED` or
`SCHNABEL-ESKOW-PRECONDITIONED-CG-SOLVER-USED` or
`USERS-PRECONDITIONED-CG-SOLVER-USED` or
`BANDSOLVER-PRECONDITIONED-CG-SOLVER-USED (integer)` or
`MULTIFRONTAL-SOLVER-USED`
`DIRECT-MODIFIED-MULTIFRONTAL-SOLVER-USED`

In the second stage of the process to find an (approximate) optimum value of the model, the variables which lie on their bounds at the Cauchy point are fixed and the optimum value of the model with respect to the remaining variables sought. This optimization is equivalent to the solution of one or more systems of linear equations. The coefficient matrix of each system is the Hessian matrix of the merit function, taken with respect to the variables that are not fixed at the Cauchy point. The user may choose between a number of appropriate linear equation solvers. The best choice will depend on the structure and conditioning of the coefficient matrix.

If the keyword `CG-METHOD-USED` is included, the conjugate gradient method without preconditioning is used. Conjugate gradient methods with preconditioning are also available. These range from the simple use of diagonal scalings, to preconditioners suitable for matrices with a band structure, through incomplete factorization preconditioners, to full factorization preconditioners.

The diagonal scaling preconditioner is invoked with the `DIAGONAL-PRECONDITIONED-CG-SOLVER-USED` keyword. The band matrix preconditioner is called with the keyword `BANDSOLVER-PRECONDITIONED-CG-SOLVER-USED`; in this case, the semi-bandwidth must be specified as an non-negative integer following the keyword[3].

Two incomplete factorization preconditioners are available, The Munksgaard preconditioner may be specified with the keywords `MUNKSGAARDS-PRECONDITIONED-CG-SOLVER-USED`. Alternatively, a so-called expanding band pre-

[3]If the semi-bandwidth m is specified, a banded preconditioner with $2m + 1$ nonzero diagonals will be used. Hence $m = 1$, say, results in a tridiagonal preconditioner.

conditioner, invoked with the keywords `EXPANDING-BAND-PRECONDITIONED--CG-SOLVER-USED`, is also available.

Finally, full factorization preconditioners may be used. The modification method due to Gill, Murray, Poncelèon and Saunders may be called by one of the keywords `FULL-MATRIX-PRECONDITIONED-CG-SOLVER-USED` or `GILL-MURRAY-PONCELEON-SAUNDERS-PRECONDITIONED-CG-SOLVER-USED`. The alternative modification proposed by Schnabel and Eskow is invoked with either the `MODIFIED-MA27-PRECONDITIONED-CG-SOLVER-USED` or `SCHNABEL--ESKOW-PRECONDITIONED-CG-SOLVER-USED` keywords.

The systems may also be solved using direct methods. Options available are a multifrontal method, in which negative curvature is exploited if encountered, called with the keyword `MULTIFRONTAL-SOLVER-USED`, and a modified Cholesky method, which is included when the keyword `DIRECT-MODIFIED--MULTIFRONTAL-SOLVER-USED` is present. A description of all these techniques is given in the methods section. The default solver

```
BANDSOLVER-PRECONDITIONED-CG-SOLVER-USED 5
```

is used.

4.1.19 Restarting the Calculation

keyword:

```
RESTART-FROM-PREVIOUS-POINT
```

At the end of a call to LANCELOT, the best estimate of the solution found, together with the related Lagrange multipliers (if any) are written to a file, the saved data file. The name of this file will be installation dependent but, for instance, on a UNIX system the file will be called `SAVEDATA.d` in the current working directory. The user may wish to restart the optimization with these solution values rather than those given on the original SIF file.

For instance, it may be considered desirable to stop the optimization every ten iterations of a large calculation to see how the algorithm is proceeding. If all is well, a restart is appropriate. Alternatively, the user may wish to increase the accuracy required by the calculation and a restart may significantly reduce the time for the second optimization. Finally, the user may wish to perturb the solution values, by editing the saved data file and then recommence the calculation from the perturbed point. (The file is self explanatory. It contains the name of the problem, the number of problem variables and groups followed by the current value of the penalty parameter, the values and names of the variables and then the values of the Lagrange multipliers together with the names of the constraints to which they are associated. Only the *values* of the penalty parameter, variables and Lagrange multipliers may be changed. The format of the data, together with its order within the file must *not* be changed.) If, for whatever reason, a restart is required, the RESTART-FROM-PREVIOUS-POINT keyword should be specified.

4.2 An Example

As an example, consider the following spec file.

```
BEGIN
   PRINT-LEVEL 2
   ITERATIONS-BETWEEN-PRINTING 5
   MAXIMUM-NUMBER-OF-ITERATIONS 50
   CHECK-DERIVATIVES
   EXACT-SECOND-DERIVATIVES-USED
   BANDSOLVER-PRECONDITIONED-CG-SOLVER-USED 10
END
```

Then we wish to:

- Minimize the objective function.
- Print an expanded summary every fifth iteration.
- Stop after 50 iterations.
- Save the solution data only at the end of the optimization.
- Check the derivatives of the group and nonlinear element functions, stopping if the check reveals possible errors.
- Use exact first and second derivatives.
- Forgo automatic problem scaling.
- Use default stopping tolerances.
- Use a "box" shaped trust region, whose initial radius is chosen internally by LANCELOT.
- Accept an inaccurate solution of the inner iteration subproblem.
- Find the exact Cauchy point at each iteration.
- Use a preconditioned conjugate gradient linear equation solver. The preconditioner is based on ignoring all elements which lie outside a band of width 11 (semibandwidth 10) from the diagonal of the coefficient matrix. The remaining band matrix is factorized and its diagonals modified to ensure that the modified band matrix is positive definite.
- Start from the initial point specified in the SIF file.

Chapter 5. A Description of how LANCELOT Works

The purpose of the LANCELOT package is to solve continuous optimization problems. This is achieved in several successive steps. The first part of this chapter briefly describes these steps. The second part additionally provides a description of the functionality of the supplied commands "sdlan" and "lan" and of the options available with these commands.

5.1 General Organization of the LANCELOT Solution Process

A general flowchart of the LANCELOT solution process is shown in Figure 5.1, page 145. In this figure, the system supplied tools are shown in dashed-boxes[1]. We have also represented the SDIF, SEIF and SGIF files separately, but they would typically be concatenated in a single file with a `.SIF` suffix.

The purpose of the following subsections is to expand upon the processes summarized in this figure.

5.1.1 Decoding the Problem-Input File

In the first step, a specialized LANCELOT module (called the "SIF decoder") reads a description of the problem from a file. The problem is expressed using the SIF syntax discussed in Chapters 2 and 7. This module interprets the statements found in the file and produces a few Fortran subroutines and a data file:

`ELFUNS.f` contains a Fortran subroutine that evaluates the numerical functions corresponding to the nonlinear element function types occurring in the problem, as well as their derivatives,

`GROUPS.f` contains a Fortran subroutine that evaluates the numerical functions corresponding to the group function types occurring in the problem, as well as their derivatives,

[1]We have chosen a Unix-like convention for the naming of the intermediate files in our description.

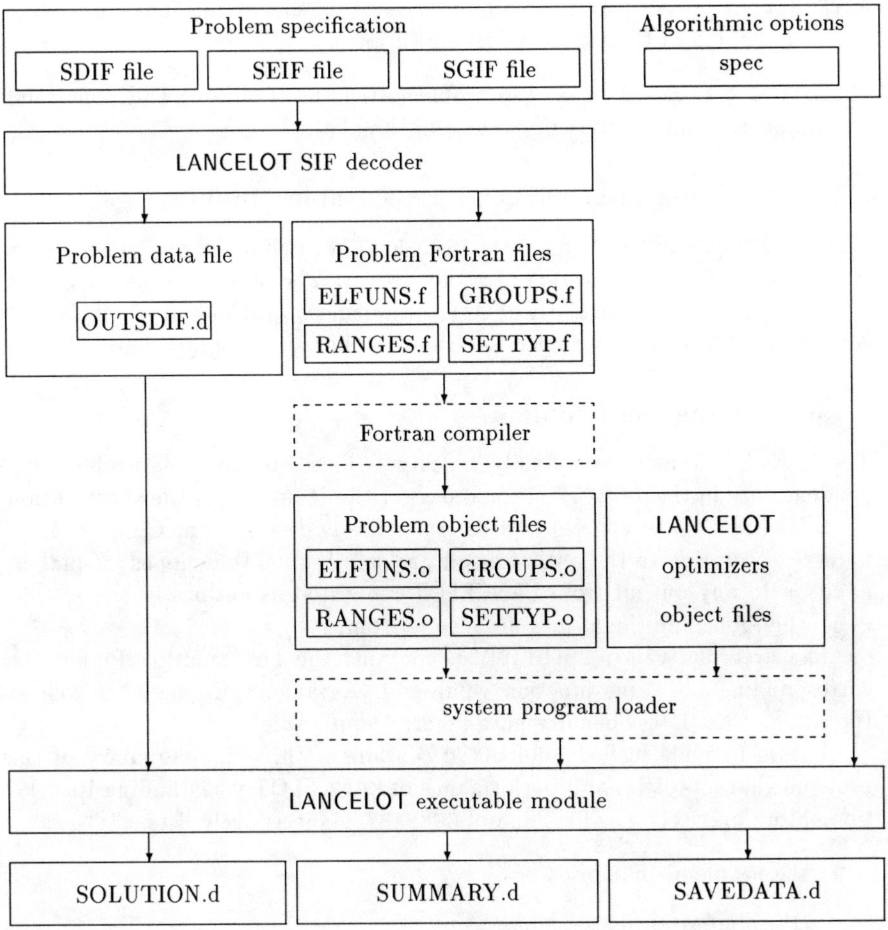

Figure 5.1: How LANCELOT works

`RANGES.f` contains a Fortran subroutine that computes the transformations from elemental to internal variables for all element types which use a nontrivial transformation,

`SETTYP.f` contains a Fortran subroutine that assigns the correct type to the nonlinear elements for the problem,

`EXTERN.f` (if present) contains all the user supplied Fortran functions found at the end of the problem SIF file (if any),

`OUTSDIF.d` contains data on the problem structure (variable and constraint names, scalings, starting point and the like).

5.1.2 Building the LANCELOT Executable Module

The files `ELFUNS.f`, `GROUPS.f`, `RANGES.f`, `EXTERN.f` (if present) and `SETTYP.f` are then compiled at the next step of the process. The resulting object modules are then loaded with the LANCELOT "optimizing object modules". This step simply uses the system-provided Fortran compiler and program loader/linker.

5.1.3 Solving the Problem

The LANCELOT module is finally executed. This module reads problem dependent data in the `OUTSDIF.d` file and algorithmic options in the specification file SPEC.SPC. The content of this latter file is described in Chapter 4. A numerical solution to the optimization problem is then (one hopes) found. In addition to any output from LANCELOT requested as options in the specification file, four additional files are also written.

The first file, called `SOLUTION.d`, contains the best solution found, the corresponding objective function value and Lagrange multipliers, as well as the value of the latest penalty parameter, if applicable.

The second file, called `SUMMARY.d` contains a three-line summary of the problem characteristics and performance of LANCELOT when finding its solution. More precisely, the first line of `SUMMARY.d` successively lists

1. the problem's name,
2. the number of free variables,
3. the number of fixed variables,
4. the number of variables that are only bounded below,
5. the number of variables that are only bounded above,
6. the number of variables bounded both above and below,
7. the number of slack variables,

8. the number of linear objective groups,
9. the number of nonlinear objective groups,
10. the number of linear equality groups,
11. the number of nonlinear equality groups,
12. the number of linear inequality groups,
13. the number of nonlinear inequality groups,

The second line of `SUMMARY.d` successively lists

1. the problem's name,
2. the total number of variables in the problem,
3. "MIN" or "MAX", depending on whether a minimizer or maximizer was sought,
4. the name of the optimizer used by LANCELOT ("SBMIN" or "AUGLG"),
5. a list of algorithmic options used (within brackets), namely
   (a) the type of trust region:
   1: hyperspherical (ℓ_2-norm),
   2: box-shaped (ℓ_∞-norm),
   (b) the method used to solve the linear systems of equations which arise at each iteration of the minimization algorithm:
   1: conjugate gradients without preconditioning,
   2: conjugate gradients with diagonal preconditioning,
   3: conjugate gradients with user supplied preconditioner,
   4: conjugate gradients with an expanding band incomplete Cholesky preconditioner,
   5: conjugate gradients with Munksgaard's preconditioner,
   6: conjugate gradients with Schnabel-Eskow modified Cholesky preconditioner,
   7: conjugate gradients with Gill-Murray-Ponceleon-Saunders preconditioner,
   8: conjugate gradients with a banded incomplete Cholesky preconditioner,
   11: multifrontal factorization,
   12: Schnabel-Eskow modified Cholesky factorization,
   (c) the type of first derivative information:

0: exact first derivatives,

1: finite-difference approximation,

(d) the type of second derivative information:

0: exact second derivatives,

1: BFGS quasi-Newton approximation,

2: DFP quasi-Newton approximation,

3: PSB quasi-Newton approximation,

4: SR1 quasi-Newton approximation,

(e) the type of generalized Cauchy point calculation:

1: first minimizer along the projected gradient path,

2: inexact projected search,

(f) the accuracy requirement on the minimization of the quadratic model within the box:

1: accurate minimization,

2: approximate minimization,

6. the number of iterations and function evaluations,

7. the number of derivative evaluations,

8. the number of conjugate gradient iterations,

9. the exit condition:

0: no error occurred,

1: the maximum number of iterations has been reached,

2: the trust region radius has become too small,

3: the step taken is too small,

4: the integer workspace is too small,

5: the real (or double precision) workspace is too small,

6: the logical workspace is too small,

10: the file `ALIVE.d` has been removed from the current directory by the user (see page 149 below).

The third line of `SUMMARY.d` successively lists

1. the problem's name,

2. the total CPU time spent in solving the problem,

3. the final value of the objective function.

The third file, named `SAVEDATA.d`, contains information written by LANCELOT that allows the optimization to be restarted from an approximation to the problem solution obtained during a previous run of the package. More specifically, the file successively contains

1. the problem's name,
2. the number of the problem's variables,
3. the number of groups,
4. the value of the penalty parameter last used by LANCELOT,
5. the values of each of the problem's variables and their names,
6. the values of each of the problem's Lagrange multipliers and their names,

Finally, a file named `ALIVE.d` is created by LANCELOT on the current directory before the first iteration of the optimization algorithm. This file contains the message

```
LANCELOT rampages onwards
```

whose precise cryptic meaning is entirely irrelevant. LANCELOT checks the presence of this file before starting each new iteration of the optimization. If the file still exists, LANCELOT continues the calculation normally. If, on the other hand, `ALIVE.d` has disappeared from the current directory, the computation stops[2] after LANCELOT has saved the current iterate, Lagrange multipliers and penalty parameter (if applicable) in the file `SAVEDATA.d` described above. This feature can be used to interactively stop the calculation when the user judges that no further effort should be spent on the considered problem with the current algorithmic settings[3]: this is achieved by simply deleting the `ALIVE.d` file. If the calculation terminates normally (that is if the `ALIVE.d` file is not removed from the current directory by the user), the file is actually deleted by LANCELOT itself before it stops.

5.2 The sdlan and lan Commands

The solution process described in the first part of this chapter can of course be automated and therefore largely hidden from the user of the LANCELOT package. This is the function of the supplied sdlan and lan commands, whose functionality and options (for supported systems) are now described.

[2]With the exit condition equal to 10 (see page 148).

[3]As specified in the `SPEC.SPC` file.

5.2.1 Functionality of the sdlan Command

The function of the sdlan command is to automate the entire solution process, from the decoding of the problem SIF file to the execution of the final LANCELOT module that produces the solution.

The main steps executed by the sdlan command are as follows.

1. Check the argument of the command for inconsistencies and interpret them. Also check that the problem specified has an associated SIF file.
2. Apply the SIF decoder to the problem SIF file, in order to produce the problem dependent data file and Fortran functions. Stop the process if any error is uncovered in the SIF file.
3. Call the lan command (described below) in order to continue the process.

5.2.2 Functionality of the lan Command

The function of the lan command is to execute the later stages in the LANCELOT solution process already described, omitting the decoding of the problem SIF file (which is presumed to have already taken place).

The main steps executed by the lan command are as follows.

1. Check the arguments of the command for inconsistencies and interpret them.
2. If no local specfile is present, copy the default version from the main LANCELOT directory/disk.
3. If a new executable module is to be built, compile the problem dependent Fortran subroutines.
4. Load together the compiled problem dependent subroutines and the LANCELOT optimization modules to produce the LANCELOT executable file.
5. Execute the LANCELOT executable file in order to solve the problem.

5.2.3 The sdlan Command on AIX, UNICOS, Unix and Ultrix Systems

The format of the command is:

```
sdlan [-s] [-h] [-k] [-m i] [-o j] [-l secs] probname
```

where optional arguments are within square brackets. The command arguments have the following meaning:

`-s` runs the single precision version of the LANCELOT package (default: double precision),

`-h` prints a simple description of the possible options for the sdlan command,

`-k` does not delete the LANCELOT executable module after execution (default: the module is deleted),

`-m i` if `i` = 0, LANCELOT chooses the minimization method used, if `i` = 1, SBMIN is requested while AUGLG is requested if `i` = 2 (default: LANCELOT chooses),

`-o j` if `j` = 0, LANCELOT runs in silent mode, with very little output added to that requested within the SPEC file, while details of the problem decoding are printed if `j` = 1 (default: silent mode),

`-l secs` an upper limit of `secs` seconds is set on the cputime used by LANCELOT in the numerical solution of the problem (default: 99999999 seconds),

`probname` is the name (without extension) of the file containing the SIF description of the problem to solve.

The use of the `[-m]` option is *not* recommended.

The sdlan command creates some intermediate files and the LANCELOT final executable module. The directory on which these files reside is defined in the sdlan command by the variable `TMP` and `EXEC` respectively. By default, sdlan uses `TMP=/tmp` and `EXEC=.` (the current directory), but these can be changed by editing the command and resetting these variables appropriately.

5.2.4 The lan Command on AIX, UNICOS, Unix and Ultrix Systems

The format of the command is:

```
lan [-s] [-h] [-k] [-n] [-o j] [-l secs]
```

where optional arguments are within square brackets. The command arguments have the following meaning:

`-s` runs the single precision version of the LANCELOT package (default: double precision),

`-h` prints a simple description of the possible options for the lan command,

`-k` does not delete the LANCELOT executable module after execution (default: the module is deleted),

`-n` reconstruct the LANCELOT executable module from the files output at the SIF decoding stage (default: run the current executable module without reconstructing it),

`-o j` if j = 0, LANCELOT runs in silent mode, with very little output added to that requested within the SPEC file, while details of the processing stages are printed if j = 1 (default: silent mode),

`-l secs` an upper limit of `secs` seconds is set on the cputime used by LANCELOT in the numerical solution of the problem (default: 99999999 seconds),

The `lan` command creates some intermediate files and the LANCELOT final executable module. The directory on which these files reside is defined in the lan command by the variable `TMP` and `EXEC` respectively. By default, lan uses `TMP=/tmp` and `EXEC=.` (the current directory), but these can be changed by editing the command and resetting these variables appropriately.

5.2.5 The sdlan Command on VAX/VMS Systems

The format of the command is:

```
sdlan [-s] [-h] [-k] [-m=i] [-o=j] probname
```

where optional arguments are within square brackets. The command arguments have the following meaning:

`-s` runs the single precision version of the LANCELOT package (default: double precision),

`-h` prints a simple description of the possible options for the sdlan command,

`-k` does not delete the LANCELOT executable module after execution (default: the module is deleted),

`-m=i` if i = 0, LANCELOT chooses the minimization method used, if i = 1, SBMIN is requested while AUGLG is requested if i = 2 (default: LANCELOT chooses),

`-o=j` if j = 0, LANCELOT runs in silent mode, with very little output added to that requested within the SPEC file, while details of the problem decoding are printed if j = 1 (default: silent mode),

`probname` is the name (without extension) of the file containing the SIF description of the problem to solve.

The use of the `[-m]` option is *not* recommended.

The sdlan command creates some intermediate files and the LANCELOT final executable module. The directory on which these files reside is the directory from which the command is executed.

5.2.6 The lan Command on VAX/VMS Systems

The format of the command is:

```
lan [-s] [-h] [-k] [-n] [-o=j]
```

where optional arguments are within square brackets. The command arguments have the following meaning:

`-s` runs the single precision version of the LANCELOT package (default: double precision),

`-h` prints a simple description of the possible options for the lan command,

`-k` does not delete the LANCELOT executable module after execution (default: the module is deleted),

`-n` reconstruct the LANCELOT executable module from the files output at the SIF decoding stage (default: run the current executable module without reconstructing it),

`-o=j` if `j` = 0, LANCELOT runs in silent mode, with very little output added to that requested within the SPEC file, while details of the processing stages are printed if `j` = 1 (default: silent mode),

The `lan` command creates some intermediate files and the LANCELOT final executable module. The directory on which these files reside is the directory from which the command is executed.

5.2.7 The SDLAN Command on VM/CMS Systems

The format of the command is:

```
SDLAN [-S] [-H] [-Mi] [-Oj] probname
```

where optional arguments are within square brackets. The command arguments have the following meaning:

`-S` runs the single precision version of the LANCELOT package (default: double precision),

`-H` prints a simple description of the possible options for the sdlan command,

`-Mi` if `i` = 0, LANCELOT chooses the minimization method used, if `i` = 1, SBMIN is requested while AUGLG is requested if `i` = 2 (default: LANCELOT chooses),

`-Oj` if `j` = 0, LANCELOT runs in silent mode, with very little output added to that requested within the SPEC file, while details of the problem decoding are printed if `j` = 1 (default: silent mode), `probname` is the name (without extension) of the file containing the SIF description of the problem to solve.

The use of the `[-M]` option is *not* recommended.

The SDLAN command creates some intermediate files and the LANCELOT final executable module. The disk on which these files reside is the disk from which the command is executed. Note that, before using the SDLAN command, the global variables

```
GLOBALV SELECT LANCELOT SETL PROBLEMS (problem_mode)
```

must have been set to point to the filemode (`problem_mode`) of the disk in which the current problem occurs.

5.2.8 The LAN Command on VM/CMS Systems

The format of the command is:

```
LAN [-S] [-H] [-Oj]
```

where optional arguments are within square brackets. The command arguments have the following meaning:

`-S` runs the single precision version of the LANCELOT package (default: double precision),

`-H` prints a simple description of the possible options for the lan command,

`-Oj` if `j` = 0, LANCELOT runs in silent mode, with very little output added to that requested within the SPEC file, while details of the processing stages are printed if `j` = 1 (default: silent mode),

The LAN command creates some intermediate files and the LANCELOT final executable module. The disk on which these files reside is the disk from which the command is executed.

As fo the SDLAN command, before using the LAN command, the global variables

```
GLOBALV SELECT LANCELOT SETL PROBLEMS (problem_mode)
```

must have been set to point to the filemode (`problem_mode`) of the disk in which the current problem occurs.

Chapter 6. Installing LANCELOT on your System

The purpose of this chapter is to describe the LANCELOT installation procedure, i.e. the sequence of steps that you should take if you wish to install the LANCELOT package on your system.

The package is available with automated installation procedures for the following range of operating systems:

- Cray: UNICOS,
- Digital: Ultrix and VMS,
- IBM: AIX and VM/CMS,
- SUN: Unix.

The following sections describe the successive steps to follow when installing LANCELOT on systems running these operating systems. The sections of this chapter that you should read, and follow, for the installation are indicated in Table 6.1.

Operating system	Sections
Cray/UNICOS	6.1, 6.2
Digital/Ultrix	6.1, 6.2
Digital/VMS	6.1, 6.3
IBM/AIX	6.1, 6.2
IBM/CMS	6.4
SUN/Unix	6.1, 6.2

Table 6.1: Which sections to read.

If the your system is different from those listed above and if you nevertheless wish to install LANCELOT on it, we recommend that you follow the suggestions made in Section 6.5.4.

6.1 Organization of the LANCELOT Directories on AIX, VMS, Unix, Ultrix and UNICOS Systems

The LANCELOT distribution comprises a set of files. All these files should be placed in the same directory, from which the installation will be performed. This directory will henceforth be called the "installation directory". The installation will, amongst other things, create a subdirectory structure suitable for the machine on which the package is being installed. This structure is shown in Figure 6.1.

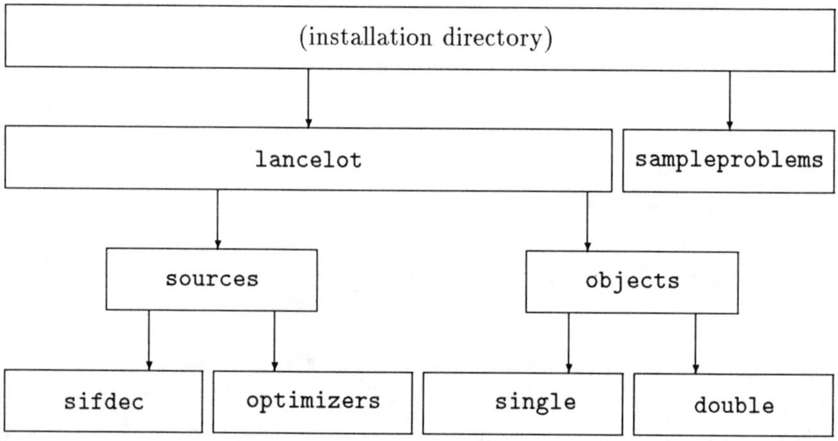

Figure 6.1: Organization of the LANCELOT directories

It is assumed that the C-shell is available in the file `/bin/csh`.

After installation, the installation directory will contain the following files:

`READ.ME`: a text similar to this installation description.

`unwrap`: a procedure to set up the LANCELOT subdirectories and copy the relevant files to their appropriate destinations[1],

`instll`: a procedure to build the LANCELOT executable modules and object files,

Going down one level, the `lancelot` directory, hereafter called the "main LANCELOT directory", contains the following files:

[1]This file comes in the distribution as one of the "unfold" procedures, but the installation instructions insist that the user rename it "unwrap".

`sdlan`: a command to decode and then solve a problem using the LANCELOT package,

`lan`: a command to run the package on a problem which has already been decoded,[2]

`compil`: an auxiliary makefile (except for VAX/VMS),

`SPEC.SPC`: the default specification file[3] for the LANCELOT algorithms.

`SAMPLE.SPC`: a template containing a complete set of algorithmic specification options.

`sifdec_d` and/or `sifdec_s`: the executable program allowing interpretation of the problem when stated in SIF format (see Chapter 2).

The `sampleproblems` directory contains two small sample problems in SIF format, namely `ALLIN.SIF` and `HS65.SIF` and a template containing a complete set of the possible statements in the SIF description of a problem, `SAMPLE.SIF`.

The `sources` directory contains the Fortran sources of the package. They are separated into two sets corresponding to two distinct modules in the package. The functions and relations of these modules are described in Chapter 5.

1. The sources for the first module, called the "SIF decoder", are found in the directory `sifdec`. This directory contains the following Fortran files[4]:

   ```
   decode.f  gps.f    inlanc.f  makefn.f  makegr.f  printp.f
   rest.f    runsd.f  utils.f
   ```

   These files contain both the single and double precision versions of the codes. The directory also contains a "makefile" whose name depends upon the particular system.

2. The sources for the second module, called the "optimizing module", are contained in the directory `optimizers`. This directory holds the following Fortran files:

   ```
   asmbl.f   assl.f    auglag.f  bndsl.f   cauch.f  cg.f
   datal.f   drche.f   drchg.f   elgrd.f   frntl.f  getpgr.f
   dumbr.f   dumhsl.f  hsprd.f   initw.f   inxac.f  lance.f
   linpac.f  modchl.f  misc.f    others.f  precn.f  runlan.f
   sbmin.f   scaln.f   sort.f    speci.f
   ```

---

[2] See Chapter 5.

[3] See Chapter 4.

[4] Note that Fortran files are denoted with a `.f` suffix, as is usually the case under Unix. The VMS operating system uses the suffix `.for` instead.

As above, the directory also contains a "makefile" whose name depends upon the particular system.

The `sources` directory additionally contains some files that are relevant for both sets:

`local.f`: the local timing and machine arithmetic dependent routines.

`tobig.f`: a program to process the LANCELOT codes to produce a version of the package capable of solving "large" problems.

`tomed.f`: a program to process the LANCELOT codes to produce a version of the package capable of solving "medium-sized" problems.

`totoy.f`: a program to process the LANCELOT codes to produce a version of the package capable of solving "small" problems.

`todble.f`: a program to process the LANCELOT codes to produce a double precision version of the package.

`tosngl.f`: a program to process the LANCELOT codes to produce a single precision version of the package.

Finally, the subdirectories of `objects` contain the single and double precision object modules that are needed for the LANCELOT optimizers. Note that only the subdirectories corresponding to installed versions of the package will be present.

6.2 Installing LANCELOT on AIX, Unix, Ultrix and UNICOS Systems

6.2.1 Running the Installation Script

The installation process requires space for temporary files. By default, these files are written to (and subsequently removed from) the directory `/tmp`. If you wish to write these files elsewhere, you must edit the file `instll` (with the suffix corresponding to your machine) at the outset and change the shell variable `TMP` to point to an appropriate directory. We recommend using the default directory `/tmp` whenever possible. Do *not* assign the current directory to `TMP`, as this will cause havoc as files try to overwrite themselves.

The installation process uses the Fortran compiler, f77, xlf or cf77, as appropriate. The compiler and linker/loader flags have been set to default values appropriate for either a SUN Sparc 1 workstation, a Digital DECstation 5000/200, an IBM RS6000 workstation or a Cray X/MP 416 as appropriate. These values may be inappropriate for other machines and will then have to be changed. The flags are set in the files

- `lancelot/lan`,
- `lancelot/compil`,
- `sifdec/makefile`,
- `optimizers/makefile`

in the variables prefaced by FFLAGS.

Finally, LANCELOT uses a number of machine dependent constants and functions. These values are set within the Fortran function **sources/local.f**. Default values are provided for the above-mentioned machines.

The first thing you must do to install the package is to "unwrap" it and create a suitable directory structure. This is achieved by issuing the commands

```
mv unfold.machine unwrap
chmod a+x unwrap
unwrap
```

where *machine* is `sun` for a SUN workstation, `dec` for a Digital DECstation, `rs6` for an IBM RS6000 or `cry` for a Cray X/MP. Note that the name `unwrap` is *essential*, as it prevents self-destruction of the installation script!

You now have the choice between single and double precision[5] versions (or both). You also have to choose a size for the package executables. Three sizes are available with preset options: small, medium and large. Which size is appropriate on your system depends on the memory available to the user and the typical sorts of problems which will subsequently be solved. If in doubt, or if you wish to customize your version of LANCELOT, we recommend that you first install the small version, and then consult Section 6.5.2.

To install the single precision version of LANCELOT, simply type

```
instll single small
```

or

```
instll single medium
```

or

```
instll single large
```

depending on which preassigned size you choose.

To install the double precision version of LANCELOT, simply type

```
instll double small
```

or

[5] On a Cray, this means an "extended" 128 bit wordlength.

```
instll double medium
```

or

```
instll double large
```

again depending on which preassigned size you choose.

One such installation takes about 3 minutes on a Cray X/MP 416, 4 minutes on an IBM RS6000, 5 minutes on a DECstation 5000/200 and 20 minutes on a SUN Sparc 1.

6.2.2 Preparing LANCELOT

The environment variable LANDIR is used by the LANCELOT commands. This variable should be set to point to the main LANCELOT directory, that is `lancelot`. We suggest this variable is set in the user's `.cshrc` file by inserting in this file the command

```
setenv LANDIR (appropriate lancelot directory)
```

where `(appropriate lancelot directory)` is the full path name of the main LANCELOT directory.

The user may need to be able to run the scripts `sdlan` and `lan` from any directory. This can be achieved in a number of ways. The simplest is to include the lines

```
alias sdlan $LANDIR/sdlan
alias lan $LANDIR/lan
```

anywhere after the line setting the `LANDIR` environment variable in the user's `.cshrc` file.

The new version of the file `.cshrc` will subsequently be active in all new shells and can be made active in the current shell by issuing the command

```
source (.cshrc)
```

where `(.cshrc)` is the absolute path name of the user's `.cshrc` file.

LANCELOT requires space for temporary files. By default, these files are written to (and subsequently removed from) the directory `/tmp`. If you wish to write these files elsewhere, you must edit the files `lancelot/sdlan` and `lancelot/lan` and change the shell variable `TMP` to point at an appropriate directory.

Finally, the LANCELOT executable module must reside (at least during execution) on a specified directory. By default, this directory is the directory from which the `sdlan` or `lan` command has been issued. If you wish the executable module to reside elsewhere, you must edit the file

```
lancelot/lan
```

and change the shell variable `EXEC` to point to an appropriate directory.

6.2.3 Running LANCELOT on an Example

To run LANCELOT on a new optimization problem, move into the directory in which the SIF file for the problem, `sif_file_name.SIF`, resides. Now issue the command

```
sdlan -s sif_file_name
```

to run the single precision package or

```
sdlan sif_file_name
```

to run the double precision version. LANCELOT will decode the SIF file, print a short summary of problem characteristics and attempt to solve the associated problem.

For instance, move into the directory `./sampleproblems` and issue the command

```
sdlan ALLIN
```

LANCELOT will then solve the associated simple problem using the double precision version of the package.

This run uses a default LANCELOT specification file. This file is copied from the `lancelot` directory `LANDIR` to the file `SPEC.SPC` in the current directory. The user may change the specification options by editing this file as explained in Chapter 4.

Once the problem has been decoded, it may be resolved with different specification options by using the `lan` script.

6.3 Installing LANCELOT on VAX/VMS Systems

6.3.1 Running the Installation Script

The installation process requires space for temporary files. By default, these files are written to (and subsequently removed from) the current directory.

The installation process uses the Fortran compiler, the linker and some machine dependent constants and functions. These, as well as the compiler flags, have been set to default values appropriate for a VAX/VMS system. Note that it is possible to change the compiler flags and/or machine dependent constants and functions after installing LANCELOT. You should read Section 6.5.3 for more detail.

The first thing you must do to install the package is to “unwrap” the package and create the suitable directory structure. This is achieved by issuing the commands

```
RENAME UNFOLD.VAX UNWRAP.COM
@UNWRAP
```

Note that the name UNWRAP is *essential*, as it prevents self-destruction of the installation procedure!

You now have the choice between single and double precision versions (or both). You also have to choose a size for the package executables. Three sizes are available with preset options: small, medium and large. Which size is appropriate on your system depends on the memory available to the user. If in doubt or for customized sizing, we recommend that you install the small size initially, and then consult Section 6.5.2.

To install the single precision version of LANCELOT, simply type

```
@INSTLL SINGLE SMALL
```

or

```
@INSTLL SINGLE MEDIUM
```

or

```
@INSTLL SINGLE LARGE
```

depending on which preassigned size you choose.

To install the double precision version of LANCELOT, simply type

```
@INSTLL DOUBLE SMALL
```

or

```
@INSTLL DOUBLE MEDIUM
```

or

```
@INSTLL DOUBLE LARGE
```

again depending on which preassigned size you choose.

One such installation takes about 5 minutes on a VAX 3500.

6.3.2 Preparing LANCELOT

The environment variable LAN_ROOT is used by the LANCELOT commands. This variable should be set to point to the installation directory. This directory and its subdirectories must be readable by the user. We suggest the logical name LAN_ROOT is assigned in the user's LOGIN.COM file by inserting in this file the command

```
ASSIGN/JOB/TRANSLATION_ATTRIBUTES=CONCEALED dv:[i.] LAN_ROOT
```

where "i" is the installation directory and where "dv" is the device where this directory resides.

The user may need to be able to run the command files SDLAN and LAN from any directory. This can be achieved in a number of ways. The simplest is to include the symbol definitions

```
SDLAN=="@LAN_ROOT:[LANCELOT]SDLAN"
LAN=="@LAN_ROOT:[LANCELOT]LAN"
```

in the user's LOGIN.COM file.

The new version of the LOGIN.COM will subsequently be active in all future sessions and can be made active in the current session by issuing the command

```
@LOGIN
```

in the user's home directory.

6.3.3 Running LANCELOT

To run LANCELOT on a new optimization problem, change the default directory to the directory in which the SIF file for the problem, sif_file_name.SIF, resides. Now issue the command

```
SDLAN -S sif_file_name
```

to run the single precision package or

```
SDLAN sif_file_name
```

to run the double precision version. LANCELOT will decode the SIF file, print a short summary of problem characteristics and attempt to solve the associated problem.

For instance, change the default directory to [.SAMPLEPROBLEMS] and issue the command

```
SDLAN ALLIN
```

LANCELOT will then solve the associated simple problem using the double precision version of the package.

This run uses a default LANCELOT specification file. This file is copied from the lancelot directory LAN_ROOT:[LANCELOT] to the file SPEC.SPC in the current directory. The user may change the specification options by editing this file as explained in Chapter 4.

Once the problem has been decoded, it may be resolved with different specification options by using the LAN command file.

6.4 Installing LANCELOT on VM/CMS Systems

6.4.1 Organization of the LANCELOT Disk

The LANCELOT distribution comprises a set of files. All these files should be placed in the same disk, from which the installation will be performed. This disk will be called henceforth the "installation disk" and its filemode will be denoted by (lancelot_mode). The installation will, amongst other things, create a MACLIB structure suitable for an easy use of the package.

After installation, the installation disk will contain the following files:

`READ ME`: a text similar to this installation description.

`UNWRAP EXEC`: an exec file to set up the LANCELOT MACLIBs and copy the relevant files where they belong[6],

`INSTLL EXEC`: an exec file to build the LANCELOT executable modules and object files,

`SDLAN EXEC`: an exec file to decode and then solve a problem using the LANCELOT package,

`LAN EXEC`: a exec file to run the package on a problem which has already been decoded,[7],

`SPEC SPC`: the default specification file[8] for the LANCELOT algorithms,

`SAMPLE SPC`: a template containing a complete set of algorithmic specification options,

`SAMPLE SIF`: a template containing a complete set of the possible statements in the SIF description of a problem,

`HS65 SIF`: a small test problem expressed in SIF,

`ALLIN SIF`: another small test problem expressed in SIF.

The installation disk also contains three MACLIBs, namely

- `SIFDEC`,
- `OPTIMIZE`,
- `SOURCES`,

whose content is now described.

1. The `SIFDEC` MACLIB contains the Fortran sources for the SIF/LANCELOT interface (the SIF "decoder"), consisting of the following files:

   ```
   DECODE F  GPS F    INLANC F  MAKEFN F  MAKEGR F  PRINTP F
   REST F    RUNSD F  UTILS F
   ```

   These files contain both the single and double precision versions of the codes.

---

[6] This file comes in the distribution as one of the "UNFOLD" procedures, but the installation instructions insist that the user rename it to "UNWRAP".

[7] See Chapter 5.

[8] See Chapter 4.

The `SIFDEC` MACLIB also contains the corresponding compiled objects in two libraries, named `SINGLE` and `DOUBLE` respectively. It also contains the exec file `MAKESD EXEC` which is used to compile and build the SIF/LANCELOT interface during the installation.

2. The `OPTIMIZE` MACLIB contains the Fortran source files for both the single and double precision versions of the optimization algorithms. These are:

ASMBL F	ASSL F	AUGLAG F	BNDSL F	CAUCH F	CG F
DATAL F	DRCHE F	DRCHG F	ELGRD F	FRNTL F	GETPGR F
DUMBR F	DUMHSL F	HSPRD F	INITW F	INXAC F	LANCE F
LINPAC F	MODCHL F	MISC F	OTHERS F	PRECN F	RUNLAN F
SBMIN F	SCALN F	SORT F	SPECI F		

As for `SIFDEC`, the `OPTIMIZE` MACLIB also contains the compiled objects corresponding to the single and double precision versions in two libraries, named `SINGLE` and `DOUBLE` respectively. It also contains the exec file `MAKEOP EXEC` which is used to compile the optimizers' code during the installation.

3. The `SOURCES` MACLIB additionally contains some files that are relevant for both the optimization algorithms and the SIF decoder:

`LOCAL F`: the local timing and machine arithmetic dependent routines.

`TOBIG F`: a program to process the LANCELOT codes to produce a version of the package capable of solving "large" problems.

`TOMED F`: a program to process the LANCELOT codes to produce a version of the package capable of solving "medium-sized" problems.

`TOTOY F`: a program to process the LANCELOT codes to produce a version of the package capable of solving "small" problems.

`TODBLE F`: a program to process the LANCELOT codes to produce a double precision version of the package.

`TOSNGL F`: a program to process the LANCELOT codes to produce a single precision version of the package.

6.4.2 Running the Installation Script

The installation process requires space for temporary files. By default, these files are written to (and subsequently removed from) the disk with filemode `(lancelot_mode)`.

The installation process uses the Fortran compiler, the loader and some machine dependent constants and functions. These as well as the compiler flags have been set to default values appropriate for a VM/CMS system. Note that it is possible to change the compiler flags and/or machine dependent constants and functions after installing LANCELOT. You should read Chapter 5 for more detail.

The first thing you must do is to declare the disk with filemode (lancelot_mode) to be the LANCELOT disk. You achieve this by issuing the command

```
GLOBALV SELECT LANCELOT SETL PROBLEMS (lancelot_mode)
```

You then "unwrap" the package and create the suitable MACLIBs. This is achieved by issuing the commands

```
RENAME UNFOLD CMS (lancelot_mode) UNWRAP EXEC (lancelot_mode)
UNWRAP
```

Note that the name UNWRAP is *essential*, in order to avoid deletion of the UNFOLD file during the process.

You now have the choice between single and double precision versions (or both). You also have to choose a size for the package executables. Three sizes are available with preset options: small, medium and large. Which size is appropriate on your system depends on the memory available to the user. If in doubt or for customized sizing, we recommend that you initially install the small version, and then consult Section 6.5.2.

To install the single precision version of LANCELOT, simply type

```
INSTLL SINGLE SMALL
```

or

```
INSTLL SINGLE MEDIUM
```

or

```
INSTLL SINGLE LARGE
```

depending on which preassigned size you choose.

To install the double precision version of LANCELOT, simply type

```
INSTLL DOUBLE SMALL
```

or

```
INSTLL DOUBLE MEDIUM
```

or

```
INSTLL DOUBLE LARGE
```

depending on which preassigned size you choose.

One such installation takes about 5 minutes on a moderately busy IBM 3090-600S.

6.4.3 Preparing LANCELOT

Other users may need to be able to run the command files SDLAN and LAN from any Virtual Machine. This can be achieved in a number of ways, the idea being that they obtain link access to the disk with filemode (lancelot_mode). How you actually do this depends on your local security rules, and is therefore left unspecified here.

6.4.4 Running LANCELOT on an Example

To run LANCELOT on a new optimization problem in a user's disk, this user must have link access to the disk with filemode (lancelot_mode) (see above) and should also issue the command

```
GLOBALV SELECT LANCELOT SETL PROBLEMS (problem_mode)
```

in order to set up the global variable LANCELOT_PROBLEMS to correspond to the disk in which the SIF file for the problem, sif_file_name SIF, resides and whose filemode is (problem_mode). Now,the command

```
SDLAN -S sif_file_name
```

can be issued to run the single precision package or

```
SDLAN sif_file_name
```

to run the double precision version. LANCELOT will decode the SIF file, print a short summary of problem characteristics and attempt to solve the associated problem.

For instance, from the disk with filemode (lancelot_mode), you can issue the commands

```
GLOBALV SELECT LANCELOT SETL PROBLEMS (lancelot_mode)
SDLAN ALLIN
```

LANCELOT will then solve the associated simple problem using the double precision version of the package.

Running the package requires a default LANCELOT specification file. This file is first copied from SPEC SPC (lancelot_mode) to the file SPEC SPC-(problem_mode). The user may change the specification options by editing this file as explained in Chapter 4.

Once the problem has been decoded, it may be resolved with different specification options by using the LAN command file.

6.5 Further Installation Issues

6.5.1 The Interface between LANCELOT and the Harwell Subroutine Library

Some of the algorithmic options (automatic scaling, several linear solver choices) in the LANCELOT package make use of the Fortran subroutines MA27,

MA31 and MC19, available from the Harwell Subroutine Library[9]. Since the LANCELOT licence does not include a licence for this Library, the LANCELOT distribution includes a "dummy Harwell Subroutine Library". Its only functionality is to print a message telling the user to use the proper library for the selected option and then to stop execution. This is admittedly frustrating behaviour if you really wish to use the incriminating option.

If you have access to the Harwell Subroutine Library, you can use the library codes within LANCELOT instead of the supplied default dummy versions.

6.5.1.1 Using the Fortran Sources of the Library

We now describe how to achieve integration of LANCELOT and the Harwell Subroutine Library in the case where you have access to the sources of the latter, and, in particular, to the Fortran source code for MA27, MA31 and MC19.

1. Determine the location of the "dummy Harwell Subroutine Library" Fortran file within your LANCELOT distribution.

   On AIX, Unix, Ultrix and UNICOS systems: this file is called `dumhsl.f` and resides in the directory **`LANDIR/sources/optimizers`**.

   On VAX/VMS systems: this file is called `dumhsl.for` and resides in the directory `@LAN_ROOT:[LANCELOT.SOURCES.OPTIMIZERS]`.

   On VM/CMS systems: this file is called `DUMHSL F` and can be found in the `OPTIMIZE` MACLIB on the main `LANCELOT` disk.

2. Make a backup copy of this dummy file (just in case) under a different name.

3. Replace the dummy HSL file by another file with the same name, but containing the Fortran code for all three subroutines MA27, MA31 and MC19 from the Harwell Subroutine Library.

---

[9] For information on the Harwell Subroutine Library, please contact

The Harwell Subroutine Library Liaison Officer ,
AEA Industrial Technology,
Building 424.4,
Harwell Laboratory,
Harwell,
Oxfordshire OX11 0RA,
England

4. Reinstall LANCELOT by again issuing the `instll` command with arguments suitable for your system (see above).

All options should now be available within the package, including those depending on the Harwell Subroutine Library.

6.5.1.2 Using the Object Modules from the Library

If you only have access to the object modules of the Library, namely the objects for MA27, MA31 and MC19, here is what you should do to integrate those with LANCELOT:

1. Find the lan command corresponding to your system (it should reside in the main LANCELOT directory/disk).

2. Modify both the single and double precision linking/loading commands[10] that actually build the LANCELOT executable module to replace the dependence of the `dumhsl` object by a dependence of your local Harwell Subroutine Library objects.

Again, this should be sufficient to enable the use of all algorithmic options in LANCELOT.

6.5.2 Changing the Size of LANCELOT

As described above, the LANCELOT package is distributed with three predetermined "sizes": large, medium and small. These sizes refer to the size of the memory available for problem decoding and solution, and hence are directly related to the size (the amount of data) of the problems that LANCELOT can tackle. It may happen that the predetermined sizes do not fit your favorite problem or your machine well, and that you wish to specify your own custom size. Typically, you would try to run too large a problem, and LANCELOT would then complain that one of the sizing parameters is too small and stop. You then have to increase this parameter (if your machine allows) in order to handle the problem. This modification of the LANCELOT array sizes is indeed possible, and it is the purpose of this section to explain how it should be done.

We first note that the dependency on problem size occurs in both the decoding of the problem SIF file into data structures and subroutines and in the numerical problem solution itself. Indeed, the problem dependent data is fully specified in its associated SIF file and must be taken into account in all stages up to the numerical solution process. It is therefore clear that the size of both the SIF decoder and the optimizers must be adequate for the problem.

[10] They are near the end of the file.

The actual choice of one of the predetermined sizes at installation time is made when running the `install` command (see above), whose second argument is the desired size. In fact, `install` calls upon two other supplied commands, `makesd` and `makeop`, that build up the SIF decoder and optimizers respectively. Both these commands accept an argument which specifies a particular "sizing program" to be applied to the source code in order to obtain the desired size. The three sizing programs are called `tobig`, `tomed` and `totoy` for large, medium and small size respectively; their source is also part of the LANCELOT distribution. All three operate in a very simple way: they select in the source code the appropriate parameter assignment statements, which in turn determine the dimensions of the various arrays used in the code. These assignment statements are differentiated by their first four characters:

- `CBIG` specifies the large size,
- `CMED` the medium size,
- `CTOY` the small size.

The process of customizing the size of LANCELOT is thus rather simple. The values assigned to the dimension parameters are first modified to suit the user's need (by modifying one of the existing sizes) and LANCELOT is then re-installed with this size. We now examine this procedure in more detail.

6.5.2.1 Sizing the Optimizers

We first consider increasing the size of the LANCELOT optimizers. This size is governed by five parameters:

`LIWK` is the size of the integer workspace array used by the algorithms;

`LWK` is the size of the single or double precision workspace array used by the algorithms, according to the precision of the version installed;

`LLOGIC` is the size of the logical workspace array used by the algorithms;

`LCHARA` is the number of ten character strings used as workspace by the algorithms;

`LFUVAL` is the size of the array used to store the problem's function and derivative values.

These five parameters are assigned values at the beginning of the workspace partitioning routine, itself the first routine in the Fortran `lance` source file[11]. The user should then

[11]This is the file `lance.f` for AIX, Unix, Ultrix and Unicos systems, the file `lance.for` for VAX/VMS systems and the file `LANCE F` for VM/CMS systems.

1. select the size that is closest to the problem needs (large, say),

2. modify the value of as many of the corresponding sizing parameters to suit the problem size as is necessary,

3. re-install LANCELOT with the selected size (large, in our case) by re-issuing the `instll` command with an adequate second argument (large, in our example).

6.5.2.2 Sizing the SIF Decoder

The modification of the SIF decoder size is entirely similar, although the number of sizing parameters is larger. These are:

NMAX: the maximum number of variables in a problem,

NGMAX: the maximum number of groups in a problem,

NGPMAX: the maximum number of group parameters in a problem,

NELMAX: the maximum number of element functions in a problem,

NEVMAX: the maximum total number of elemental variables in a problem,

NINMAX: the maximum total number of internal variables in a problem,

NSETVC: the maximum number of nonzero entries in an element Hessian,

NEPMAX: the maximum number of real parameters associated with nonlinear elements,

LA: the maximum number of nonzero entries in the linear elements,

NINDEX: the maximum number of integer parameters in a problem SIF description,

NRLNDX: the maximum number of real paremeters in a problem SIF description,

NBMAX: the maximum number of different sets of bounds on the variables which may be specified,

NSMAX: the maximum number of different sets of starting points which may be specified,

NOBMAX: the maximum number of different sets of bounds on the objective function which may be specified,

All these parameters are assigned a value at the beginning of the `SDLANC` subroutine, the second subroutine in the `runsd` Fortran source file[12]. As above, the user should then

1. select the size that is closest to the problem needs (large, say),
2. modify the value of one or more of the corresponding sizing parameters to suit the problem size,
3. re-install LANCELOT with the selected size (large, in our case) by re-issuing the `instll` command with an adequate second argument (large, in our example).

Note: The *same* size must be chosen for modification in both the SIF decoder and the optimizers!

6.5.3 Changing Compiler Flags and System Dependent Constants

In some circumstances, it might be useful to alter the predetermined compiler flags. An example might be when some new level of code optimization becomes available on your machine[13] and you wish to use it. Some operating system revisions might also require that you change machine dependent constants or procedures (such as the timer). It is the purpose of this section to describe how these modifications can be done.

6.5.3.1 Compiler and Linker Flags

Compilation of Fortran code occurs in three particular instances in the package. Flags are set (and can be modified) separately for each of these instances, as we now detail.

1. The first occurrence is when the SIF decoder is compiled and linked during the installation procedure. The corresponding compilation flags are set in the file `makesd.`*system* of the distribution.

   - For AIX, Unix, Ultrix and Unicos systems, this file is copied into

     `LANDIR/sources/sifdec/makefile`

[12] That is the file `runsd.f` for AIX, Unix, Ultrix and Unicos systems, the file `runsd.for` for VAX/VMS systems and the file `RUNSD F` for VM/CMS systems.

[13] Note that care should be exercised with code optimizers: we know of cases where the optimizers introduce real bugs into the code... As a consequence, it might be a good idea to turn optimization off before deciding that some strange behaviour of the package is anomalous and worth reporting. This is another reason why modifying compiler flags might be useful.

The flags used for compiling the SIF decoder are set in the variable `FFLAGS` while `FFLAGSLD` holds the flags used for linking the compiled objects into an executable module. These are the flags that you should modify, if you wish to.

- For VAX/VMS systems, this file is copied into

  `@LAN_ROOT:[LANCELOT.SOURCES.SIFDEC]makesd.com`

  The compiler flags are set in the string variable **`fflags`** which is defined at the beginning of the file. The linker flags are set in the string variable **`fflagsld`**. You should modify these flags if you wish to modify the compiling/linking default.

- For VM/CMS systems, this file is copied into

  `MAKESD EXEC`

  on the LANCELOT disk. The compiler flags are set in the string variable `FFLAGS` which is defined at the beginning of the file. The linker flags are set in the string variable `FFLAGSLD`. You should modify these flags if you wish to modify the compiling/linking default.

2. The second occurrence of compilation flags is when the optimization algorithms are compiled during installation. The associated compilation flags are set in the distribution file `makeop.`*system*.

   - For AIX, Unix, Ultrix and Unicos systems, this file is copied into

     `LANDIR/sources/optimizers/makefile`

     The flags used for compiling the SIF decoder are set in the variable `FFLAGS`. These are the flags that you should modify, if you wish to.

   - For VAX/VMS systems, this file is copied into

     `@LAN_ROOT:[LANCELOT.SOURCES.OPTIMIZERS]makeop.com`

     The flags used for compiling the optimizers are set in the variable `FFLAGS`. These are the flags that you should modify, if you wish to change the default.

   - For VM/CMS systems, this file is copied into

     `MAKEOP EXEC`

     on the LANCELOT disk. The compiler flags are set in the string variable `FFLAGS` which is defined at the beginning of the file. You should modify these flags if you wish to modify the compiling/linking default.

3. The last, and more frequent, case is when the problem dependent subroutines (those which calculate values and derivatives of groups and nonlinear elements) are compiled and linked with the optimizers during a normal execution of the sdlan or lan commands.

- For AIX, Unix, Ultrix and Unicos systems, the compiler flags are set in the variable `FFLAGS` at the beginning of the file

  `LANDIR/lancelot/compil`

  This is where they can be changed if you wish to alter the default.

- For VAX/VMS systems, the flags used to compile the problem dependent functions are set in the string variable `fflags` within the file

  `LANDIR/lancelot/lan`

  while the flags used for linking the compiled optimizers with the compiled problem dependent functions are set in the string variable `lflags` within the same file. You should modify these flags if you wish to modify the compiling/linking default.

- For VM/CMS systems, the flags used to compile the problem dependent functions are set in the string variable `FFLAGS` within the file

  `LAN EXEC`

  in the main LANCELOT disk, while the flags used for linking the compiled optimizers with the compiled problem dependent functions are set in the string variable `LFLAGS` within the same file. You should modify these flags if you wish to modify the compiling/linking default.

6.5.3.2 System Dependent Constants and Functions

In this section, we indicate where the system dependent constants and functions are present in the LANCELOT package. In fact, they are all gathered in a single Fortran source file. On AIX, Unix, Ultrix and Unicos systems, this file is

`LANDIR/lancelot/sources/local.f`

On VAX/VMS, this file is

`@LAN_ROOT:[LANCELOT.SOURCES]local.for`

On VM/CMS, this file is

`LOCAL F`

within the `SOURCES MACLIB` in the main LANCELOT disk. It contains the following items.

A set of hashing routines The routines `HASHA`, `HASHB`, `HASHC` and `HASHE` provide a Fortran hashing tool. They are system dependent in that they rely on the number of bytes used to represent an integer within the particular Fortran dialect used. This number of bytes is set in the parameter NBYTES both in HASHB and HASHC. If your Fortran compiler uses an "unorthodox" number of bytes for its integers, you will have to change the value of NBYTES.

A definition of the arithmetic constants The supplied functions SMACHR and DMACHR return values for various machine dependent constants, for the single and double precision arithmetic respectively. These are

RC(1): the smallest positive number such that $1 + \text{RC}(1) > 1$;

RC(2): the smallest positive number such that $1 - \text{RC}(2) < 1$;

RC(3): the smallest nonzero positive number;

RC(4): the smallest full precision positive number;

RC(5): the largest finite positive number.

Each of these numbers should be modified, when necessary.

A cpu timer This is the real function CPUTIM, which returns the current cpu-time used by the package, expressed in seconds. This timer is highly system dependent, as can be seen from the diversity of supplied options.

6.5.4 Installing LANCELOT on an Unsupported System

This final section is intended to help if you wish to install LANCELOT on a system which is not in our list of supported systems (page 155).

Broadly speaking, you should follow the steps that we now describe.

6.5.4.1 Recommended Reading

The first thing to do is to become familiar with the ways in which LANCELOT works, which is the object of most of this book. In particular, we recommend that you read the chapters detailed in Table 6.2, page 176, carefully. They are listed in the Table by order of decreasing importance[14].

6.5.4.2 The Code-Specialization Mechanism

After reading the recommended material, you might still wonder about the mechanism that is being used by the installation in order to customize the code for a given machine/system and for a given precision, a question that we now clarify.

This mechanism is as follows. Each source file contains statements for both precisions, and for all supported machines and systems. In order to allow the

---

[14] For the purpose of a new installation, of course!

Chapter or Section	Content
Chapter 5	describes the way in which the package works
parts of this Chapter (6)	describes the installation of the package on the system you are most familiar with[15]
the entire Section 6.5	describes machine dependent issues in the package
Chapter 4	describes the available algorithmic options
Chapter 3	describes the algorithms used in the package

Table 6.2: Recommended reading before an unsupported installation

necessary customization, the first five characters of each of these precision or machine/system dependent statements within the LANCELOT programs contains a code. These codes or keywords are used in the following way.

- The first character is always "C", indicating that the statement should be considered as a Fortran comment, at least if the statement is not modified.
- The characters 2 to 5 are used to indicate the particular dependency on precision, machine or system. Table 6.3 describes the keywords defined at the date of printing.

Keyword	Meaning
CS	for the single precision version
CD	for the double precision version
CSUN	for SUN workstations
CIBM	for IBM machines (excluding the RISC6000)
CRS6	for IBM RISC6000 workstations
CVAXD	for Digital VAX machines using standard arithmetic
CVAXG	for Digital VAX machines using arithmtic with generalized exponent
CDEC	for Digital RISC workstations
CUNIX	for Unix-like systems
CCRAY	for Cray machines
CIEEE	for standard IEEE arithmetic

Table 6.3: Code keywords

The actual statement selection is performed by applying system dependent programs to the code, which "uncomment[16]" the suitable lines. These programs are written in Fortran and named `TOSNGL` and `TODBLE` for the single precision and for the double precision versions respectively.

[15] You should select the parts you prefer to read from Table 6.1.

[16] Replace the statement's first five characters by five blanks.

Note that the code will need further processing to pick out an appropriate size, as is explained in Section 6.5.2.2.

6.5.4.3 Designing a new Installation

At this point, you probably know enough about LANCELOT and you could tackle the installation itself. We next outline what we feel should be the main steps of this process.

1. You should first copy all the LANCELOT installation files to the same "place", whatever that means on your system[17]. It might also be wise to take a backup copy, just in case.
2. You then have to choose one or more new keyword(s) (always starting with "`C`") that will specify your installation,
3. You should insert system dependent statements suitable for your system in the package source files. The files you should process in this way are

   `runsd, runlan, local`

   The first two contain file opening and closing statements, which might be system dependent, the content of the third being already described in Section 6.5.3.2.
4. You should next write your own versions of the `TOSNGL` and `TODBLE` programs, that would select the statements with your newly defined keywords. This is best done by adapting existing versions to your system.
5. You should at this point decide on the structure of the installation on your machine, that is decide on "places" where to store the various files that together constitute a working LANCELOT installation. We suggest that you consider separating
   - the sources files for the SIF decoder, with the procedure to build the decoder (`makesd`, yet to be written),
   - the sources files for the optimizers, with the procedure to compile them (`makeop`, yet to be written),
   - the sources files for the system dependent tools (the precision casting and sizing programs as well as the `local` Fortran source file),
   - the LANCELOT commands and executable SIF decoder,
   - the libraries of single and double precision compiled optimization algorithms.

[17] Directory, disk, ...

An example of such an organization is presented (for Unix-like systems) in Section 6.1.

6. Once the organization is decided, you should write (in the command language of your system), a procedure (`unfold`) that would create that structure and transfer all the distribution files where they belong. Again, using existing `unfold` files will help.

7. The next step is to write a procedure that will compile all the Fortran source files for the SIF decoder and link them into an executable file. In our terminology, this is the **`makesd`** file. The relevant source files are

   ```
   decode  gps    inlanc  makefn  makegr  printp
   rest    runsd  utils   local
   ```

   Remember that your local version of the precision casting program (i.e. `tosngl` or `todble`) must be applied to all source files first. The adequate size casting program (`tobig`, `tomed` or `totoy`) must also be applied to `runsd` and `local` before compilation. It is good practice to let **`makesd`** have arguments allowing the user to specify the desired precision and program size. Examples are available in existing **`makesd`** files.

   We suggest you store the resulting executable file in the main LANCELOT directory/disk.

8. After the SIF decoder, you should now write your local **`makeop`** procedure, whose purpose is to compile (but not link) the Fortran source files for the optimization package itself. The relevant source files are

   ```
   asmbl   assl    auglag  bndsl   cauch  cg
   datal   drche   drchg   elgrd   frntl  getpgr
   dumbr   dumhsl  hsprd   initw   inxac  lance
   linpac  modchl  misc    others  precn  runlan
   sbmin   scaln   sort    speci   local
   ```

   As before, remember that your local version of the precision casting program (`tosngl` or `todble`) must be applied to all source files first. The adequate size casting program (`tobig`, `tomed` or `totoy`) must also be applied to `lance` and `local` before compilation. It is good practice to let **`makeop`** have arguments allowing the user to specify the desired precision and program size. Examples are available in existing **`makeop`** files.

   We recommend you store the compiled objects in a dedicated library/directory/disk, depending on what is appropriate for your system.

9. You are now ready to write the final installation procedure itself, `instll`. The tasks of the installation procedure are

   - check its arguments specifying the desired precision and program size,

- compile and link the necessary size and precision casting program, i.e. `tosngl` or `todble` and one of `tobig`, `tomed` or `totoy`,
- build the SIF decoder by calling the `makesd` procedure with suitable arguments,
- compile the optimizers by calling the `makeop` procedure with suitable arguments.

10. The next step is to write the sdlan and lan commands. The functionality of these commands is described in Sections 5.2.1 and 5.2.2 respectively. Existing sdlan and lan files should also provide help.

11. Comprehensive testing of the installation procedures and the commands should take place at this stage. Remember to test both the single and the double precision versions of the installation, as well as the different sizing options. The sdlan and lan commands can be tested by running LANCELOT on the two test problems supplied in the distribution: `ALLIN` and `HS65`.

12. When the installation and package are working, you should then write your local version of the `README` document, whose purpose is to help in further installations on the same system. Templates for this file are provided by existing `README` files.

13. If you reach this stage, you are indeed judged worthy of entering the selective circle of LANCELOT wizards! We would be *very* pleased to welcome you to this prestigeous company. To formally qualify for this exclusive honour, please send us your complete system dependent installation files, that is your local versions of

    ```
    instll lan    makeop makesd README sdlan
    todble tosngl unfold
    ```

    and any other file needed in your installation. Other users will then benefit from your work. We genuinely appreciate this type of collaboration and thank you very much in advance.

Chapter 7. The SIF Reference Report

At this stage the reader will presumably have already digested the primer in Chapter 2. In the present chapter, we provide a definitive description of the standard input format. It is intended to be used primarily as a reference document.

7.1 Introduction

The mathematical modelling of many real-world applications involves the minimization or maximization of a function of unknown parameters or variables. Frequently these parameters have known bounds; sometimes there are more general relationships between the parameters. When the number of variables is modest, say up to ten, the input of such a problem to an optimization procedure is usually fairly straightforward. Unfortunately many application areas now require the solution of optimization problems with thousands of variables; in this case merely the input of the problem data is extremely time-consuming and prone to error. Moreover, the mathematical programming community is only now designing algorithms for solving problems of this scale.

The work detailed in this chapter (and already introduced in Chapter 2) was motivated directly by the difficulties the authors were experiencing entering test examples to the LANCELOT large-scale nonlinear optimization package. It soon became apparent that if others were to be encouraged to carry out similar tests and even enticed to use our software, the process of specifying problems had to be considerably simplified. Thus we were inevitably drawn to provide a preliminary version of what is described in the present chapter: a standard input format (SIF) for nonlinear programming problems, together with an appropriate translator from the input file to the form required by the authors' minimization software. While understandably reflecting our views and experience, the present proposal is intended to be broadly applicable.

During the subsequent (and successive) stages of development of these preliminary ideas, various important considerations were discussed. These strongly influenced the present proposal.

- There are many reasons for proposing a standard input format. The most obvious one is the increased consistency in coding nonlinear programming problems, and the resulting improvement in code reliability.

As every problem is treated in a similar and standardized way, it is more difficult to overlook certain aspects of the problem definition. The provision of a SIF file for a given problem also allows some elementary (and very often helpful) automatic error and consistency checking.

- A further advantage of having a standard input format is the long awaited possibility of having a portable testbed of meaningful problems. Moreover, such a testbed that can be expected to grow. The authors soon experienced the daunting difficulties associated with specifying large scale problems — not only the difficulty of writing down the specification correctly but also the actual coding (and frequent re-coding) of a particular problem which often results in non-trivial differences between the initial and final data. These differences could be a major obstacle to valid comparisons between competing optimization codes. By contrast, having a SIF file allows simple and unambiguous data transfer via diskette, tape or electronic mail. The success of the NETLIB and Harwell/Boeing problem collections for linear programming and sparse linear algebra (see [26], and [19]) is a good recommendation for such flexibility. The formality required by the SIF approach may admittedly appear formidable for very simple problems, but is soon repaid when dealing with more complex ones.

- Of course, the SIF format should cover a large part of the practical optimization problems that users may want to specify. Explicit provision should be made not only for unconstrained problems, but also for constraints of different types and complexity: simple bounds on the variables, linear and/or nonlinear equations and inequalities should be handled without trouble. Special structure of the problem at hand is also a mandatory part of an SIF file. For example, the structure of least-squares problems must be described in an exploitable form. Sparsity of relevant matrices and partial separability of involved nonlinear functions must be included in the standard problem description when they are known. Finally, the special case of systems of nonlinear equations should also be covered.

- The existence and worldwide success of the MPS standard input format for linear programming must be considered as a *de facto* basis for any attempt to define an SIF for nonlinear problems. The number of problems already available in this format is large, and many nonlinear problems arise as a refinement of existing linear ones whose linear part and sparsity structure are expected to be described in the MPS format. It therefore seems reasonable to require that an SIF for nonlinear programming problems should conform to the MPS format. We were thus led to choose a standard that corresponds to MPS, augmented with additional constructs and structures, thus allowing nonlinearity, and the general features that we wished, to be described properly.

- The requirement of compatibility with the MPS format has a number of consequences, not all of which are pleasant. The first one is that the new SIF must be based on fixed format for the SIF file. Indeed, blanks are significant characters in MPS, when they appear in the right data fields, and cannot be used as general separators for free format input in any compatible system. The second one is the *a priori* existence of a "style" for keywords and overall layout of the problem description, a format which is not always ideally suited to nonlinear problems. Our present proposal accepts these limitations.

- The SIF should not be dependent on a specific operating system and/or manufacturer. In this respect, it must avoid relying on tools that may be excellent but are too specific (yacc and lex, for example). This of course does not prevent any implementation of an SIF interpreter using whatever facilities are locally available.

- In principle, the SIF should not be dependent on a particular high level programming language. However, as the intention is that SIF files may be converted into executable programs, restrictions on the symbolic names allowed by different programming languages may influence the choice of names within the problem description itself. For instance, in Fortran, symbolic variables may only contain up to six characters from a restricted set. We have chosen to base the present proposal, where necessary, on Fortran as this appears to be the most restrictive of the more popular high level languages. This dependence has been isolated as much as possible.

The authors are very well aware of the shortcomings of the SIF approach when compared to more elaborate modelling languages (see, for example, GAMS [2], AMPL [24] and OMP [17]). These probably remain the best way to allow easy and error free input of large problems. However, we contend that there is at present no language in the public domain which satisfactorily handles the nonlinear aspects of mathematical programming problems. While the advent of a tool of this nature is very much hoped for, it nevertheless seems necessary to provide something like the SIF now. This (we hope, intermediate) step is indeed crucial for the development and comparison of algorithms for solving large scale nonlinear problems, without which a more elaborate tool would be meaningless anyway. The SIF for nonlinear problems may also be considered as a first attempt to specify the minimal structures that should be present in a true modelling language for such problems. It is also of interest to develop a relatively simple input format, given that researchers developing new optimization methods may have to implement their own code for translating the SIF file into a form suitable for their algorithms. At this level, some compromise between completeness and simplicity seems necessary. Finally, the existence and availability of modelling languages for linear programming for a number of years has not yet made the MPS format irrelevant.

Hence, the reader should be aware that what sometimes appear as unnecessarily restrictive "features" of the proposed standard are often direct consequences of the considerations outlined above.

At this point, the reader should probably return to Section 1.2 and make sure the notions explained in this section are well understood. One further reason to return to this section is that we will use the three examples presented in Sections 1.2.3–1.2.5 as support for the present chapter, whose outline is now presented.

In the remainder of this section, we explain how we propose to exploit the structure in problems of the form (1.2.1)–(1.2.4) . We do this both in general and for a number of examples. Details of the way such structure may be expressed in a standard data input format follow in Section 2. The input of nonlinear information for element and group functions is covered in Section 3 and Section 4 respectively. The formats proposed in Sections 2–4 are quite rigid. A more flexible, free-form, input is considered in Section 5. The relationship to existing work is presented in Section 6 and conclusions drawn in Section 7.

7.2 The Standard Data Input Format

We now consider how to pass the data for optimization problems to an optimization procedure. In our description, we will concentrate on our third example, as presented in Section 1.2.5; we will show how the input file might be specified for this example to motivate the overall structure of such a file and then follow this with the general syntax allowed.

The data which defines a particular problem is written in a file in a standard format. It is intended that this data file is interpreted by an appropriate decoding program and converted into a format useful for input to an optimization package or program. The content of the file is specified line by line. As our format is intended to be compatible with the MPS linear programming format [14], we preserve the MPS terminology and call these lines *cards*.

A SIF comprises one or more files. The first of these files is known as the Standard *Data* Input Format (SDIF). As its name suggests, data which describes how the parts of the optimization problem are related, together with all fixed constants, are given in this file. Indeed, a SIF for linear programming problems can be completely specified by an MPS file; the SIF comprises a single section, the SDIF file, and that section is merely the MPS file.

7.2.1 Introduction to the Standard Data Input Format

As we have just said, the data format is designed to be compatible with the MPS linear programming format. There are, however, extensions to allow the user to input nonlinear problems. The user must prepare an input file consisting of three types of cards:

- Indicator cards, which specify the type of data to follow.
- Data cards, which contain the actual data.
- Comment cards.

Indicator cards contain a simple keyword to specify the type of data that follows. The first character of such cards must be in column 1; indicator cards are the only cards, with the exception of comment cards, which start in column 1. Possible indicator cards are given in Table 7.1.

Keyword	Comments	Presence	Described in §
`NAME`		mandatory	7.2.2.1
	either		
`GROUPS`		mandatory	7.2.2.6
`ROWS`	synonym for `GROUPS`		7.2.2.6
`CONSTRAINTS`	synonym for `GROUPS`		7.2.2.6
`VARIABLES`		mandatory	7.2.2.7
`COLUMNS`	synonym for `VARIABLES`		7.2.2.7
	or		
`VARIABLES`		mandatory	7.2.2.8
`COLUMNS`	synonym for `VARIABLES`		7.2.2.8
`GROUPS`		mandatory	7.2.2.9
`ROWS`	synonym for `GROUPS`		7.2.2.9
`CONSTRAINTS`	synonym for `GROUPS`		7.2.2.9
`CONSTANTS`		optional	7.2.2.10
`RHS`	synonym for `CONSTANTS`		7.2.2.10
`RHS'`	synonym for `CONSTANTS`		7.2.2.10
`RANGES`		optional	7.2.2.11
`BOUNDS`		optional	7.2.2.12
`START POINT`		optional	7.2.2.13
`ELEMENT TYPE`		optional	7.2.2.14
`ELEMENT USES`		optional	7.2.2.15
`GROUP TYPE`		optional	7.2.2.16
`GROUP USES`		optional	7.2.2.17
`OBJECT BOUND`		optional	7.2.2.18
`ENDATA`		mandatory	7.2.2.2

Table 7.1: Possible indicator card

Indicator cards must appear in the order shown, except that the `GROUPS` and `VARIABLES` sections may be interchanged to allow specification of the linear terms by rows or columns. The cards `CONSTANTS`, `RHS'`, `RHS`, `RANGES`, `BOUNDS`, `START POINT`, `ELEMENT TYPE`, `ELEMENT USES`, `GROUP TYPE`, `GROUP USES` and `OBJECT BOUND` are optional.

The data cards are divided into six fields. The content of each field varies with each type of data card as described in Section 7.2.2. Those in fields 1, 2, 3 and 5 must always be left justified within the field. Field 1, which appears in columns 2 and 3, may contain a *code* (that is, a one or two character string which defines the expected contents of the remaining fields on the card), fields 2, 3 and 5 may hold *names* and fields 4 and 6 might store *numerical values*. The numerical values are defined by up to 12 characters which may include a decimal point and an optional sign (a positive number is assumed unless a – is given). The value may be followed by a decimal exponent, written as an `E` or `D`, followed by a signed or unsigned one or two digit integer; the first blank after the `E` or `D` terminates the field. The names of variables, nonlinear elements or groups may be up to ten characters long. These names may include integer indices (see Section 7.2.1.1).

Any card with the character * in column 1 is a comment card; the remaining contents of the card are ignored. Such a card may appear anywhere in the data file. In addition, completely blank cards are ignored when scanning the input file and may thus be used to space the data. Finally, the presence of a $ as the first character in fields 3 or 5 of a data card indicates that the content of the remaining part of the card is a comment and will be ignored.

7.2.1.1 Names

One of the positive features of the MPS standard is the ability to give meaningful names to problem constraints and variables. As our proposal is intended to be MPS compatible, we too have this option. However, one of the less convenient features of the MPS standard is the cumbersome way that repetitious structure is handled. In particular, the name of each variable and constraint must be defined on a separate line, and structure within constraints is effectively ignored when setting up the constraint matrix. We consider it important to overcome this deficiency of the MPS standard when formulating large-scale examples. One way is to allow variable, group and nonlinear element names to have indices and to have syntactic devices which enable the user to define many items at once.

Unless otherwise indicated, we allow any name which uses up to ten valid characters. A *valid* character is any ASCII character whose decimal code lies in the range 32 to 126 (binary 0100000 to 11111110, hex 20 to 7E) (see, for instance [49]). This includes lower and upper case roman alphabetic characters, the digits 0 to 9, the blank character and other mathematical and grammatical symbols. A name can be one of the following:

1. a scalar name of the form $\mathcal{LLLLLLLLLL}$ where each $\mathcal{L}$ is a valid character type excepting that the first $\mathcal{L}$ cannot be a `$`. A completely blank

string is also not allowed. Futhermore, the strings 'SCALE', 'MARKER', 'DEFAULT', 'INTEGER' and 'ZERO-ONE' are reserved for special operations.

2. an array name of the form name(index), where index is a list of integer index names, name is a list of valid characters (the first character may not be a $) and the maximum possible size of the *expanded name* does not exceed ten characters. The list of index names must be of the form list1, list2, list3, where list1, list2 and list3 are predefined index (parameter) names (see Section 7.2.2.3, below) and all three indices are optional. The indices are only allowed to take on integer values. Commas are only required as separators; the presence of an open bracket "(" announces a list of indices and a close bracket ")" terminates the list. An array *name* is *expanded* as namenumber1, number2, number3, where numberi, i = 1, 2, 3 is the integer value allocated to the index listi at the time of use.

As an example, the expanded form of the array name X(I,J,K) when I, J and K have the values 3, 4 and 6 respectively would be X3,4,6, while it would take the form X-6,0,3 if I, J and K have the values -6, 0 and 3 respectively. However, X(I,J,K) could not be expanded if I, J and K were each allowed to be as large as 100 as, for instance, X100,100,100 is over ten characters long and thus not a valid expanded name.

An array item may be referred to by either its array name (so long as the index lists have been specified) or by its expanded name. Thus, if I, J and K have been specified as 2, 7 and 9 respectively, X(I,J,K) and X2,7,9 are identical.

If two separators (opening or closing brackets and commas) are adjacent in an array name, the intervening index is deemed not to exist and is ignored when the name is expanded. Thus, the expanded name of Y() is just Y, while that of Z(I,,K) is Z3,4 if I is 3 and K is 4. Furthermore, any name which does not include the characters "(", ")" or "," may be used as an array name and is its own expanded name. Thus the name X may be a scalar or array name whereas W(and V, can only be scalar names.

We defer the definition of integer indices until Section 7.2.2.3.

Note that blanks are considered to be significant characters. Thus if _ denotes a blank, the names _x and x_ are different. It is recommended that all names are left-shifted within their relevant data fields to avoid possible user-instigated name recognition errors.

7.2.1.2 Fortran Names

A notable exception to the above are Fortran names. A *Fortran* name takes the form of a sequence of one to six upper case letters or digits, the first of which must not be a digit. These names are used in Sections 7.2.2.14–7.2.2.17, 7.3.4.1–7.3.4.3 and 7.4.2.1.

7.2.1.3 Numerical Values

The definition of a specific problem normally requires the use of numerical (integer or real) data values. Such values can be specified in two ways. Firstly, the values may simply occur as integer or floating-point numbers in data fields 4 and 6. Secondly, values may be allocated to named parameters, known as integer or real parameters, and a value subsequently used by reference to a particular integer or real parameter name. This second method may only be used to allocate values on certain cards; when this facility is used, the first character in field 1 on the relevant data card will be an `X` or a `Z`. This latter approach is particularly useful when a value is to be used repeatedly or if a value is to be changed within a do-loop (see Section 7.2.2.4).

We defer the definition of integer and real parameters until Section 7.2.2.3.

7.2.1.4 An Example

Before we give the complete syntax for an SDIF file, we give an illustrative example. In order to exhibit as many constructs as possible, we consider how we might encode the example in Section 1.2.5. We urge the reader to study this section in detail. As always, there are many possible ways of specifying a particular problem; we give one in Figures 7.1 and 7.2, pages 192 and 193. The horizontal and vertical lines are merely included to indicate the extent of data fields. The actual widths of the fields are given at the top of the figure, and the column numbers given at its foot.

The SDIF file naturally divides into two parts. In the first part, lines 2 to 39 of the figure, we specify information regarding linear functions used in the example. In the second part, lines 40 to 93, we specify nonlinear information. The first part is merely an extension of the MPS input format; the second part is new.

The file must always start with a `NAME` card, on which a name (in this case `EG3`) for the example may be given (line 1), and must end with an `ENDATA` card (line 93). A comment is inserted at the end of line 1 as to the source of the example. The character `$` identifies the remainder of the line as a comment; the comment is ignored when interpreting the input file.

We next specify names of parameters which will occur frequently in specifying the example (lines 2 to 5). In our case the integer and real parameters `1` and `ONE` are given along with `N`, a problem dimension — here `N` is set to 100, but it would be trivial to change the example in 6 to allow variables $x_1, \ldots, x_n$ for any n. We make a comment to this effect on line 4; any card with the character `*` in column 1 is a comment card and its content is ignored when interpreting the input file.

We now name the problem variables and groups (in our example objective function and constraints) used. The groups may be specified before or after the variables. We choose here to name the groups first. The objective function will be known as `OBJ` (line 7); the character `N` in field 1 specifies that this is an objective function group. The inequality constraints (1.2.19) and (1.2.20) are named `CONLE1`, ..., `CONLE99` and `CONGE1`, ..., `CONGE100` respectively. Rather than specify them individually, a do-loop is used to make an array definition. Thus the constraints `CONLE1`, ..., `CONLE99` are defined *en masse* on lines 9 to 11 with the do-loop index `I` running from the previously defined value 1 to the value `NM1`. The integer parameter, `NM1`, is defined on line 8 to be the sum of `N` and the value `-1` and in our case will be 99. The characters `XL` in field 1 of line 10 indicate that an array definition is being made (the `X`) and that the groups are less-than-or-equal-to constraints (the `L`). The do-loop introduced on line 9 with the characters `DO` in field 1 is terminated on line 11 with the characters `ND` in its first field. In a similar way, the constraints `CONGE1`, ..., `CONGE99` are defined all together on lines 12 to 14; that these constraints involve bounds on both sides is taken care of by considering them to be greater-than-or-equal-to constraints (`XG`) on line 13 and later specifying the additional upper bounds in the `RANGES` section (lines 26 to 29). Finally, the equality constraint (1.2.21) is to be called `CONEQ` (line 15); the character `E` in field 1 specifies that this is an equality constraint group.

Having named the groups, we next name the problem variables. At the same time, we include the coefficients of all the linear elements used. The variables are named `X1`, ..., `X100` and `Y`; an array declaration is made for the former set on lines 17 to 19 and `Y` is defined on line 20. The character `X` in field 1 of line 18 indicates that an array definition is used. Only the objective function (1.2.18), inequality constraint (1.2.19) and equality constraint groups (1.2.21) contain linear elements. As well as introducing `Y`, line 20 also specifies that the linear element associated with group `OBJ` (field 3) involves variable `Y`, and that `Y` 's coefficient in the linear element is 1.0 (field 4). A do-loop is now used in lines 21 to 23 to show that the linear elements for constraints (1.2.19) also use the variable `Y`. It is assumed that unless a variable is explicitly identified with a linear element, the element is independent of that variable. Thus, although (1.2.21) uses a linear element, the element is constant and need not be specified in the `VARIABLES` section.

The only remaining part of the linear elements which must be specified is the constant term. Again, only nonzero constants need be given. For our example, only the equality constraint group (1.2.21) has a nonzero constant term and this data is specified on lines 24 and 25. The string `C1` in field 2 of line 25 is the name given to a specific set of constants. In general, more than one set of constants may be specified in the SDIF file and the relevant one selected in a postprocessing stage. Here, of course, we only have one set.

As we have seen, the inequality constraint groups (1.2.20) are bounded from above as well as from below. In the `RANGES` section (lines 26 to 30) we specify these upper bounds (or range constraints as they are sometimes known). The numerical values $\frac{1}{2}$ are specified for each bound for the relevant groups in an array definition on line 28; the string `R1` in field 2 is once again a name given to a specific set of range values as it is possible to define more than one set in the `RANGES` section.

We now turn to the simple bounds (1.2.22) which are specified in lines 30 to 36 of the example. All problem variables are assumed to have lower bounds of zero and no upper bounds unless otherwise specified. All but one of the variables for our example have lower bounds of -1. We thus change the default value for the value of the lower bound on line 31 - the set of bounds is named `BND1`. The character `L` specifies that it is the lower bound default that is to be changed. The string `'DEFAULT'` in field 3 indicates that the default is being changed. The variable x_i is given an upper bound of i. We encode that in a do-loop on lines 32 to 35 of the figure. The do-loop index `I` is an integer. We change its current value to a real on line 33 and assign that value as the upper bound on line 34. The character `Z` in field 1 of this line indicates that an array definition is being made and that the data is taken from a parameter in field 5 (as opposed to a specified numerical value in field 4) and the character `U` specifies that the upper bound value is to be assigned. The variable `y` is unbounded or, as it is often known, free. This is specified on line 36, the string `FR` in field 1 indicating that `Y` is free.

The final "linear" piece of information given is an estimate of the solution to the problem (if known) or at least a set of values from which to start a minimization algorithm. This information is given on lines 37 to 39. For our problem, we choose the values $x_i = \frac{1}{2}, 1 \leq i \leq 100$ and $y = 0$. Unless otherwise specified, all starting values take a default of zero. We change that default on line 38 to $\frac{1}{2}$ — the set of starting values are named `START1` — and then specify the individual value for the variable `Y` on line 39.

We now specify the nonlinear information. We saw in Section 1.2.5 that there are four element types for the problem, being of the form (i) $(v_1 - v_2)v_3$, (ii) $p_1 v_1 v_2$, (iii) $\sin v_1$ and (iv) $(v_1 + v_2)^2$. In the `ELEMENT TYPE` section on lines 40 to 48, we record details of these types. We name the four types (i)–(iv) `3PROD, 2PROD, SINE` and `SQUARE` respectively. For `3PROD`, we define the elemental variables (lines 41 and 42) to be `V1, V2` and `V3` and the internal variables (line 43) to be `U1` and `U2`. Elemental variables may be defined, two to a line, on lines for which field 1 is `EV`. Internal variables, on the other hand, are defined on lines with `IV` in field 1. Similar definitions are made for `2PROD` (line 44), `SINE` (line 46) and `SQUARE` (line 47). The type `2PROD` also makes use of a parameter p_1. This is named `P1` on line 45 for which field 1 reads `EP`.

Having specified the element types, we next specify individual nonlinear elements in the `ELEMENT USES` section. As we have seen, the objective function group uses a single nonlinear element of type `3PROD`. We name this particular element `OBJ1`. On line 50, the character `T` in field 1 indicates that the `OBJ1` is of type `3PROD`. The assignment of problem to elemental variables is made on lines 51 to 53. Problem variables X1 and X2 are assigned to elemental variables V1 and V3; the assignment is indicated by the character V in field 1. In order to assign x_{100} (or in general x_n) to v_2, we assign the array entry X(N) to V2. Notice that as an array element is being used, this must be specially flagged (ZV in field 1) as otherwise the wrong variable (called X(N) rather than X100, which is the expanded form of X(N)) would be assigned. There are two nonlinear elements for each inequality constraint group (1.2.19), each being of the same type 2PROD. We name these elements CLEA1,...,CLEA99 and CLEB1,...,CLEB99. The assignments are made on lines 54 to 67 within a do-loop. On lines 56 and 60 the elements are named and their types assigned.

As array assignments are being used, field 1 for both lines contains the string XT. The elemental variables are then associated with problem variables on lines 57–58 and 61–62 respectively. Again array assignments are used and field 1 contains the string ZV. Notice that on line 58 v_2 is assigned the problem variable x_{i+1}, where the index IP1 is defined as the sum of the index I and the integer value 1 on line 55. It remains to assign values for the parameter p_1 for each element. This is straightforward for the elements CLEA1,...,CLEA99 as the required value is always 1 and the assignment is made on line 59 on a card with first field XP. The remaining elements have varying parameter values $1 + 2/i$. This value is calculated on lines 63 to 65 and assigned on line 66. Line 63 assigns REALI to have the floating point value of the index I. This new value is then divided into the value 2 on line 64 and the value assigned to ONE is added to the resulting value on the final line. Thus the parameter 2OVAI+1 holds the required value of p_1 and the array assignment is made on line 66. On this line the string ZP indicates that an array assignment is being made, taking its value from the parameter 2OVAI+1 in field 4 (the Z) and that the elemental parameter P1 in field 3 is to be assigned (the P). The definition of the nonlinear elements for the remaining constraint groups is straightforward. The inequality constraints (1.2.20) each use a single element, named CGE1,...,CGE100, of type SINE and the appropriate array assignments are made on lines 68 to 70. Finally, the equality constraint (1.2.21) is named CEQ1 and typed SQUARE with appropriate elemental variable assignments on lines 72 to 74.

We next need to specify the nontrivial group types. This is done in the GROUP TYPE section on lines 75 to 77. We saw in Section 1.2.5 that a single nontrivial group, $p_1\alpha^2$, is required. On line 76, the name PSQUARE is given for the type and the group type variable α is named ALPHA. The string GV pin field 1 indicates that a type and its variable are to be defined. On the following line field 1 is GP and this is used to announce that the group type parameter p_1 is named P1.

Finally, we need to allocate nonlinear elements to groups and specify what type the resulting groups are to be. This takes place within the GROUP USES section which runs from line 78 to 90. The objective function group is nontrivial and its type is announced on line 79. The group uses the single nonlinear element `OBJ1` specified on line 80 and the group-type parameter p_1 is set to the value $\frac{1}{2}$ on the next line. The characters `T`, `E` and `P` in the first fields of these three cards announce their purposes. The inequality groups (1.2.19) each use two nonlinear elements, but the groups themselves are trivial (and thus their types do not have to be made explicit). The assignment of the elements to each group is made in an array definition on lines 82 to 84; line 83 is flagged as assigning elements to a group with the string `XE` in field 1. The second set of inequality constraints (1.2.20) use the nontrivial group type `PSQUARE` with parameter value 1. Each group uses a single nonlinear element and the appropriate array assignments are contained on lines 85 to 89. Lastly the trivial equality constraint group (1.2.21) is assigned the nonlinear element `CEQ1` on line 90.

The definition of the problem is now complete. However, it often helps the intended minimization program if known lower and upper bounds on the possible values of the objective function can be given. For our example, the objective function (1.2.18) cannot be smaller than zero. This data is specified on lines 91 and 92. The string `LO` in field 1 of line 92 indicates that a lower bound is known for the value of (1.2.18). The string `OBOUND` in field 2 of this line is a name given to this known bound. The value of the lower bound now follows in field 4. No upper bound need be specified as the function is initially assumed to lie between plus and minus infinity.

```
 <> <---10---> <---10---><----12---->   <---10---><----12---->
 F.1 Field 2   Field 3    Field 4        Field 5   Field 6
NAME          EG3                      $ The example of §1.6
 IE 1                   1
 IE N                   100
*Variants of §1.6 obtained by choice of N on previous card
 RE ONE                 1.0
GROUPS
 N  OBJ
 IA NM1       N         -1
 DO I         1                        NM1
 XL CONLE(I)
 ND
 DO I         1                        N
 XG CONGE(I)
 ND
 E  CONEQ
VARIABLES
 DO I         1                        N
 X  X(I)
 ND
    Y         OBJ       1.0
 DO I         1                        NM1
 X  Y         CONLE(I)  1.0
 ND
CONSTANTS
    C1        CONEQ     1.0
RANGES
 DO I         1                        NM1
 X  R1        CONGE(I)  0.5
 ND
BOUNDS
 LO BND1      'DEFAULT' -1.0
 DO I         1                        N
 RI REALI     I
 ZU BND1      X(I)                     REALI
 ND
 FR BND1      Y
START POINT
    START1    'DEFAULT' 0.5
    START1    Y         0.0
ELEMENT TYPE
 EV 3PROD     V1                       V2
 EV 3PROD     V3
 IV 3PROD     U1                       U2
 EV 2PROD     V1                       V2
 EP 2PROD     P1
 EV SINE      V1
 EV SQUARE    V1                       V2
 IV SQUARE    U1

1 2 3 5    10   1415      2425       36   40      4950       61
```

Figure 7.1: SDIF file (part 1) for the example of Section 1.2.5

<> F.1	<---10---> Field 2	<---10---> Field 3	<----12----> Field 4	<---10---> Field 5	<----12----> Field 6
ELEMENT USES					
T	OBJ1	3PROD			
V	OBJ1	V1		X1	
ZV	OBJ1	V2		X(N)	
V	OBJ1	V3		X2	
DO	I	1		NM1	
IA	IP1	I	1		
XT	CLEA(I)	2PROD			
ZV	CLEA(I)	V1		X(1)	
ZV	CLEA(I)	V2		X(IP1)	
XP	CLEA(I)	P1	1.0		
XT	CLEB(I)	2PROD			
ZV	CLEB(I)	V1		X(I)	
ZV	CLEB(I)	V2		X(N)	
RI	REALI	I			
RD	2OVERI	REALI	2.0		
R+	2OVAI+1	2OVERI		ONE	
ZP	CLEB(I)	P1		2OVAI+1	
ND					
DO	I	1		N	
XT	CGE(I)	SINE			
ZV	CGE(I)	V1		X(I)	
ND					
T	CEQ1	SQUARE			
V	CEQ1	V1		X1	
ZV	CEQ1	V2		X(N)	
GROUP TYPE					
GV	PSQUARE				
GP	PSQUARE	P1			
GROUP USES					
T	OBJ	SQUARE			
E	OBJ	OBJ1			
P	OBJ	P1	0.5		
DO	I	1		NM1	
XE	CONLE(I)	CLEA(I)		CLEB(I)	
ND					
DO	I	1		N	
XT	CONGE(I)	PSQUARE			
XE	CONGE(I)	CGE(I)			
XP	CONGE(I)	P1	1.0		
ND					
E	CONEQ	CEQ1			
OBJECT BOUND					
LO	OBOUND		0.0		
ENDATA					
1 2 3	5 10 14	15 24	25 36	40 49	50 61

Figure 7.2: SDIF file (part 2) for the example of Section 1.2.5

7.2.2 Indicator and Data Cards

We now give details of the indicator cards and the data cards which follow them.

7.2.2.1 The NAME Indicator Card

The `NAME` indicator card is used to announce the start of the input data for a particular problem. The user may specify a name for the problem; this name is entered on the indicator card in field 3 and may be at most 8 characters long. The syntax for the `NAME` card is given in Figure 7.3.

```
                <---8-->
                 Field 3
NAME            prob-nam
                ↑       ↑
                15      24
```

Figure 7.3: The indicator card `NAME`

7.2.2.2 The ENDATA Indicator Card

The `ENDATA` indicator card simply announces the end of the input data. The data for a particular problem, in the form of indicator and data cards, must lie between a `NAME` and an `ENDATA` card. The syntax for the `ENDATA` card is given in Figure 7.4.

```
ENDATA
```

Figure 7.4: The indicator card `ENDATA`

7.2.2.3 Integer and Real Parameters

We shall use the word *parameter* to mean the name given to any quantity which is associated with a specified numerical value. The numerical value will be known as the *parameter* value. Integer and real values may be associated with parameters in two ways. The easiest way is simply to set a parameter to a specified parameter value, or to obtain a parameter from a previously defined parameter by simple arithmetic operations (addition, subtraction, multiplication and division). The second way is to have a parameter value specified in a do-loop, or to obtain a parameter from one specified in a do-loop (see Section 7.2.2.4 below).

The syntax for associating a parameter with a specific value is given in Figure 7.5.

The two character string in data field 1 (F.1) specifies the way in which the parameter value is to be assigned. If the first of these characters is a `I`, the assigned value is an integer; the parameter will be referred to as an *integer parameter* or *integer index.* Alternatively, if the first of these characters is an `R` or an `A`, the assigned value is a real and the parameter will be called a *real parameter.*

```
<> <---10---><---10---><----12---->   <---10--->
F.1  Field 2   Field 3     Field 4       Field 5
 IE int-p-name          numerical-vl
 IR int-p-namerl--p-name
 IA int-p-nameint-p-namenumerical-vl
 IS int-p-nameint-p-namenumerical-vl
 IM int-p-nameint-p-namenumerical-vl
 ID int-p-nameint-p-namenumerical-vl
 I= int-p-nameint-p-name
 I+ int-p-nameint-p-name               int-p-name
 I- int-p-nameint-p-name               int-p-name
 I* int-p-nameint-p-name               int-p-name
 I/ int-p-nameint-p-name               int-p-name
 RE rl--p-name          numerical-vl
 RI rl--p-nameint-p-name
 RA rl--p-namerl--p-namenumerical-vl
 RS rl--p-namerl--p-namenumerical-vl
 RM rl--p-namerl--p-namenumerical-vl
 RD rl--p-namerl--p-namenumerical-vl
 RF rl--p-namefunct-namenumerical-vl
 R= rl--p-namerl--p-name
 R+ rl--p-namerl--p-name               rl--p-name
 R- rl--p-namerl--p-name               rl--p-name
 R* rl--p-namerl--p-name               rl--p-name
 R/ rl--p-namerl--p-name               rl--p-name
 R( rl--p-namefunct-name               rl--p-name
 AE r-p-a-name          numerical-vl
 AI r-p-a-nameint-p-name
 AA r-p-a-namer-p-a-namenumerical-vl
 AS r-p-a-namer-p-a-namenumerical-vl
 AM r-p-a-namer-p-a-namenumerical-vl
 AD r-p-a-namer-p-a-namenumerical-vl
 AF r-p-a-namefunct-namenumerical-vl
 A= r-p-a-namer-p-a-name
 A+ r-p-a-namer-p-a-name               r-p-a-name
 A- r-p-a-namer-p-a-name               r-p-a-name
 A* r-p-a-namer-p-a-name               r-p-a-name
 A/ r-p-a-namer-p-a-name               r-p-a-name
 A( r-p-a-namefunct-name               r-p-a-name
 ↑↑ ↑        ↑↑        ↑↑          ↑  ↑        ↑
 2 3 5      1415      2425        36 40       49
```

Figure 7.5: Possible cards for specifying parameter values

If the string is `IE`, the integer parameter `int-p-name` named in field 2 is to be given the integer value specified in field 4. The parameter may be up to ten characters long, and the integer value can occupy up to twelve positions.

If the string is `IR`, the integer parameter value named in field 2 is to be assigned the value of the nearest integer (closer to zero) to the value of the real parameter `rl--p-name` specified in field 3. The parameter appearing in field 3 must have already been assigned a value.

If the string is `IA`, the integer parameter named in field 2 is to be formed by adding the value of the parameter `int-p-name` referred to in field 3 to the integer value specified in field 4. The parameter appearing in field 3 must have already been assigned a value.

If the string is `IS`, the integer parameter named in field 2 is to be formed by subtracting the value of the parameter `int-p-name` referred to in field 3 from the integer value specified in field 4. The parameter appearing in field 3 must have already been assigned a value.

If the string is `IM`, the value of the integer parameter named in field 2 is to be obtained by multiplying the value already specified for the parameter in field 3 by the integer value specified in field 4. Once again, the parameter appearing in field 3 must have already been assigned a value.

If the string is `ID`, the value of the integer parameter named in field 2 is to be obtained by dividing the integer value specified in field 4 by the value already specified for the parameter in field 3. Once again, the parameter appearing in field 3 must have already been assigned a value.

If the string is `I=`, the value of the integer parameter named in field 2 is to be set to the integer value specified for the parameter in field 3. The parameter appearing in field 3 must have already been assigned a value.

If the string is `I+`, the value of the integer parameter named in field 2 is to be calculated by adding the values of the integer parameters `int-p-name` referred to in fields 3 and 5. The parameters appearing in fields 3 and 5 must have already been assigned values.

If the string is `I-`, the value of the integer parameter named in field 2 is to be calculated by subtracting the value of the integer parameters `int-p-name` referred to in field 5 from that in field 3. The parameters appearing in fields 3 and 5 must have already been assigned values.

If the string is `I*`, the value of the integer parameter named in field 2 is to be formed as the product of the values already specified for the integer parameters in fields 3 and 5. The parameters appearing in fields 3 and 5 must have already been assigned values.

Finally, if the string is `I/`, the value of the integer parameter named in field 2 is to be formed by dividing the value specified for the integer parameters in field 3 by that specified for the integer parameters in field 5. Once again, the parameters appearing in fields 3 and 5 must have already been assigned values.

Note that, as an array name can only be a maximum of 10 characters long, any integer parameter which is to be the index of an array can only be at most seven characters in length. Furthermore, such a parameter name may not include the characters "(", ")" or ",".

If the string is `RE`, the real parameter `rl--p-name` named in field 2 is to be given the real value specified in field 4. The parameter may be up to ten characters long, and the real value can occupy up to twelve positions.

If the string is `RI`, the real parameter value named in field 2 is to be assigned the equivalent floating point value of the integer parameter `int-p-name` specified in field 3. The parameter appearing in field 3 must have already been assigned a value.

If the string is `RA`, the value of the real parameter named in field 2 is to be formed by adding the value of the real parameter `rl--p-name` referred to in field 3 to the real value specified in field 4. The parameter appearing in field 3 must have already been assigned a value.

If the string is `RS`, the value of the real parameter named in field 2 is to be formed by subtracting the value of the real parameter `rl--p-name` referred to in field 3 from the real value specified in field 4. The parameter appearing in field 3 must have already been assigned a value.

If the string is `RM`, the value of the parameter named in field 2 is to be formed by multiplying the value specified for the real parameter in field 3 by the real value specified in field 4. Once again, the parameter appearing in field 3 must have already been assigned a value.

If the string is `RD`, the value of the parameter named in field 2 is to be formed by dividing the real value specified in field 4 by the value specified for the real parameter in field 3. The parameter appearing in field 3 must have already been assigned a value.

If the string is `RF`, the value of the parameter named in field 2 is to be formed by evaluating the function named in field 3 at the real value specified in field 4. The function `funct-name` — and its mathematical equivalent $f(x)$ — may be one of: `ABS` ($f(x) = |x|$), `SQRT` ($f(x) = \sqrt{x}$), `EXP` ($f(x) = e^x$), `LOG` ($f(x) = \log_e x$), `LOG10` ($f(x) = \log_{10} x$), `SIN` ($f(x) = \sin x$), `COS` ($f(x) = \cos x$), `TAN` ($f(x) = \tan x$), `ARCSIN` ($f(x) = \sin^{-1} x$), `ARCCOS` ($f(x) = \cos^{-1} x$), `ARCTAN` ($f(x) = \tan^{-1} x$), `HYPSIN` ($f(x) = \sinh x$), `HYPCOS` ($f(x) = \cosh x$) or `HYPTAN` ($f(x) = \tanh x$). Certain of the functions may only be evaluated for arguments lying within restricted ranges. The argument for `SQRT` must be non-negative, those for `LOG` and `LOG10` must be strictly positive, and those for `ARCSIN` and `ARCCOS` must be no larger than one in absolute value.

If the string is `R=`, the parameter value named in field 2 is to be assigned the value of the real parameter `rl--p-name` referred to in field 3. The parameter appearing in field 3 must have already been assigned a value.

If the string is `R+`, the parameter value named in field 2 is to be formed as the sum of the values of the real parameters `rl--p-name` referred to in fields 3 and 5. The parameters appearing in fields 3 and 5 must have already been assigned values.

If the string is `R-`, the parameter value named in field 2 is to be formed by subtracting the value of the real parameter `rl--p-name` referred to in field 5 from the value of that referred to in field 3. The parameters appearing in fields 3 and 5 must have already been assigned values.

If the string is `R*`, the value of the real parameter named in field 2 is to be formed as the product of the values already specified for the real parameters in fields 3 and 5. Once again, the parameters appearing in fields 3 and 5 must have already been assigned values.

If the string is `R/`, the parameter value named in field 2 is to be formed by dividing the value of the real parameter `rl--p-name` referred to in field 3 by the value of that referred to in field 5. The parameters appearing in fields 3 and 5 must have already been assigned values.

Finally, if the string is `R(`, the value of the parameter named in field 2 is to be formed by evaluating the function named in field 3 at the value of the real parameter `rl--p-name` specified in field 5. The function (and its mathematical equivalent) may be any of those named in the `RF` paragraph and the restrictions on the allowed argument ranges given above still apply.

If the first character in field 1 is an `A`, an array of real parameters is to be defined. The particular type of definition is as for the `R` cards, excepting that any name, `r-p-a-name`, referred to in fields 2, 3 or 5, with the exception of integer parameters named in field 3 of `AI` cards and functions named in the same field of `AF` and `A(` cards, must be a real parameter array name with a valid index.

Parameter assignments may be made at any point within the SDIF file between the `NAME` and `ENDATA` indicator cards. It is anticipated that parameters will be used to store values such as the total number of variables and groups, which are used later in array definitions, and to allow a user to enter regular and repetitious data in a straightforward and compact way.

7.2.2.4 Do-loops

A do-loop may occur at any point in the `GROUPS`, `VARIABLES`, `CONSTANTS`, `RANGES`, `BOUNDS`, `START POINT`, `ELEMENT USES` or `GROUP USES` sections. Do-loops are used to make array definitions, that is, to make compact definitions of several quantities at once. The syntax required for do-loops is given in Figure 7.6.

The two-character string in data field 1 specifies either the start or the end of a do-loop. The start of a loop is indicated by the string `DO`. In this case an integer parameter named in field 2 is defined to take values starting from the integer parameter value given in field 3 and ending with the last value before the integer parameter value given in field 5 has been surpassed. The parameters named in fields 3 and 5 must have been defined on previous data cards. The parameter name defined in field 2 can occupy up to ten locations. If the next data card does not have the characters `DI` as its first

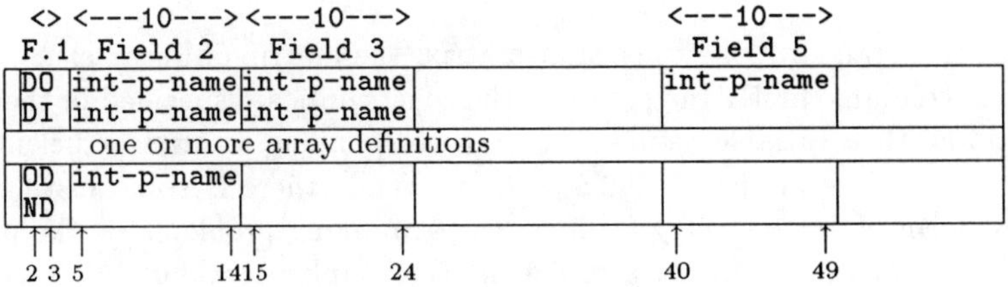

Figure 7.6: Syntax for do-loops

field, the parameter defined on the `DO` card, `iloop` say, will take all integer values starting from that given in field 3, say `istart`, and ending on that in field 5, `iend` say. If `istart` is larger than `iend`, the loop will be skipped.

If the data card following a `DO` card has the string `DI` in field 1, the do-loop parameter named in field 2 is to be incremented by the amount, `incr` say, specified for the integer parameter given in field 3. Once again, the parameter in field 3 must have been previously defined. The index `iloop` will now take values

$$\texttt{iloop} = \texttt{istart} + j \cdot \texttt{incr}$$

for all positive j for which `iloop` lies between (and including) `istart` and `iend`. If `incr` is negative and `istart` is larger than `iend`, the parameter specifies a decreasing sequence of values. If `incr` is positive and `istart` is larger than `iend`, or if `incr` is negative and `istart` is smaller than `iend`, the loop will be skipped.

Once a do-loop has been started, any array definitions which use its do-loop index specify that the definition is to be made for all values of the integer parameter specified in the loop. Loops can be nested up to three deep; this corresponds to the maximum number of allowed indices in an array index list.

A do-loop must be terminated. A particular loop can be terminated on a data card in which field 1 contains the characters `OD`; the name of the loop parameter must appear in field 2. Alternatively, all loops may be terminated at once using a data card in which field 1 contains the characters `ND`.

In addition, parameter assignments with the syntax given in Figure 7.5 — that is, cards whose first field are `IE`, `IR`, `IA`, `IS`, `IM`, `ID`, `I=`, `I+`, `I-`, `I*`, `I/`, `RE`, `RI`, `RA`, `RS`, `RM`, `RD`, `RF`, `R=`, `R+`, `R-`, `R*`, `R/`, `R(`, `AE`, `AI`, `AA`, `AS`, `AM`, `AD`, `AF`, `A=`, `A+`, `A-`, `A*`, `A/` or `R(` — may be inserted at any point in a do-loop; it is only necessary that a parameter is defined prior to its use.

Note that array definitions may occur both within and outside do-loops; all that is required for a successful array definition is that the integer indices used have defined values when they are needed. The use of do-loops is illustrated in Section 7.2.4.

7.2.2.5 The Definition of Variables and Groups

In the MPS standard, the constraint matrix, the matrix of linear elements, is input by columns; firstly the names of the constraints are specified in the `ROWS` section and then variable names and the corresponding matrix coefficients are set one at a time in the `COLUMNS` section. While there is some justification for this form of matrix entry for linear programming problems — the principal solution algorithm for such problems, the simplex method [16], is usually column oriented — there seems no good reason why the coefficients of linear elements might not also be input by rows. After all, it is more natural to think of specifying the constraints for a problem one at a time. Furthermore, requiring that a complete row or column has been specified before the next may be processed is unnecessarily restrictive.

We thus allow the data to be input in a either a group-wise (row-wise) or variable-wise (column-wise) fashion. In a group/row-wise scheme, one or two coefficients and their variable/column names are specified for a given group/row; for a variable/column-wise scheme, one or two coefficients and their group/row names are specified for a given variable/column. We do, however, still require that in a group/row-wise storage scheme, the names of all the variables/rows *which appear in linear elements* are completely specified before the coefficients are input. Similarly, in a variable/column-wise storage scheme, the names of all the groups/rows *which have a linear element* must be completely specified before the coefficients are input. This allows for some checking of the input data. If the groups/rows are specified first, there is no requirement that variables/columns are input one at a time (but of course they may be). When processing the data file, variable/column names should be inspected to see if they are new or where they have appeared before. Likewise, if the variables/columns are specified first, there is no requirement that groups/rows are ordered on input. The coordinates of new data values can then be stored as a linked triple (group/row, variable/column, value). Conversion from such a component-wise input scheme to a row or column based storage scheme may be performed very efficiently if desired (see [20, pp. 30–31] and subroutine MC39 in the Harwell Subroutine Library). If a variable/column-wise input scheme is to be adopted, the data file will contain a `GROUPS/ROWS/CONSTRAINTS` indicator card and section followed by a `VARIABLES/COLUMNS` card and section. The allowed data cards are discussed in Section 7.2.2.6 and Section 7.2.2.7. If a group/row-wise input scheme is to be adopted, the data file will contain a `VARIABLES/COLUMNS` indicator card and section followed by a `GROUPS/ROWS/CONSTRAINTS` card and section. The data cards for this scheme are discussed in Section 7.2.2.8 and Section 7.2.2.9.

7.2.2.6 The GROUPS, ROWS or CONSTRAINTS Data Cards (variable/column-wise)

The GROUPS, ROWS and CONSTRAINTS indicator cards are used interchangeably to announce the names of the groups which make up the objective function or, for constrained problems, the names of the constraints (or rows, as they are often known in linear programming applications). The user may give a scaling factor for the groups or constraints. In addition, groups which are linear combinations of previous groups may be specified. The syntax for the data cards which follow these indicator cards is given in Figure 7.7.

```
<> <---10---><---10---><----12---->  <---10---><----12---->
F.1  Field 2   Field 3      Field 4      Field 5     Field 6
GROUPS or
ROWS or
CONSTRAINTS
 N  group-name$$$$$$$$$$numerical-vl
 G  group-name$$$$$$$$$$numerical-vl
 L  group-name$$$$$$$$$$numerical-vl
 E  group-name$$$$$$$$$$numerical-vl
 XN group-name$$$$$$$$$$numerical-vl
 XG group-name$$$$$$$$$$numerical-vl
 XL group-name$$$$$$$$$$numerical-vl
 XE group-name$$$$$$$$$$numerical-vl
 ZN group-name$$$$$$$$$$               r-p-a-name
 ZG group-name$$$$$$$$$$               r-p-a-name
 ZL group-name$$$$$$$$$$               r-p-a-name
 ZE group-name$$$$$$$$$$               r-p-a-name
 DN group-name$$$$$$$$$$numerical-vl   $$$$$$$$$$numerical-vl
 DG group-name$$$$$$$$$$numerical-vl   $$$$$$$$$$numerical-vl
 DL group-name$$$$$$$$$$numerical-vl   $$$$$$$$$$numerical-vl
 DE group-name$$$$$$$$$$numerical-vl   $$$$$$$$$$numerical-vl
 ↑↑ ↑        ↑↑        ↑↑          ↑   ↑        ↑↑          ↑
 2 3 5      1415      2425        36  40       4950        61
```

Figure 7.7: Possible data cards for GROUPS, ROWS or CONSTRAINTS (column-wise)

The one- or two-character string in data field 1 specifies the type of group, row or constraint to be input. Possible values for the first character are:

N : the group is to be specially marked (for constrained problems, the group/-row is an objective function group/row).

G : the group is to use an extra "artificial" variable; this variable will only occur in this particular group, will be non-negative and its value will be subtracted from the group function. For constrained problems, this is equivalent to requiring the constraint/row be non-negative; the extra variable is then a surplus variable and whether it is used explicitly (considered as a problem variable) or implicitly will depend upon the optimization technique to be used. Thus, if the problem variables are x, and the k-th group has a linear element $a_k^T x - b_k$, the linear element that will be passed to the optimization procedure could be $a_k^T x - y_k - b_k$, for some non-negative variable y_k.

`L` : the group is to use an extra "artificial" variable; this variable will only occur in this particular group, will be non-negative and its value will be added to the group function. For constrained problems, this is equivalent to requiring the constraint/row be non-positive; the extra variable is then a slack variable and may be used explicitly or implicitly by the optimization procedure. Thus, if the linear element is as specified above, the linear element that will be passed to the optimization procedure could be $a_k^T x + y_k - b_k$, for some non-negative variable y_k.

`E` : the group is a normal one (for constrained problems, the row/constraint is an equality),

`X` and `Z` : an array of groups are to be defined at once. When the first character is an `X` or `Z`, the second character may be one of `N`, `G`, `L` or `E`. The resulting array of groups then each has the characteristics of an `N`, `G`, `L` or `E` group as just described.

`D` : the group is to be formed as a linear combination of two previous groups. When the first character is a `D`, the second character may be one of `N`, `G`, `L` or `E`. The resulting group then has the characteristics of an `N`, `G`, `L` or `E` group as just described.

The string `group-name` in data field 2 gives the name of the group (or row or constraint) under consideration. This name may be up to ten characters long, excepting that the name `'SCALE'` is not allowed. For `X` data cards, the expanded array name must be valid and the integer indices must have been defined in a parameter assignment (see Section 7.2.2.3).

The string $$$$$$$$$$ in data field 3 may be blank; this happens when field 2 is merely announcing the name of a group. If it is not blank, it is used for two purposes.

- It may be used to announce that all the entries (if any) in the linear element for the group under consideration are to be scaled, that is *divided* by a constant scale factor; in this case field 3 will contain the string `'SCALE'`. If the first character in field 1 is a `Z`, the string in data field 5 gives the name of a previously defined real parameter and the numerical value associated with this parameter gives the scale factor. Otherwise, the string `numerical-vl`, occupying up to 12 locations in data field 4, contains the scale factor. Fields 5 and 6 are not then used.

- If the first character in field 1 is a `D`, the current group is to be formed as a linear combination of the groups mentioned in fields 3 and 5; the multiplication factors are then recorded in fields 4 and 6 respectively. Thus we will have

$$\text{group in field 2} = \text{group in field 3} * \text{field 4} + \text{group in field 5} * \text{field 6}.$$

In this case, the names of the groups in fields 3 and 5 must have already been defined. The multiplication factors may occupy up to 12 locations in fields 4 and 6.

7.2.2.7 The `VARIABLES` or `COLUMNS` Data Cards (Variable/Column-Wise)

The `VARIABLES` or `COLUMNS` indicator cards are used interchangeably to announce the (problem) variables for the minimization. In addition, the entries for the linear elements are input here. The user may also give a scaling factor for the entries in any column. The syntax for data following this indicator card is given in Figure 7.8.

```
<> <---10---><---10---><----12---->  <---10---><----12---->
F.1  Field 2    Field 3     Field 4      Field 5     Field 6
VARIABLES or
COLUMNS
    varbl-name$$$$$$$$$$numerical-vl  $$$$$$$$$$numerical-vl
 X  varbl-name$$$$$$$$$$numerical-vl  $$$$$$$$$$numerical-vl
 Z  varbl-name$$$$$$$$$$              r-p-a-name
 ↑↑ ↑        ↑↑        ↑↑          ↑   ↑         ↑↑          ↑
 2 3 5       1415      2425        36  40        4950        61
```

Figure 7.8: Possible data cards for `VARIABLES` or `COLUMNS` (column-wise)

The string `varbl-name` in data field 2 gives the name of the variable (or column) under consideration. This name may be up to ten characters long excepting that the name '`SCALE`' is not allowed. If data field 1 holds the character `X` or `Z`, an array of variables is to be defined. In this case, the expanded array name of the variables (or columns) must be valid and the integer indices must have been defined in a parameter assignment (see Section 7.2.2.3).

The string `$$$$$$$$$$` in data field 3 is used for five purposes.

- If the string is empty, the card is just defining the name of a problem variable.
- It may be used to specify that the variable mentioned in field 2 occurs in the linear element for the group given in field 3. In this case, the string in field 3 must have been defined in the `GROUPS` section. If an array definition is being made, the string in field 3 must be an array name.
- It may be used to announce that all the entries in the linear elements for the variable under consideration are to be scaled; in this case field 3 will contain the string '`SCALE`'.

- It may be used to specify that the variable(s) mentioned in field 2 is(are) only allowed to take integer values. In this case field 3 will contain the string `'INTEGER'`.
- It may be used to specify that the variable(s) mentioned in field 2 is(are) only allowed to take the values 0 or 1. In this case field 3 will contain the string `'ZERO-ONE'`.

A numerical value, whose purpose depends on the string in the previous field, is now specified. On `Z` cards, the value is that previously associated with the real parameter `r-p-a-name` in field 5. On other cards, the actual numerical value `numerical-vl` may occupy up to 12 characters in data field 4.

If field 3 indicates that an entry for the linear element for a group is to be defined, the specified numerical value gives the coefficient of that entry. If, on the other hand, field 3 indicates that all entries for the variable in field 2 are to be scaled, the specified value gives the scale factor, that is the factor by which each entry is to be divided.

On non `Z` cards, the strings in fields 5 and 6 are optional and are used exactly as for strings 3 and 4 to define further entries or a scale factor.

7.2.2.8 The `VARIABLES` or `COLUMNS` Data Cards (Group/Row-Wise)

The `VARIABLES` or `COLUMNS` indicator cards are used interchangeably to announce the (problem) variables for the minimization. The user may also give a scaling factor for the entries in the column. The syntax for data following this indicator card is given in Figure 7.9.

```
 <><---10---><---10---><----12---->  <---10---><----12---->
 F.1 Field 2   Field 3    Field 4       Field 5     Field 6
VARIABLES or
COLUMNS
   varbl-name$$$$$$$$$$numerical-vl
 X varbl-name$$$$$$$$$$numerical-vl
 Z varbl-name$$$$$$$$$$            r-p-a-name
 ↑↑ ↑        ↑↑        ↑↑        ↑ ↑         ↑↑          ↑
 2 3 5       1415      2425      36 40       4950        61
```

Figure 7.9: Possible data cards for `VARIABLES` or `COLUMNS` (row-wise)

The string `varbl-name` in data field 2 gives the name of the variable (or column) under consideration. This name may be up to ten characters long excepting that the name `'SCALE'` is not allowed. If data field 1 holds the character `X` or `Z`, an array definition is to be made. In this case, the expanded array name of the variables (or columns) must be valid and the integer indices must have been defined in a parameter assignment (see Section 7.2.2.3).

The string `$$$$$$$$$$` in data field 3 is used for four purposes.

- If the string is empty, the card is just defining the name of a problem variable. Such a card must be inserted for all variables that only appear in nonlinear elements.

- It may be used to announce that all the entries in the linear elements for the variable under consideration are to be scaled. On `Z` cards, the numerical value of this scale factor, the amount by which each entry is to be divided, is that previously associated with the real parameter `r-p-a-name` given in field 5. On other cards, the actual scale factor `numerical-vl` occupies up to 12 characters in data field 4.

- It may be used to specify that the variable(s) mentioned in field 2 is(are) only allowed to take integer values. In this case field 3 will contain the string `'INTEGER'`.

- It may be used to specify that the variable(s) mentioned in field 2 is(are) only allowed to take the values 0 or 1. In this case field 3 will contain the string `'ZERO-ONE'`.

7.2.2.9 The `GROUPS`, `ROWS` or `CONSTRAINTS` Data Cards (Group/Row-Wise)

The `GROUPS`, `ROWS` and `CONSTRAINTS` indicator cards are used interchangeably to announce the names of the groups which make up the objective function and, for constrained problems, the names of the constraints (or rows, as they are often known in linear programming applications). In addition, the entries for the linear elements are input here. The user may give a scaling factor for the groups or constraints. Furthermore, groups which are linear combinations of previous groups may be specified. The syntax for the data cards which follow these indicator cards is given in Figure 7.10.

The one- or two-character string in data field 1 specifies the type of group, row or constraint to be input. Possible values for the first character and their interpretations are exactly as in Section 7.2.2.6.

The string `group-name` in data field 2 gives the name of the group (or row or constraint) under consideration. This name may be up to ten characters long excepting that the name `'SCALE'` is not allowed. For `X` and `Z` data cards, the expanded array name must be valid and the integer indices must have been defined in a parameter assignment (see Section 7.2.2.3). The kind of group (`N`, `L`, `G` or `E`) will be taken to be that which is defined on the *first* occurrence of a data card for that group. Subsequent contradictory information will be ignored.

The string `$$$$$$$$$$` in data field 3 is used for three purposes.

- It may be used to specify that the group mentioned in field 2 has a linear element involving the variable given in field 3. In this case, the string in field 3 must have been defined in the `VARIABLES` section. If an array

```
 <><---10---><---10---><----12---->  <---10---><----12---->
F.1 Field 2   Field 3      Field 4      Field 5      Field 6
GROUPS or
ROWS or
CONSTRAINTS
 N  group-name$$$$$$$$$$numerical-vl  $$$$$$$$$$numerical-vl
 G  group-name$$$$$$$$$$numerical-vl  $$$$$$$$$$numerical-vl
 L  group-name$$$$$$$$$$numerical-vl  $$$$$$$$$$numerical-vl
 E  group-name$$$$$$$$$$numerical-vl  $$$$$$$$$$numerical-vl
 XN group-name$$$$$$$$$$numerical-vl  $$$$$$$$$$numerical-vl
 XG group-name$$$$$$$$$$numerical-vl  $$$$$$$$$$numerical-vl
 XL group-name$$$$$$$$$$numerical-vl  $$$$$$$$$$numerical-vl
 XE group-name$$$$$$$$$$numerical-vl  $$$$$$$$$$numerical-vl
 ZN group-name$$$$$$$$$$              r-p-a-name
 ZG group-name$$$$$$$$$$              r-p-a-name
 ZL group-name$$$$$$$$$$              r-p-a-name
 ZE group-name$$$$$$$$$$              r-p-a-name
 DN group-name$$$$$$$$$$numerical-vl  $$$$$$$$$$numerical-vl
 DG group-name$$$$$$$$$$numerical-vl  $$$$$$$$$$numerical-vl
 DL group-name$$$$$$$$$$numerical-vl  $$$$$$$$$$numerical-vl
 DE group-name$$$$$$$$$$numerical-vl  $$$$$$$$$$numerical-vl
 ↑↑ ↑         ↑↑        ↑↑          ↑ ↑         ↑↑          ↑
 2 3 5       1415      2425        36 40       4950        61
```

Figure 7.10: Possible data cards for GROUPS, ROWS, or CONSTRAINTS (row-wise)

definition is being made, the string in field 3 must be an array name. The numerical value of the coefficient of the linear term corresponding to the variable must now be specified. On Z cards, the value is that previously associated with the real parameter `r-p-a-name` given in field 5. On other cards, the actual numerical value `numerical-vl` may occupy up to 12 characters in data field 4.

- It may be used to announce that all the entries (if any) in the linear element for the group under consideration are to be scaled; in this case field 3 will contain the string 'SCALE'. The numerical value of the scale factor, that is the factor by which the group is to be divided, is now specified exactly as above.

  In these first two cases, fields 5 and 6 may be used to define further coefficients or a scale factor for non Z cards.

- If the first character in field 1 is a D, the current group is to be formed as a linear combination of the groups mentioned in fields 3 and 5; the multiplication factors are then recorded in fields 4 and 6 respectively. Thus we will have

  $$\text{group in field } 2 = \text{group in field } 3 * \text{field } 4 + \text{group in field } 5 * \text{field } 6.$$

  In this case, the names of the groups in fields 3 and 5 must have already been defined. The multiplication factors may occupy up to 12 locations in fields 4 and 6.

7.2.2.10 The CONSTANTS, RHS or RHS' Data Cards

The `CONSTANTS`, `RHS` or `RHS'` indicator cards are used interchangeably to announce the definition of a vector of the constant terms b_i (in the constrained case, the right-hand-sides) for each linear element. The syntax for data following this indicator card is given in Figure 7.11.

```
<> <---10---><---10---><----12---->  <---10---><----12---->
F.1  Field 2    Field 3     Field 4       Field 5     Field 6
CONSTANTS or
RHS or
RHS'
    rhs--name $$$$$$$$$$numerical-vl  group-namenumerical-vl
 X  rhs--name $$$$$$$$$$numerical-vl  group-namenumerical-vl
 Z  rhs--name $$$$$$$$$$              r-p-a-name
 2 3 5      1415      2425        36  40      4950        61
```

Figure 7.11: Possible data cards for `CONSTANTS`, `RHS` or `RHS'`

The string `rhs--name` in data field 2 gives the name of the vector of group constants/ right-hand-sides. This name may be up to ten characters long. More than one vector of group constants may be defined.

The strings `$$$$$$$$$$` is used for two purposes.

- It may be used to assign a default value to all the constants in a particular vector. In this case field 3 will contain the string `'DEFAULT'`.
- It may contain the name of a group/row/constraint for which the constant term/right-hand-side is to be specified. Such a string must have been defined in the `GROUPS` section.

The string `numerical-vl` in data field 4 and (optionally) 6 now contains the numerical value of the constant/right-hand-side and may occupy up to 12 locations.

Constants for an array of groups may also be defined on cards in which field 1 contains the character `X` or `Z`. On such cards, the expanded array name in field 3 and (as an option on `X` cards) 5 must be valid and the integer indices must have been defined in a parameter assignment (see Section 7.2.2.3). On `Z` cards, the numerical value of the constant/right-hand-side is that previously associated with the real parameter array, `r-p-a-name`, given in field 5. On `X` cards, the actual numerical value `numerical-vl` may occupy up to 12 characters in data fields 4 and (optionally) 6.

Any constants not specified take a default value. The default value for the components of each vector is initially zero. This default may be changed using a card whose third field contains the string `'DEFAULT'` as mentioned above. On such a card, the default value for the vector in field 2 is given in field 4 whenever field 1 is blank or contains the character `X`. If field 1 contains the character `Z`, the default value is that associated with the real parameter, `rl--p-name`, named in field 5. The default value applies to each constant not explicitly specified; if the default is to be changed, the change must be made on the first card naming a particular vector of constants.

7.2.2.11 The `RANGES` Data Cards

The `RANGES` indicator card is used to announce the definition of a vector of additional bounds on the artificial variables introduced in the `GROUPS` section (in the constrained case, this corresponds to saying that specified inequality constraints/rows have both lower and upper bounds). The syntax for data following this indicator card is given in Figure 7.12.

```
 <> <---10---><---10---><----12---->  <---10---><----12---->
 F.1  Field 2    Field 3     Field 4      Field 5      Field 6
RANGES
 | range-name$$$$$$$$$$numerical-vl  group-namenumerical-vl
 |X range-name$$$$$$$$$$numerical-vl  group-namenumerical-vl
 |Z range-name$$$$$$$$$$              r-p-a-name
 ↑↑ ↑          ↑↑        ↑↑          ↑  ↑         ↑↑          ↑
 2 3 5        1415      2425        36 40        4950        61
```

Figure 7.12: Possible data cards for `RANGES`

The string `range-name` in data field 2 gives the name of the vector of range values. This name may be up to ten characters long. More than one vector of range values may be defined.

The string `$$$$$$$$$$` is used for two purposes.

- It may be used to assign a default value to all the range values in a particular vector. In this case field 3 will contain the string `'DEFAULT'`.
- It may contain the name of a group/row/constraint for which the range value is to be specified. Such a string must have been defined in the `GROUPS` section.

In addition, the (optional) string `group-name` in data field 5 may also define the name of a group/row/constraint for which the range value is to be specified.

The string `numerical-vl` in data field 4 and (optionally) 6 now contains the numerical value of the relevant range value and may occupy up to 12 locations. Only groups initially specified with a `G` or `L` in columns 1 or 2 of field 1 in the `GROUPS` section use range values and therefore only these groups may be specified.

Range values for an array of groups may also be defined on cards on which field 1 is the character `X` or `Z`. On such cards, the expanded array name in field 3 and (as an option on `X` cards) 5 must be valid and the integer indices must have been defined in a parameter assignment (see Section 7.2.2.3). On `Z` cards, the range value is that previously associated with the real parameter, `r-p-a-name`, given in field 5. On `X` cards, the actual numerical value `numerical-vl` may occupy up to 12 characters in data fields 4 and (optionally) 6. Using the terminology of Section 7.2.2.6, the extra bound is taken to imply the inequality $0 \leq y_k \leq |\text{field 4 or 6}|$ on the artificial variable y_k.

Any component in a range vector not specified takes a default value. The default value for the components of each vector is initially infinite. This default may be changed using a card whose third field contains the string `'DEFAULT'` as mentioned above. On such a card, the default value for the vector in field 2 is given in field 4 whenever field 1 is blank or contains the character `X`. If field 1 contains the character `Z`, the default value is that associated with the real parameter, `rl--p-name`, named in field 5. The default value applies to each range value not explicitly specified. If the default is to be changed, the change must be made on the first card naming a particular vector of range values.

7.2.2.12 The `BOUNDS` Data Cards

The `BOUNDS` indicator card is used to announce a vector of data giving lower and upper bounds on the unknown variables.The syntax for data following this indicator card is given in Figure 7.13.

```
 <> <---10---><---10---><----12---->  <---10---><----12---->
 F.1 Field 2   Field 3     Field 4      Field 5
BOUNDS
 LO bound-name$$$$$$$$$$numerical-vl
 UP bound-name$$$$$$$$$$numerical-vl
 FX bound-name$$$$$$$$$$numerical-vl
 FR bound-name$$$$$$$$$$
 MI bound-name$$$$$$$$$$
 PL bound-name$$$$$$$$$$
 XL bound-name$$$$$$$$$$numerical-vl
 XU bound-name$$$$$$$$$$numerical-vl
 XX bound-name$$$$$$$$$$numerical-vl
 XR bound-name$$$$$$$$$$
 XM bound-name$$$$$$$$$$
 XP bound-name$$$$$$$$$$
 ZL bound-name$$$$$$$$$$              r-p-a-name
 ZU bound-name$$$$$$$$$$              r-p-a-name
 ZX bound-name$$$$$$$$$$              r-p-a-name
 ↑↑ ↑         ↑↑        ↑↑          ↑ ↑        ↑
 2 3 5        1415      2425        36 40      49
```

Figure 7.13: Possible data cards for `BOUNDS`

The two-character string in data field 1 specifies the type of bound to be input. Possible values are: LO, XL or ZL, in the case of a lower bound, UP, XU, ZU,in the case of an upper bound, FX, XX, ZX,in the case of a fixed variable, i.e., the lower and upper bounds are equal, FR or XR if the variable is free, i.e., the lower and upper bounds are infinite, MI or XM, if there is no lower bound, and PL or XP, if there is no upper bound. The string bound-name in data field 2 gives the name of the bound vector under consideration. This name may be up to ten characters long. Several different bound vectors may be defined in the BOUNDS section.

The string $$$$$$$$$$ is used for two purposes.

- It may contain the name of a variable/column for which a bound is to be specified. This name may be up to ten characters long and must refer to a variable defined in the VARIABLE data.

  If the card is of type LO, UP, FX, FR, MI, or PL, the string in data field 3 specifies to which variable the bound is applied. If the card is of type XL, ZL, XU, ZU, XX, ZX, XR, XM or XP, this string specifies an array of variables which are to be bounded. On such cards, the expanded array name of this string must be valid and the integer indices must have been defined in a parameter assignment (see Section 7.2.2.3).

  For bounds of type LO, UP, FX, XL, XU or XX, the numerical value of the bound or array of bounds is given as the string numerical-vl, using at most 12 characters, in data field 4. For bounds of type ZL, ZU or ZX, the numerical value of the array of bounds is that previously associated with the real parameter array r-p-a-name specified in field 5. When both lower and upper bounds on a variable are required, they must be specified on separate cards. Possible combinations are LO–UP, LO–PL, MI–UP, XL–XU, XL–XP, XM–XU, ZL–XU, XL–ZU, ZL–ZU, ZL–XP and XM–ZU.

- It may be used to assign a default value to all the lower and/or upper bounds in a particular vector. In this case field 3 will contain the string 'DEFAULT'. Each bound vector is given default lower and upper bounds on every variable. The value of the default lower bound is initially zero and the upper bound is initially infinite. These default values may be changed. A new default value for the vector in field 2 is then given in field 4 whenever field 1 is LO, UP, FX, FR, MI, PL or starts with the character X. If field 1 starts with the character Z, the default value is that associated with the real parameter, rl--p-name, named in field 5. The appropriate default value applies to each bound not explicitly specified. If the default is to be changed, the change must be made on the first card naming a particular vector of bounds.

  If default lower and upper bounds of zero and infinity, respectively, are in effect and a card with MI or XM in field 1 is encountered, the relevant lower bound is changed to minus infinity *and* the upper bound becomes zero. If the same defaults are in effect and an upper bound of zero is specified on a UP, XU or ZU card, the relevant lower bound becomes minus infinity. These two features are necessary for MPS compatibility.

7.2.2.13 The START POINT Data Cards

The START POINT indicator card is used to announce a vector of initial estimates of the values of the unknown variables and, in the case of problems with general constraints, Lagrange multipliers. The Lagrangian function associated with (1.2.1)–(1.2.4) is the function

$$l(x,\lambda) = \sum_{i\in I_0} g_i\left(\sum_{j\in J_i} w_{i,j} f_j(x_j) + a_i^T x - b_i\right) + \sum_{i\in I_E\cup I_I} \lambda_i g_i\left(\sum_{j\in J_i} w_{i,j} f_j(x_j) + a_i^T x - b_i\right)$$

where the scalars λ_i are known as Lagrange multipliers. Good estimates of these parameters can sometimes be useful for optimization procedures (see, for example, [31]). The syntax for data following this indicator card is given in Figure 7.14.

```
<> <---10---><---10---><----12---->  <---10---><----12---->
F.1  Field 2   Field 3     Field 4     Field 5     Field 6
START POINT
 V  start-name$$$$$$$$$$numerical-vl  varbl-namenumerical-vl
 XV start-name$$$$$$$$$$numerical-vl  varbl-namenumerical-vl
 ZV start-name$$$$$$$$$$              rl--p-name
 M  start-name$$$$$$$$$$numerical-vl  multp-namenumerical-vl
 XM start-name$$$$$$$$$$numerical-vl  multp-namenumerical-vl
 ZM start-name$$$$$$$$$$              rl--p-name
    start-name$$$$$$$$$$numerical-vl  varbl-namenumerical-vl
    start-name$$$$$$$$$$numerical-vl  multp-namenumerical-vl
 X  start-name$$$$$$$$$$numerical-vl  varbl-namenumerical-vl
 X  start-name$$$$$$$$$$numerical-vl  multp-namenumerical-vl
 Z  start-name$$$$$$$$$$              rl--p-name
 ↑↑ ↑         ↑↑        ↑↑          ↑ ↑         ↑↑          ↑
 2 3 5        1415      2425       36 40        4950        61
```

Figure 7.14: Possible data cards for START POINT

The V, XV and ZV cards are used to define the starting value for variables. In any of these cards, the string **start-name** in data field 2 gives the name of a starting vector and may be up to ten characters long. Several different starting vectors may be defined in the START POINT section. The string $$$$$$$$$$ in field 3 is used for two purposes.

- It must contain the name of a variable defined in the VARIABLES section, when a starting value is to be assigned to that variable. If field 1 does not contain ZV, the optional string `varbl-name` in data field 5 may also contain the name of such a variable whose starting value is to be assigned.

- It may be used to assign a default value to all the starting values for variables in a particular vector. In this case field 3 must contain the string 'DEFAULT'.

  Each starting point vector is given a default value for every variable. This value is initially zero, but it may be changed. The appropriate default value applies to each variable not explicitly specified.

The M, XM and ZM cards are used to define the starting value for Lagrange multipliers. In any of those cards, the string `start-name` in data field 2 gives the name of a starting vector and may be up to ten characters long. Several different starting vectors may be defined in the START POINT section. The string $$$$$$$$$$ in field 3 is used for two purposes.

- It must contain the name of a group defined in the GROUPS section which is not an objective function group, when a starting value is to be assigned to the corresponding Lagrange multiplier. If field 1 does not contain ZV, the optional string `multp-name` in data field 5 may also contain the name of such a multiplier whose starting value is to be assigned.

- It may be used to assign a default value to all the starting values for Lagrange multipliers in a particular vector. In this case field 3 must contain the string 'DEFAULT'.

  Each starting point vector is given a default value for every Lagrange multiplier. This value is initially zero, but it may be changed. The appropriate default value applies to each multiplier not explicitly specified.

The effect of a card whose field 1 is blank or contains X or Z is similar to that of V, XV, ZV, M, XM and ZM cards, except that

- variables and Lagrange multipliers may be mixed up on cards whose field 3 does not contain 'DEFAULT',

- default values are assigned to both variables and Lagrange multipliers on card whose field 3 contains 'DEFAULT'. The default value for both variables and multipliers is initially zero.

If the defaults are to be changed, the change must be made on the first card naming a particular starting point vector.

Starting values for an array of variables or Lagrange multipliers may only be defined on cards whose field 1 begins with the character X or Z; on X cards two arrays may be defined on a single card. On such cards, the expanded array name in field 3 (and field 5 for X cards) must be valid and the integer indices must have been defined in a parameter assignment (see Section 7.2.2.3).

It remains to specify the numerical value of the default or individual starting point as appropriate. On cards whose field 1 starts with `Z`, the value is that previously associated with the real parameter **`rl--p-name`** or array of real parameters **`rl--p-name`** (respectively) given in field 5. On other cards, the numerical value is (or values are) specified using up to twelve characters in the string(s) **`numerical-vl`** in data field 4 (and if required field 6).

7.2.2.14 The `ELEMENT TYPE` Data Cards

The `ELEMENT TYPE` indicator card is used to announce the data for the different types of nonlinear elements to be used. The names of the elemental and, optionally, internal variables and parameters for each element type are specified in this section. The syntax for data cards following the indicator card is given in Figure 7.15.

```
<> <---10---><--6->                          <--6->
F.1  Field 2      Field 3                     Field 5
ELEMENT TYPE
 EV etype-nameev-nam                          ev-nam
 IV etype-nameiv-nam                          iv-nam
 EP etype-nameep-nam                          ep-nam
 ↑↑ ↑         ↑↑    ↑                         ↑    ↑
 2 3 5        1415  20                        40   45
```

Figure 7.15: Possible data cards for `ELEMENT TYPE`

The string in field 1 may be one of `EV`, `IV` or `EP`. This indicates whether the names of elemental variables (`EV`), internal variables (`IV`) or elemental parameters (`EP`) are to be specified on the given data card. If no cards with the string `IV` in field 1 are found for a particular element type, the element is assumed to have no useful internal variables; the internal variables are then allocated the same names as the elemental ones. Likewise, if no cards with the string `EP` in field 1 are found for a particular element type, the element is assumed not to depend on parameter values.

The string **`etype-name`** in data field 2 gives the name of the element type under consideration. This name may be up to ten characters long. The data for a particular element must be specified on consecutive data cards.

The strings in data fields 3 and (optionally) 5 give the names of elemental variables (field 1 = `EV`), internal variables (field 1 = `IV`) or parameters (field 1 = `EP`) for the element type specified in field 2. These strings must be valid Fortran names (see Section 7.2.1.2). The names of the variables for different element types may be the same; the names of the elemental variables, internal variables and parameters (if the latter two are given) for a specific element type must all be different.

7.2.2.15 The ELEMENT USES Data Cards

The ELEMENT USES indicator card is used to specify the names and types of the nonlinear element functions. The element types may be selected from among those defined in the ELEMENT TYPE section. Associations are made between the problem variables and the elemental variables for the elements used and parameter values are assigned. The syntax for data following this indicator card is given in Figure 7.16.

```
 <> <---10---> <--6->      <----12---->  <--6->      <----12---->
 F.1  Field 2    Field 3      Field 4       Field 5       Field 6
ELEMENT USES
 T  $$$$$$$$$$etype-nam
 XT $$$$$$$$$$etype-nam
 V  elmnt-nameev-nam                     varbl-name
 ZV elmnt-nameev-nam                     varbl-name
 P  elmnt-nameep-nam     numerical-vl    ep-nam       numerical-vl
 XP elmnt-nameep-nam     numerical-vl    ep-nam       numerical-vl
 ZP elmnt-nameep-nam                     r-p-a-nam
 ↑↑ ↑        ↑↑    ↑    ↑↑          ↑   ↑    ↑   ↑↑          ↑
 2 3 5      1415   20   2425        36  40   45  4950        61
```

Figure 7.16: Possible data cards for ELEMENT USES

There are three sorts of data cards in the ELEMENT USES section. For cards of the second and third kinds, the string elmnt-name in data field 2 gives the name, or an array of names, of a nonlinear element function. This name may be up to ten characters long and each nonlinear element name must be unique. On array cards (those prefixed by X or Z), the expanded element array name in field 2 must be valid and the integer indices must have been defined in a parameter assignment (see Section 7.2.2.3).

The first kind of cards, identified by the characters T or XT in field 1, give the name, or an array of names, of an element and its type. On such cards, the string $$$$$$$$$$ in field 2 is used for two purposes.

- It may contain the name, or an array of names, of a nonlinear element function whose type is to be defined. This name may be up to ten characters long and each nonlinear element name must be unique. On XT cards the expanded element array name in field 2 must be valid and the integer indices must have been defined in a parameter assignment (see Section 7.2.2.3).

- It may be used to assign a default type to all the nonlinear element functions. In this case field 2 must contain the string 'DEFAULT'. Any element not explicitly typed is assumed to belong to a default type. If a default is to be used, it must be specified on a card and such a card may only appear before the first T or XT card in the ELEMENT USES section. The string etype-name in data field 3 gives the name of the element type to be used. This name may be up to ten characters long and must have appeared in the `ELEMENT TYPE` section.

The second kind of data card, identified by the characters `V` or `ZV` in field 1, is used to assign problem variables to the elemental variables appropriate for the element type. On this data card, the string `ev-nam` in data field 3 gives the name of one of the elemental variables for the given element type. This name must have been set in the `ELEMENT TYPE` section and be a valid Fortran name (see Section 7.2.1.2). The string `varbl-name` in data field 5 then gives the name of the problem variable that is to be assigned to the specified elemental variable. The name of this variable may have been set in the `VARIABLES/COLUMNS` section or may be a new variable (often known as a *nonlinear variable*) introduced here and can be up to ten characters long. On a `ZV` card, the name of the variable must be an element of an array of variables, with a valid name and index.

The last kind of data card, identified by the characters `P`, `XP` or `ZP` in field 1, is used to assign numerical values to the parameters for the element functions (`P`) or array of element functions (`XP` and `ZP`). On this data card, the string `ep-nam` in data field 3 (and, for `P` and `XP` cards, optionally 5) must give the name of a parameter. This name must have been set in the `ELEMENT TYPE` section and be a valid Fortran name, see Section 7.2.1.2. On `P` and `XP` cards, the strings `numerical-vl` in data fields 4 and (optionally) 6 contain the numerical value of the parameter. These values may each occupy up to 12 locations within their field. On `ZP` cards, the string `r-p-a-name` in data field 5 gives a real parameter array name. This name must have been previously defined and its associated value then gives the numerical value of the parameter.

7.2.2.16 The `GROUP TYPE` Data Cards

The `GROUP TYPE` indicator card is used to announce the data for the different types of nontrivial groups which are to be used. The names of the group-type variable and, optionally, of group parameters for each group type are specified in this section. The syntax for data cards following the indicator card is given in Figure 7.2.2.16.

The string in field 1 may be either `GV` or `GP`. This indicates whether the name of a group-type variable (`GV`) or one or more group parameters (`GP`) are to be specified on the given data card. The data for a particular group type must be specified on consecutive data cards. The string `gtype-name` in data field 2 gives the name of a nontrivial group type and may be up to ten characters long. If data field 1 holds `GV`, the string `gv-nam` in data field 3 then gives the name of the group-type variable for this group type. This string may be up to 6 characters long. This string must be a valid Fortran name (see Section 7.2.1.2). The names of the variables for different group types may be the same. Alternatively, if data field 1 holds `GP`, the strings `gp-nam` in data

```
<> <---10---><--6->                     <--6->
F.1  Field 2      Field 3                 Field 5
GROUP TYPE
 GV gtype-namegv-nam
 GP gtype-namegp-nam                      gp-nam
  ↑↑ ↑          ↑↑    ↑                    ↑      ↑
  2 3 5        1415   20                   40     45
```

Figure 7.17: Possible data cards for GROUP TYPE

fields 3 and (optionally) 5 give the names of parameters for the group type. These strings must again be valid Fortran names. The names of parameters for different group types may be the same; the names of the group-type variable and parameters (if the latter appear) for a specific group type must all be different.

7.2.2.17 The GROUP USES Data Cards

The GROUP USES indicator card is used to announce which of the nonlinear elements appear in each group and the type of group function involved. The group types may be selected from among those defined in the GROUP TYPE section of the data while the elements may be selected from among the types defined in the ELEMENT USES section. In addition, group parameter values are assigned. The syntax for data following this indicator card is given in Figure 7.18.

```
<> <---10---> <--6->        <----12---->  <--6->      <----12---->
F.1  Field 2     Field 3        Field 4        Field 5       Field 6
GROUP USES
 T  group-namegtype-nam
 XT group-namegtype-nam
 E  group-nameelmnt-nam  blank/num-vl   elmnt-name blank/num-vl
 XE group-nameelmnt-nam  blank/num-vl   elmnt-name blank/num-vl
 ZE group-nameelmnt-nam                 r-p-a-name
 P  group-namegp-nam     numerical-vl   gp-nam     numerical-vl
 XP group-namegp-nam     numerical-vl   gp-nam     numerical-vl
 ZP group-namegp-nam                    r-p-a-name
  ↑↑ ↑         ↑↑    ↑   ↑↑          ↑   ↑     ↑   ↑↑           ↑
  2 3 5       1415   20  2425        36  40    45  4950         61
```

Figure 7.18: Possible data cards for GROUP USES

There are three sorts of data cards in the GROUP USES section. For cards of the second and third kinds, the string group-name in data field 2 gives the name, or an array of names, of the group(s) (or row(s) or constraint(s)) under consideration. The name may be up to ten characters long and must have been defined in the GROUPS/ROWS/CONSTRAINTS section.

On array cards (those prefixed by X or Z), the expanded element array name in field 2 must be valid and the integer indices must have been defined in a parameter assignment (see Section 7.2.2.3).

The first kind of cards, identified by the characters T or XT in field 1, give the name, or an array of names, of a group function and its type. On such cards, the string $$$$$$$$$$ in field 2 is used for two purposes.

- It may contain the name, or an array of names, of a group function whose type is to be defined. The name may be up to ten characters long and must have been defined in the GROUPS/ROWS/CONSTRAINTS section. On XT cards the expanded element array name in field 2 must be valid and the integer indices must have been defined in a parameter assignment (see Section 7.2.2.3).

- It may be used to assign a default type to all the group functions. In this case field 2 must contain the string 'DEFAULT'. Such a card may only appear before the first T or XT card in the GROUP USES section. Any group not explicitly typed is assumed to belong to a default type. The initial default type is trivial but the default may be changed. The string gtype-name in data field 3 gives the name of the group type to be used. This name may be up to ten characters long and must have appeared in the GROUP TYPE section.

The second kind of data card, identified by the characters E, XE or ZE in field 1, is an indication that particular nonlinear elements a re to be included in a given group. Optionally the given elements may be multiplied by specified weights. On these data cards, the string elmnt-name in data fields 3 (and optionally 5 on E and XE cards) hold the names of nonlinear elements to be used. The names in both fields may be up to ten characters long and must have been defined in the ELEMENT USES section. On XE and ZE cards, the names of the nonlinear elements must be components of an array of nonlinear elements, with a valid name and index. The elements are multiplied by given weights. By default, each weight takes the value 1.0. Only non-unit weights need to be specified explicitly. On E and XE cards, the weights are assigned the numerical values specified in data fields 4 (and optionally 6). These values may occupy up to 12 locations of their specified field. The default value of 1.0 is taken whenever these fields are empty. On ZE cards, the string r-p-a-name in data field 5 gives a real parameter array name. This name must have been previously defined and its associated value then gives the numerical value of the weight. Any group that is not named on an E or XE card is taken to have no nonlinear elements.

The last kind of data card, identified by the characters P, XP or ZP in field 1, is used to assign numerical values to the parameters for the group functions (P) or array of group functions (XP and ZP). On this data card, the strings gp-nam in data fields 3 (and, for P and XP cards, optionally 5) give the names of parameters. These names must have been set in the GROUP TYPE section and be valid Fortran name, see Section 7.2.1.2. On P and XP cards, the strings numerical-vl in data fields 4 and (optionally) 6 contain the numerical value of the parameter. These values may each occupy up to 12 locations of their field. On ZP cards, the string r-p-a-name in data field 5 gives a real parameter array name. This name must have been previously defined and its associated value then gives the numerical value of the parameter.

The T or XT card for a particular group must appear before its E, XE, ZE, P, XP or ZP cards.

7.2.2.18 The OBJECT BOUND Data Cards

The OBJECT BOUND indicator card is used to announce known lower and upper bounds on the value of the objective function for the problem. The syntax for data following this indicator card is given in Figure 7.19.

```
 <> <---10--->             <----12---->  <---10--->
 F.1  Field 2                 Field 4      Field 5
OBJECT BOUND
 LO obbnd-name             numerical-vl
 UP obbnd-name             numerical-vl
 XL obbnd-name             numerical-vl
 XU obbnd-name             numerical-vl
 ZL obbnd-name                           rl--p-name
 ZU obbnd-name                           rl--p-name
 ↑↑ ↑         ↑            ↑          ↑  ↑         ↑
 2 3 5        14           25         36 40        49
```

Figure 7.19: Possible data cards for OBJECT BOUND

The two-character string in data field 1 specifies the type of bound to be input. Possible values are: LO, XL or ZL for a lower bound, and UP, XU or ZU for an upper bound. The string obbnd-name in data field 2 gives a name to the bounds under consideration. This name may be up to ten characters long. Several different known bounds on the objective function may be defined in the OBJECT BOUND section.

For bounds of type LO or UP, the numerical value of the bound is given as the string numerical-vl using at most 12 characters in data field 4. For bounds of type ZL or ZU, the numerical value of the bound is that previously associated with the real parameter array r-p-a-name specified in field 5. When both lower and upper bounds on the objective are known, they must be specified on separate cards.

The objective function is assumed by default to be unbounded both below and above. The values for each named bound set may only be changed on a LO, UP, XL, XU, ZL or ZU card.

7.2.3 Another Example

In Section 1.2.3, we gave an example. An SDIF file for this example is given in Figure 7.20. The problem is given the name DOC. The groups are referred to as GROUP1/2/3 and the variables are X1/2/3. The vector of bounds is called BN1 and the two types of nonlinear element are ELEMENT1/2. The elemental variables are assigned names beginning with U and the internal variables for the second nonlinear element start with V. The two group types are GTYPE1/2. Finally the nonlinear element in GROUP2 is given the name G2E1, while those in GROUP3 are G3E1/2.

7.2.4 A Further Example

In Section 1.2.4, we gave a second example. Because of its repetitious structure, this example is well suited to use array names and do-loops. An SDIF file for this example is given in Figure 7.21. The problem is given the name DOC2. The variables are referred to as X1,..., X1000 and the groups are G1, ..., G1000. The vector of bounds is called BND, the constants are CONST and the single nonlinear element type is SQUARE, with elemental variable V. Note that the BND section is necessary since the variables are unrestricted and we must override the default lower bounds of zero and upper bounds of infinity. The nonlinear elements are given the names E1,..., E1000. Finally, the single group type is SINE with group-type variable ALPHA and parameter P.

```
 <> <---10---><---10---><----12---->   <---10---><----12---->
 F.1 Field 2   Field 3    Field 4       Field 5    Field 6
NAME          DOC
GROUPS
 E  GROUP1
 E  GROUP2
 E  GROUP3
VARIABLES
    X1        GROUP1    1.0
    X2        GROUP3    1.0
    X3
BOUNDS
 FR BN1       X1
 LO BN1       X2        -1.0D+0
 LO BN1       X3        1.0D+0
 UP BN1       X2        1.0D+0
 UP BN1       X3        2.0D+0
ELEMENT TYPE
 EV ETYPE1    V1
 EV ETYPE1    V2
 EV ETYPE2    V1
 EV ETYPE2    V2
 EV ETYPE2    V3
 IV ETYPE2    U1
 IV ETYPE2    U2
ELEMENT USES
 T  G2E1      ETYPE1
 V  G2E1      V1                       X2
 V  G2E1      V2                       X3
 T  G3E1      ETYPE2
 V  G3E1      V1                       X2
 V  G3E1      V2                       X1
 V  G3E1      V3                       X3
 T  G3E2      ETYPE1
 V  G3E2      V1                       X1
 V  G3E2      V2                       X3
GROUP TYPE
 GV GTYPE1    ALPHA
 GV GTYPE2    ALPHA
GROUP USES
 T  GROUP1    GTYPE1
 T  GROUP2    GTYPE2
 E  GROUP2    G2E1
 E  GROUP3    G3E1
 E  GROUP3    G3E2
ENDATA
 ↑↑ ↑        ↑↑        ↑↑          ↑   ↑        ↑↑          ↑
 2 3 5       14 15     24 25       36  40       49 50       61
```

Figure 7.20: SDIF file for the example of Section 1.2.3

```
<> <---10---><---10---><----12----> <---10---><----12---->
F.1  Field 2   Field 3     Field 4      Field 5    Field 6
NAME          DOC2
 IE ONE                 1
 IE N                   1000
 IA NM1                 -1
VARIABLES
 DO I         ONE                      N
    X         X(I)
 ND
GROUPS
 DO I         ONE                      NM1
 XN G(I)      X(ONE)    1.0
 ND
 XN G(N)
CONSTANTS
    CONST     'DEFAULT' 1.0
 X  CONST     G(N)      0.0
BOUNDS
 FR BND       'DEFAULT'
ELEMENT TYPE
 EV SQUARE    V
ELEMENT USES
 DO I         ONE                      N
 XT E(I)      SQUARE
 ZV E(I)      V                        X(I)
 ND
GROUP TYPE
 GV SINE      ALPHA
 GP SINE      P
GROUP USES
 DO I         ONE                      NM1
 XT G(I)      SINE
 XE G(I)      E(I)                     E(N)
 XP G(I)      P         1.0
 ND
 E  G1000     E1000
 P  G1000     P         0.5
ENDATA
 2 3 5     1415      2425        36  40       4950       61
```

Figure 7.21: SDIF file for the example of Section 1.2.4

7.3 The Standard Input Format for Nonlinear Elements

In addition to the problem data described in Section 7.2, the user might also wish to specify the nonlinear element functions, and their derivatives, in a systematic way. A particular nonlinear element function is defined in terms of its problem variables and its type; both of these quantities are specified in Section 7.2. Thus, the only details which remain to be specified are the function and derivative values of the *element types* and the transformations between elemental and internal variables, if any.

In this section, we present one approach to this issue. As before, data is specified in a file. The file comprises an ordered mixture of indicator and data cards; the latter allow function and derivative definitions in appropriate high-level language statements.

7.3.1 Introduction to the Standard Element Type Input Format

7.3.1.1 The Values and Derivatives Required

It is assumed that a nonlinear element type is specified in terms of internal variables $\mathbf{u}$, whose names are those given on the `ELEMENT TYPE` data cards in an SDIF file (if the element has no useful internal variables, the internal and elemental variables are the same and the internal variables will have been named after the elementals), see Section 7.2.2.14. An optimization procedure is likely to require the values of the element functions and possibly their first and second, derivatives. These derivatives need only be given with respect to the internal variables. For if we denote the gradient and Hessian matrix of an element function f with respect to u by

$$\nabla_u f \text{ and } \nabla_{uu} f$$

respectively, the gradient and Hessian matrices with respect to the elemental variables are

$$W^T \nabla_u f \text{ and } W^T \nabla_{uu} f W,$$

where W is defined by (1.2.11).

We thus need only supply derivatives with respect to u. Formally, we must define the function value f, possibly the gradient vector $\nabla_u f$ (i.e., the vector whose i-th component is the first partial derivative with respect to the i-th internal variable) and, possibly, the Hessian matrix $\nabla_{uu} f$ (i.e., the matrix whose i,j-th entry is the second partial derivative with respect to the i-th and j-th internal variables), all evaluated at u. We now describe how to set up the data for a given problem.

7.3.2 Indicator Cards

As before, the user must prepare an input file, the SEIF (Standard Element type Input Format) file, consisting of indicator and data cards. The former contain a simple keyword to specify the type of data that follows. Possible indicator cards are given in Figure 7.22.

Indicator cards must appear in the order shown. The cards `TEMPORARIES`, `GLOBALS` and `INDIVIDUALS` are optional.

The data cards are of two kinds. The first are like those described in Section 7.2.1. The others use four fields, fields 1, 2 and 3, as before, and

Keyword	Comments	Presence	Described in §
`ELEMENTS`	same as **`NAME`**	mandatory	7.2.2.1
`TEMPORARIES`		optional	7.3.4.1
`GLOBALS`		optional	7.3.4.2
`INDIVIDUALS`		optional	7.3.4.3
`ENDATA`		mandatory	7.2.2.2

Figure 7.22: Possible indicator cards

field 7 which starts in column 25 and is 41 characters long. This last field is used to hold arithmetic expressions. An *arithmetic expression* is as defined in the Fortran programming language standard (ANSI X3.9-1978). We allow the use of any of the language's intrinsic functions in such an expression. Continuation of an expression over at most nineteen lines is also permitted.

7.3.3 An Example

Before we give the complete syntax for an SEIF file, we continue the illustrative example that we started in Section 7.2.1.4 and show how to specify an input file appropriate for the problem of Section 1.2.5. Once again, there are many possible ways of specifying a particular problem; we give one in Figure 7.23. The arithmetic expressions given are written in Fortran.

The file must always start with an `ELEMENTS` card, on which a name (in this case `EG3`) for the example may be given (line 1), and must end with an `ENDATA` card (line 40).

We next need to specify the names and attributes of any auxiliary quantities and functions that we intend to use in our high level description of the element functions. These are needed to allow for consistency checks in the subsequent high-level language statements and must always occur in the `TEMPORARIES` section of the input file. Lines 3 to 6 indicate that we shall be using temporary quantities `SINV1`, `ZERO`, `ONE` and `TWOP1`, and the character `R` in the first field for these lines states that these quantities will be associated with floating point (real) values. The character `M` in field 1 of Lines 7 and 8 indicates that we may use the intrinsic (machine) functions `SIN` and `COS`. These are of course Fortran intrinsic functions appropriate for the high-level language used here.

We now specify any numerical values which are to be used in one or more element descriptions within the `GLOBALS` section. On lines 10 and 11, we allocate the values 0 and 1 to the previously defined quantities `ZERO` and `ONE`. Note that such cards require the character `A` in field 1 - if an assignment were to take more than 41 characters (the width of field 7), it could be continued on subsequent lines for which the string `A+` is required in field 1.

```
ELEMENTS      EG3
TEMPORARIES
 R  SINV1
 R  ZERO
 R  ONE
 R  TWOP1
 M  SIN
 M  COS
GLOBALS
 A  ZERO                0.0
 A  ONE                 1.0
INDIVIDUALS
 T  3PROD
 R  U1        V1        1.0            V2        -1.0
 R  U2        V3        1.0
 F                      U1*U2
 G  U1                  U2
 G  U2                  U1
 H  U1        U1        ZERO
 H  U1        U2        ONE
 H  U2        U2        ZERO
 T  2PROD
 F                      V1*V2
 G  V1                  V2
 G  V2                  V1
 H  V1        V1        ZERO
 H  V1        V2        ONE
 H  V2        V2        ZERO
 T  SINE
 A  SINV1               SIN(V1)
 F                      SINV1
 G  V1                  COS(V1)
 H  V1        V1        -SINV1
 T  SQUARE
 R  U1        V1        1.0            V2        1.0
 A  TWO                 2.0
 F                      U1*U1
 G  U1                  TWO*U1
 H  U1        U1        TWO
ENDATA
```

Figure 7.23: SEIF file for the example of Section 1.2.5

Finally we need to make the actual definitions of the function and derivative values for the element types and specify the transformations from elemental to internal variables if they are used. Such specifications occur in the `INDIVIDUALS` section from lines 12 to 39 of the example. We recall that there are four element types `3PROD`, `2PROD`, `SINE` and `SQUARE` and that their attributes (names of elemental and internal variables and parameters) have been described in the SDIF file set up in Section 7.2.1.4. Two of the element types (`3PROD` and `SQUARE`) use internal variables so we need to describe the relevant transformation for those.

On line 13, the presence of the character `T` in field 1 announces that the data for the element type `3PROD` is to follow. All the data for this element must be specified before another element type is considered. On lines 14 and 15 we describe the transformation from elemental to internal variables that is used for `3PROD`. Recall that the transformation is $u_1 = v_1 - v_2$ and $u_2 = v_3$. On line 14, the first of these transformations is given, namely that `U1` is to be formed by adding 1.0 times `V1` to -1.0 times `V2`. The second transformation is given on the following line, namely that `U2` is formed by taking 1.0 times `V3`. Both lines are marked as defining transformations by the character `R` in field 1 — continuation lines are possible for transformations involving more than two elemental variables on lines in which the string `R+` appears in the same field.

We now specify the function and derivative values of the element type u_1u_2 with respect to its internal variables. On line 16, the code `F` in field 1 indicates that we are setting the value of the element type to `U1*U2`, the Fortran expression for multiplying `U1` and `U2`. On lines 17 and 18, we specify the first derivatives of the element type with respect to its two internal variables `U1` and `U2` - the character `G` in field 1 indicates that gradient values are to be set. On line 17, the derivative with respect to the variable `U1`, specified in field 2, is taken and expressed as `U2` in field 7. Similarly, on line 18, the derivative with respect to the variable `U2` (in field 2), `U1`, is given in field 7. Finally, on lines 19 to 21, the second partial derivatives with respect to both internal variables are given. These derivatives appear on cards whose first field contains the character `H`. On line 19, the second derivative with respect to the variables `U1` (in field 2) and `U1` (in field 3), 0.0, is given in field 7. Similarly the second derivative with respect to the variables `U1` (in field 2) and `U2` (in field 3), 1.0, occurs in field 7 of line 20 and that with respect to `U2` (in field 2) and `U2` (in field 3), 0.0, is given in field 7 of the following line.

The same principle is applied to the specification of range transformations, values and derivatives for the remaining element types. The type `2PROD` does not use a transformation to internal variables, so derivatives are taken with respect to the elemental variables `V1` and `V2` (or one might think of the internal variables being `V1` and `V2`, related to the elemental variables through the identity transformation). The values and derivatives for this element type are given on lines 22 to 28. The type `SINE` again does not use special internal variables and the required value and derivatives are given on lines 29 to 33.

Note, however, that the value and its second derivative with respect to v_1 both use the quantity $\sin v_1$; for efficiency, we set the auxiliary quantity `SINV1` to the Fortran value `SIN(V1)` on line 30 and thereafter refer to `SINV1` on lines 31 and 33. Notice that this definition of auxiliary quantities occurs on a line whose first field contains the character `A`. Finally, the type `SQUARE`, which uses a transformation from elemental to internal variables $u_1 = v_1 + v_2$, is defined on lines 34 to 39. Again notice that the value 2.0 occurs in both first and second derivatives, so the auxiliary quantity `TWO` is set on line 36 to hold this value.

7.3.4 Data Cards

The `ELEMENTS` and `ENDATA` indicator cards perform the same function as the cards `NAME` and `ENDATA` in Section 7.2.2.1 and 7.2.2.2. The problem name specified in field 3 on the `ELEMENTS` card must be the same as that given in the same field on the `NAME` card of the SDIF file.

7.3.4.1 The `TEMPORARIES` Data Card

When specifying the function and derivative values of a nonlinear element, it often happens that an expression occurs more than once. It is then convenient to define an auxiliary parameter to have the value of the common expression and henceforth to refer to the auxiliary parameter. For instance, a nonlinear element of the two internal variables u_1 and u_2 might be $u_1 e^{u_2}$. (The names of the internal variables have already been specified in the `ELEMENT TYPE` section of the SDIF and are known as *reserved* parameters.) Its gradient vector (vector of first partial derivatives) has components e^{u_2} and $u_1 e^{u_2}$. If we define the auxiliary parameter $w = e^{u_2}$, the derivatives are then w and $u_1 w$.

```
 <> <--6->
 F.1 Field 2
TEMPORARIES
 I  p-name
 R  p-name
 L  p-name
 M  p-name
 F  p-name
 ↑↑ ↑    ↑
 2 3 5   10
```

Figure 7.24: Possible data cards for `TEMPORARIES`

The `TEMPORARIES` indicator card is used to announce the names of any auxiliary parameters which are to be used in defining the function and derivative values of the nonlinear elements. This list should also include the name of any intrinsic and external functions used. The syntax for data cards following the indicator card is given in Figure 7.24.

The single-character string in field 1 specifies the type of auxiliary parameter that is to be defined. Possible types are integer (`I`), real (`R`), logical (`L`), intrinsic function (`M`) or external function (`F`). The string **`p-name`** in field 2 then gives the name of the auxiliary parameter. The name must be a valid Fortran name, see Section 7.2.1.2, but must not be a reserved one, i.e., one of the names assigned to the internal variables or parameters for the element in question in the `ELEMENT TYPE` section of the SDIF (see, Section 7.2.2.14). Any auxiliary parameter that is to be used must be defined in the `TEMPORARIES` section along with all intrinsic and external function names.

7.3.4.2 The `GLOBALS` Data Cards

The `GLOBALS` indicator card is used to announce the assignment of general parameter values. the syntax for data cards following the indicator card is given in Figure 7.25.

```
 <> <--6->   <--6->   <-------------------41------------------->
 F.1 Field 2  Field 3                  Field 7
GLOBALS
 A  p-name                &&&&&&&&&&&&&&&&&&&&&&&&&&&&&&&&&&&&&&&&&
 A+                       &&&&&&&&&&&&&&&&&&&&&&&&&&&&&&&&&&&&&&&&&
 I  l-name    p-name      &&&&&&&&&&&&&&&&&&&&&&&&&&&&&&&&&&&&&&&&&
 I+                       &&&&&&&&&&&&&&&&&&&&&&&&&&&&&&&&&&&&&&&&&
 E  l-name    p-name      &&&&&&&&&&&&&&&&&&&&&&&&&&&&&&&&&&&&&&&&&
 E+                       &&&&&&&&&&&&&&&&&&&&&&&&&&&&&&&&&&&&&&&&&
 ↑↑ ↑    ↑    ↑    ↑      ↑                                        ↑
 2 3 5   10   15   20     25                                       65
```

Figure 7.25: Possible data cards for `GLOBALS`

The one or two character string in field 1 specifies the type of assignment that is to be made from the card. Possible values for the first character of the string are:

`A` This card announces that an auxiliary parameter is to be assigned a value. The string `p-name` in field 2 gives the name of the auxiliary parameter that is to be defined; this name must be a valid Fortran name, see Section 7.2.1.2, and must have been previously defined in the `TEMPORARIES` section. The string in field 7 is an arithmetic expression. The assignment

$$\text{auxiliary variable named in field 2} \leftarrow \text{field 7}$$

is made, where again $\leftarrow$ means "is given the value"; any variable mentioned in the arithmetic expression must either be reserved (see Section 7.3.4.1), or have been defined in the `TEMPORARIES` section. If in this latter case, the variable is integer or real, it must have been allocated a value itself on a previous `GLOBALS` data card.

`I` This card announces that an auxiliary parameter is to be assigned a value whenever a second logical auxiliary parameter has the value `.TRUE.` The string `p-name` in field 3 gives the name of the auxiliary parameter that is to be defined; this name must be a valid Fortran name, see Section 7.2.1.2, and must have been previously defined in the `TEMPORARIES` section. The string in field 7 is an arithmetic expression. The assignment

$$\text{auxiliary variable named in field 3} \leftarrow \text{field 7}$$

will be made if and only if the logical auxiliary parameter `l-name` specified in field 2 has the value `.TRUE.`; the logical parameter must have been previously defined in the `TEMPORARIES` section and allocated a value in the `GLOBALS` section. The arithmetic expression must obey the rules set out in the `A` section above.

`E` This card announces that an auxiliary parameter is to be assigned a value whenever a second logical auxiliary parameter has the value `.FALSE.` The string `p-name` in field 3 gives the name of the auxiliary parameter that is to be defined; this name must be a valid Fortran name, see Section 7.2.1.2, and must have been previously defined in the `TEMPORARIES` section. The string in field 7 is an arithmetic expression. The assignment

$$\text{auxiliary variable named in field 3} \leftarrow \text{field 7}$$

will be made if and only if the logical auxiliary parameter, `l-name`, specified in field 2 has the value `.FALSE.`; the logical parameter must have been previously defined in the `TEMPORARIES` section and allocated a value in the `GLOBALS` section. The arithmetic expression must obey the rules set out in the `A` section above.

The data started on an `A`, `I` and `E` card may be continued on a card whose first field contains an `A+`, `I+` or `E+` respectively. Such cards contain an arithmetic expression in field 7 and no further data; the arithmetic expression must obey the rules set out in the `A` section above. At most nineteen continuations of a single assignment are allowed.

The `GLOBALS` section is intended for the definition of auxiliary variables which occur in more than one element type. If an auxiliary variable occurs in a single element type, it may be defined in the `INDIVIDUALS` section (see Section 7.3.4.3).

7.3.4.3 The `INDIVIDUALS` Data Cards

The `INDIVIDUALS` indicator card is used to announce the definition of function and derivative values and the transformation between elemental and internal variables for the types of nonlinear element functions required. The syntax for data cards following the indicator card is given in Figure 7.26.

```
                                  <--------------41----Field 7-------------->
 <> <---10---> <---10---> <----12---->  <---10---> <----12---->
F.1  Field 2    Field 3      Field 4       Field 5      Field 6
+-----------------------------------------------------------------+
|INDIVIDUALS                                                      |
| |T |etype-name|         |                                       |
| |R |iv-nam    |ev-nam   |numerical-vl|ev-nam   |numerical-vl|   |
| |A |p-name    |         |&&&&&&&&&&&&&&&&&&&&&&&&&&&&&&&&&&&&&&&&&|
| |A+|          |         |&&&&&&&&&&&&&&&&&&&&&&&&&&&&&&&&&&&&&&&&&|
| |I |l-name    |p-name   |&&&&&&&&&&&&&&&&&&&&&&&&&&&&&&&&&&&&&&&&&|
| |I+|          |         |&&&&&&&&&&&&&&&&&&&&&&&&&&&&&&&&&&&&&&&&&|
| |E |l-name    |p-name   |&&&&&&&&&&&&&&&&&&&&&&&&&&&&&&&&&&&&&&&&&|
| |E+|          |         |&&&&&&&&&&&&&&&&&&&&&&&&&&&&&&&&&&&&&&&&&|
| |F |          |         |&&&&&&&&&&&&&&&&&&&&&&&&&&&&&&&&&&&&&&&&&|
| |F+|          |         |&&&&&&&&&&&&&&&&&&&&&&&&&&&&&&&&&&&&&&&&&|
| |G |iv-nam    |         |&&&&&&&&&&&&&&&&&&&&&&&&&&&&&&&&&&&&&&&&&|
| |G+|          |         |&&&&&&&&&&&&&&&&&&&&&&&&&&&&&&&&&&&&&&&&&|
| |H |iv-nam    |iv-nam   |&&&&&&&&&&&&&&&&&&&&&&&&&&&&&&&&&&&&&&&&&|
| |H+|          |         |&&&&&&&&&&&&&&&&&&&&&&&&&&&&&&&&&&&&&&&&&|
+-----------------------------------------------------------------+
  2 3 5    10  1415   20  2425        36  40      4950       61  65
```

Figure 7.26: Possible data cards for `INDIVIDUALS`

The one- or two-character string in field 1 specifies the type of data contained on the card. Possible values for the first character of the string are:

`T` This card announces that a new element type is to be considered. The string `etype-name` in field 2 gives the name of the element type; the name may be up to ten characters long and must have been defined in the `ELEMENT TYPE` section of the SDIF file (see Section 7.2.2.14).

`R` This card announces that information concerning the transformation between the elemental and internal variables for the element type is to be given. Such information is appropriate only for element types which have been defined with internal variables in the `ELEMENT TYPE` section of the SDIF file (see Section 7.2.2.14). The transformation is specified by the matrix W of Section 1.2.2; only nonzero coefficients of W need be specified here.

The string `inv-name` in field 2 contains the name of an internal variable (i.e., row of W). The name must be a valid Fortran name, see Section 7.2.1.2, and have been defined on an `IV` data line in the `ELEMENT TYPE` section of the SDIF file. The strings `iv-nam` in fields 3 and (optionally) 5 then give the names of elemental variables (i.e., columns of W). The names must be valid Fortran names and have been defined on `EV` data lines in the `ELEMENT TYPE` section of the SDIF file. The strings in fields 4 and (optionally) 6 contain the numerical values of the coefficients of W corresponding to the row given in field 2 and the columns given in fields 3 and 5 respectively. These numerical values may each be up to 12 characters long. The entries of W may be defined in any order.

As an example, the transformation (1.2.9) could be entered with three `R` data cards. On the first, field 2 would hold the name given to the internal variable u_1; field 3 would hold the name given to the elemental variable v_1 and field 4 would contain 1.0. Similarly field 5 would hold the name given to the elemental variable v_2 and field 6 would also contain 1.0. On the second, field 2 would also hold the name given to the internal variable u_1; field 3 would now hold the name given to the elemental variable v_3 and field 4 would contain -2.0. On the third card, field 2 would hold the name given to the internal variable u_2; field 3 would hold the name given to the elemental variable v_1 and field 4 would contain 1.0. Field 5 would now hold the name given to the elemental variable v_3 and field 6 would contain -1.0.

`A` This card announces that an auxiliary parameter, specific to the current element type, is to be assigned a value. The string `p-name` in field 2 gives the name of the auxiliary parameter that is to be defined; this name must be a valid Fortran name, see Section 7.2.1.2, and have been previously defined in the `TEMPORARIES` section. The string in field 7 is an arithmetic expression. The assignment

auxiliary variable named in field 2 $\leftarrow$ field 7

is made, where again $\leftarrow$ means "is given the value"; any variable mentioned in the arithmetic expression must either be reserved (see Section 7.3.4.1), or have been defined in the `TEMPORARIES` section. If in this latter case, the variable is integer or real, it must have been allocated a value itself either on a previous `GLOBALS` data card or on a previous `A`, `E` or `I` card for the current element type in the `ELEMENTS` section.

`I` This card announces that an auxiliary parameter, specific to the current element type, is to be assigned a value whenever a second logical auxiliary parameter has the value `.TRUE.` The string, `p-name`, in field 3 gives the name of the auxiliary parameter that is to be defined; this name must be a valid Fortran name, see Section 7.2.1.2, and have been previously defined in the `TEMPORARIES` section. The string in field 7 is an arithmetic expression. The assignment

auxiliary variable named in field 3 $\leftarrow$ field 7

will be made if and only if the logical auxiliary parameter, `l-name`, specified in field 2 has the value `.TRUE.`; the logical parameter must have been previously defined in the `TEMPORARIES` section and allocated a value in the `GLOBALS` or `INDIVIDUALS` section. The arithmetic expression must obey the rules set out in the `A` section above.

`E` This card announces that an auxiliary parameter, specific to the current element type, is to be assigned a value whenever a second logical auxiliary parameter has the value `.FALSE.` The string, `p-name`, in field 3 gives the name of the auxiliary parameter that is to be defined; this name must be a valid Fortran name, see Section 7.2.1.2, and have been previously defined in the `TEMPORARIES` section. The string in field 7 is an arithmetic expression. The assignment

$$\text{auxiliary variable named in field 3} \leftarrow \text{field 7}$$

will be made if and only if the logical auxiliary parameter, `l-name`, specified in field 2 has the value `.FALSE.`; the logical parameter must have been previously defined in the `TEMPORARIES` section and allocated a value in the `GLOBALS` or `INDIVIDUALS` section. The arithmetic expression must obey the rules set out in the `A` section above.

`F` This card specifies the value of the nonlinear element. The string in field 7 is an arithmetic expression; the assignment

$$\text{nonlinear element function} \leftarrow \text{field 7}$$

is made; any variable mentioned in the expression must obey the rules set out in the `A` section above.

`G` This card specifies the value of a component of the gradient of the nonlinear element. The string, `iv-nam`, in field 2 contains the name of an internal variable. The component of the gradient specified on the card will be taken with respect to this variable. The string must be a valid Fortran name, see Section 7.2.1.2, and have been defined on an `IV` data line, for a nonlinear element defined with internal variables, or an `EV` data line, for an element without explicit internal variables, in the `ELEMENT TYPE` section of the SDIF file. The string in field 7 is an arithmetic expression; the assignment

$$\text{derivative of element w.r.t. variable in field 2} \leftarrow \text{field 7}$$

is made; any variable mentioned in the arithmetic expression must obey the rules set out in the `A` section above. `G` cards are optional. However, once the user starts to form the gradient for an element type, any component not explicitly specified will be assumed to have the value zero.

`H` This card specifies the value of a component of the Hessian matrix of the nonlinear element. The strings `iv-nam` in fields 2 and 3 contain the names of internal variables. The component of the Hessian specified on the card will be taken with respect to these variables. Either string must be a valid Fortran name, see Section 7.2.1.2, and have been defined on an `IV` data line, for a nonlinear element defined with internal variables, or an `EV` data line, for an element without explicit internal variables, in the `ELEMENT TYPE` section of the SDIF file. The string in field 7 is an arithmetic expression; the assignment

second derivative of element w.r.t. variables in fields 2 and 3 $\leftarrow$ field 7

is made; any variable mentioned in the arithmetic expression must obey the rules set out in the `A` section above. `H` cards are optional. However, once the user starts to specify the Hessian matrix for an element type, any component not specified will be assumed to have the value zero. The matrix is assumed to be symmetric and so the user needs only supply values for one of

$$\frac{\partial^2 f}{\partial u_i \partial u_j} \text{ or } \frac{\partial^2 f}{\partial u_j \partial u_i} \quad (i \neq j)$$

it does not matter which. Observe that defaulting Hessian components to zero gives a very simple way of inputing sparse matrices; however, as we stressed in the introduction, we do not generally recommend this method of specifying invariant subspaces.

The data started on an `A`, `I`, `E`, `F`, `G` and `H` card may be continued on a card whose first field contains an `A+`, `I+`, `E+`, `F+`, `G+` or `H+` respectively. Such cards contain an arithmetic expression in field 7 and no further data; the arithmetic expression must obey the rules set out in the `A` section above. At most nineteen continuations of a single assignment are allowed.

The data for a single element type must occur on consecutive cards and in the order given in Figure 7.26, excepting that `A`, `I` and `E` cards may be intermixed. A new element type is deemed to have started whenever a `T` card is encountered. The `F` card is compulsory for all element types; elements with useful transformations from elemental to internal variables must also have `R` cards. The data for a particular card type is considered to have been completed whenever another card type is encountered.

7.3.5 Two Further Examples

In Section 1.2.3, we gave an example. An SEIF file for this example is given in Figure —refF3.3.1. The problem is again given the name `DOC`. The two types of nonlinear element were assigned the names `ELEMENT1/2` by the previous SDIF file. The elemental variables were given names beginning with `V` and the internal variables for the second nonlinear element started with `U`. The constant 0.0 occurs in the derivatives of both elements, so an auxiliary variable is assigned to hold its value (although, we could have just not specified these particular components, which would then have taken their default zero value). The function value and derivatives of the second element type use both sines and cosines of u_2 and again auxiliary variables are assigned to hold these

```
                                   <--------------41----Field 7----------->
<> <---10---> <---10---> <----12---->  <---10---><----12---->
F.1  Field 2     Field 3      Field 4       Field 5     Field 6
ELEMENTS      DOC
TEMPORARIES
 R  CS
 R  SN
 R  ZERO
 R  SIN
 R  COS
GLOBALS
 G  ZERO                0.0D0
INDIVIDUALS
 T  ETYPE1
 F                      V1*V2
 G  V1                  V2
 G  V2                  V1
 H  V1        V1        ZERO
 H  V1        V2        1.0D0
 H  V2        V2        ZERO
 T  ETYPE2
 R  U1        V1        1.0D0
 R  U2        V2        1.0D0          V3        1.0D0
 A  CS                  COS(U2)
 A  SN                  SIN(U2)
 F                      U1*SN
 G  U1                  SN
 G  U2                  U1*CS
 H  U1        U1        ZERO
 H  U1        U2        CS
 H  U2        U2        -U1*SN
ENDATA
 ↑↑ ↑    ↑   ↑↑    ↑   ↑↑          ↑  ↑        ↑↑          ↑  ↑
 2 3 5   10  1415  20  2425        36 40       4950        61 65
```

Figure 7.27: SEIF file for the element types for the exampleof Section 1.2.3

values, this time as variables local to `ELEMENT2`. The second derivatives are sufficiently straightforward to compute that we provide them.

We gave a second example in Section 1.2.4. An SEIF file for this example is given in Figure 7.28 on page 234. The problem is again given the name `DOC2`. The only type of nonlinear element was assigned the name `SQUARE` in the previous SDIF file, its elemental variable was called `V` and there was no useful range transformation.

7.4 The Standard Input Format for Nontrivial Groups

In addition to the problem data and the nonlinear element types described in Section 7.2 and 7.3, the user might also wish to specify the nontrivial group functions, and their derivatives, in a systematic way. A particular nontrivial

```
                              <--------------41----Field 7------------>
 <> <---10---> <---10---><----12---->  <---10---><----12---->
 F.1  Field 2    Field 3     Field 4      Field 5     Field 6
ELEMENTS        DOC2
INDIVIDUALS
 T  SQUARE
 F                        V**2
 G  V                     2.0D0*V
 H  V           V         2.0D0
ENDATA

 2 3 5   10    1415   20   2425        36  40      4950       61  65
```

Figure 7.28: SEIF file for the element types for the exampleof Section 1.2.4

group function is defined in terms of its group type and variable; both of these quantities are specified in Section 7.2. Thus, the only details which remain to be specified are the function and derivative values of the group types.

Once again, we present an approach to this issue. As before, data is specified in a file. The file comprises an ordered mixture of indicator and data cards; the latter allow function and derivative definitions in appropriate high-level language-statements.

7.4.1 Introduction to the Standard Group Type Input Format

7.4.1.1 The Values and Derivatives Required

It is assumed that a nonlinear group type i s specified in terms of its group-type variable as described on a `GROUP TYPE` data card in an SDIF file, see Section 7.2.2.16. An optimization procedure is likely to require the values of the group functions and their first and second derivatives (taken with respect to the variable). We now describe how to set up the data for a given problem.

7.4.1.2 Indicator Cards

As before, the user must prepare an input file, the SGIF (Standard Group type Input Format) file, consisting of indicator and data cards. The former contain a simple keyword to specify the type of data that follows. Possible indicator cards are given in Figure 7.29.

Indicator cards must appear in the order shown. The cards `TEMPORARIES`, `GLOBALS` and `INDIVIDUALS` are optional.

The data cards are of a single kind, using four fields, fields 1, 2, 3 and 7, exactly as described in Section 7.3.2.

Keyword	Comments	Presence	Described in §
GROUPS	same as **NAME**	mandatory	7.2.2.1
TEMPORARIES		optional	7.3.4.1
GLOBALS		optional	7.3.4.2
INDIVIDUALS		optional	7.4.2.1
ENDATA		mandatory	7.2.2.2

Figure 7.29: Possible indicator cards

7.4.1.3 An Example

Before we give the complete syntax for an SGIF file, we finish the illustrative example that we started in Section 7.2.1.4 and Section 7.3.3 and show how to specify an input file appropriate for the problem of Section 1.2.5. The format is fairly similar to that for the SEIF file of Section 7.3. Once again, there are many possible ways of specifying a particular problem; we give one in Figure 7.30.

```
      <> <---10---> <---10---> <-------------------41------------------->
line  F.1  Field 2    Field 3                   Field 7
  1  GROUPS         EG3
  2  TEMPORARIES
  3   R  TWOP1
  4  INDIVIDUALS
  5   T  PSQUARE
  6   A  TWOP1                 2.0*P1
  7   F                        P1*ALPHA*ALPHA
  8   G                        TWOP1*ALPHA
  9   H                        TWOP1
 10  ENDATA
     12 3 5    10  1415    20  2425                                     65
```

Figure 7.30: SGIF file for the element types for the example of Section 1.2.5

The file must always start with a GROUPS card, on which a name (in this case EG3) for the example may be given (line 1), and must end with an ENDATA card (line 10).

We next need to specify the names and attributes of any auxiliary quantities and functions that we intend to use in our high level description of the group functions. These are needed to allow for consistency checks in the subsequent high-level language statements and must always occur in the TEMPORARIES section of the input file. Line 3 indicates that we shall be using temporary quantities TWOP1 and the character R in the first field of this lines states that the quantity will be associated with a floating point (real) value.

We now make the actual definitions of the function and derivative values for the nontrivial group type used; we recall that there is a single nontrivial group type `PSQUARE` and that its attributes (name of group-type variable and parameter) have been described in the SDIF file set up in Section 7.2.1.4. This definition takes place within the `INDIVIDUALS` section. The presence of the character `T` in field 1 of line 5 announces that the data for the group type `PSQUARE` is to follow. All the data for this group must be specified before another group type is considered. We note that the quantity $2p_1$ occurs in both first and second derivatives of the group type function and so the auxiliary quantity `TWOP1` is set on line 6 to hold this value. The first field of a line on which such an assignment is made contains the character `A`. The value (line 7), its first derivative (line 8) and second derivative (line 9) with respect to the group-type variable are now given. A Fortran expression for these values occurs in field 7 on each of these lines; the lines contain the characters `F`, `G` and `H` respectively in field 1 for such assignments.

If there had been more than a single group type with one or more expressions in common, these expressions could have been assigned to previously attributed quantities in a `GLOBALS` section. This section would then have appeared between the `TEMPORARIES` and `INDIVIDUALS` sections.

7.4.2 Data Cards

The `GROUPS` and `ENDATA` indicator cards perform the same function as the cards `NAME` and `ENDATA` in Section 7.2.2.1 and 7.2.2.2 Likewise, the `TEMPORARIES` and `GLOBALS` data cards have exactly the same syntax as those in Section 7.3.4.1 and 7.3.4.2, excepting that the reserved parameters are now the group-type variables specified in the `GROUP TYPE` section of the SDIF file.

7.4.2.1 The `INDIVIDUALS` Data Cards

The `INDIVIDUALS` indicator card is used to announce the definition of function and derivative values for the types of nontrivial group functions required. The syntax for data cards following the indicator card is given in Figure 7.4.2.1.

The one- or two-character string in field 1 specifies the type of data contained on the card. Possible values for the first character of the string are:

`T` This card announces that a new group type is to be considered. The string `gtype-name` in field 2 gives the name of the group type; the name may be up to ten characters long and must have been defined in the `GROUP TYPE` section of the SDIF file (see Section 7.2.2.15).

`A` This card announces that an auxiliary parameter, specific to the current group type, is to be assigned a value. The string `p-name` in field 2 gives the name of the auxiliary parameter that is to be defined; this name must be a valid Fortran name, see Section 7.2.1.2, and have been previously defined in the `TEMPORARIES` section. The string in field 7 is an arithmetic

```
 <> <---10---> <---10---> <------------------41------------------->
 F.1  Field 2    Field 3                Field 7
 T  gtype-name
 A  p-name              &&&&&&&&&&&&&&&&&&&&&&&&&&&&&&&&&&&&&&&&&
 A+                     &&&&&&&&&&&&&&&&&&&&&&&&&&&&&&&&&&&&&&&&&
 I  l-name    p-name    &&&&&&&&&&&&&&&&&&&&&&&&&&&&&&&&&&&&&&&&&
 I+                     &&&&&&&&&&&&&&&&&&&&&&&&&&&&&&&&&&&&&&&&&
 E  l-name    p-name    &&&&&&&&&&&&&&&&&&&&&&&&&&&&&&&&&&&&&&&&&
 E+                     &&&&&&&&&&&&&&&&&&&&&&&&&&&&&&&&&&&&&&&&&
 F                      &&&&&&&&&&&&&&&&&&&&&&&&&&&&&&&&&&&&&&&&&
 F+                     &&&&&&&&&&&&&&&&&&&&&&&&&&&&&&&&&&&&&&&&&
 G                      &&&&&&&&&&&&&&&&&&&&&&&&&&&&&&&&&&&&&&&&&
 G+                     &&&&&&&&&&&&&&&&&&&&&&&&&&&&&&&&&&&&&&&&&
 H                      &&&&&&&&&&&&&&&&&&&&&&&&&&&&&&&&&&&&&&&&&
 H+                     &&&&&&&&&&&&&&&&&&&&&&&&&&&&&&&&&&&&&&&&&
 ↑↑ ↑    ↑   ↑↑    ↑   ↑↑                                       ↑
 2 3 5   10  1415  20  2425                                      65
```

Figure 7.31: Possible data cards for `INDIVIDUALS`

expression. The assignment

auxiliary variable named in field 2 ← field 7

is made; any variable mentioned in the arithmetic expression must either be reserved (see Section 7.4.2), or have been defined in the `TEMPORARIES` section. If, in this latter case, the variable is integer or real, it must have been allocated a value itself either on a previous `GLOBALS` data card or on a previous `A` card for the current element type in the `INDIVIDUALS` section.

`I` This card announces that an auxiliary parameter, specific to the current group type, is to be assigned a value whenever a second logical auxiliary parameter has the value `.TRUE.`. The string, `p-name`, in field 3 gives the name of the auxiliary parameter that is to be defined; this name must be a valid Fortran name, see Section 7.2.1.2, and have been previously defined in the `TEMPORARIES` section. The string in field 7 is an arithmetic expression. The assignment

auxiliary variable named in field 3 ← field 7

will be made if and only if the logical auxiliary parameter, `l-name`, specified in field 2 has the value `.TRUE.`; the logical parameter must have been previously defined in the `TEMPORARIES` section and allocated a value in the `GLOBALS` or `INDIVIDUALS` section. The arithmetic expression must obey the rules set out in the `A` section above.

`E` This card announces that an auxiliary parameter, specific to the current group type, is to be assigned a value whenever a second logical auxiliary parameter has the value `.FALSE.` The string, `p-name`, in field 3 gives the name of the auxiliary parameter that is to be defined; this name must be a valid Fortran name, see Section 7.2.1.2, and have been previously defined in the `TEMPORARIES` section. The string in field 7 is an arithmetic expression. The assignment

auxiliary variable named in field 3 $\leftarrow$ field 7

will be made if and only if the logical auxiliary parameter, `l-name`, specified in field 2 has the value `.FALSE.`; the logical parameter must have been previously defined in the `TEMPORARIES` section and allocated a value in the `GLOBALS` or `INDIVIDUALS` section. The arithmetic expression must obey the rules set out in the `A` section above.

`F` This card specifies the value of the nontrivial group. The string in field 7 is an arithmetic expression; the assignment

nontrivial group function $\leftarrow$ field 7

is made; any variable mentioned in the expression must obey the rules set out in the `A` section above.

`G` This card specifies the value of the first derivative of the nonlinear group function with respect to its group-type variable. The string in field 7 is an arithmetic expression; the assignment

first derivative of group function $\leftarrow$ field 7

is made; any variable mentioned in the arithmetic expression must obey the rules set out in the `A` section above.

`H` This card specifies the value of the second derivative of the the nonlinear group function with respect to its group-type variable. The string in field 7 is an arithmetic expression; the assignment

second derivative of group function $\leftarrow$ field 7

is made; any variable mentioned in the arithmetic expression must obey the rules set out in the `A` section above.

The data started on an `A`, `I`, `E`, `F`, `G` and `H` card may be continued on a card whose first field contains an `A+`, `I+`, `E+`, `F+`, `G+` or `H+` respectively. Such cards contain an arithmetic expression in field 7 and no further data; the arithmetic expression must obey the rules set out in the `A` section above. At most nineteen continuations of a single assignment are allowed.

The data for a single group type must occur on consecutive cards and in the order given in Figure 7.31. A new group type is deemed to have started whenever a `T` card is encountered. The `F` card is compulsory for all group types.

7.4.3 Two Further Examples

In Section 1.2.3, we gave an example. An SGIF file for this example is given in Figure 7.32.

```
<> <---10---> <---10---> <------------------41------------------>
F.1  Field 2    Field 3                 Field 7
GROUPS        DOC
TEMPORARIES
 R  ALPHA2
 R  TWO
INDIVIDUALS
 T  GTYPE1
 A  TWO                 2.0D+0
 F                      ALPHA*ALPHA
 G                      TWO*ALPHA
 H                      TWO
 T  GTYPE2
 A  ALPHA2              ALPHA*ALPHA
 F                      ALPHA2*ALPHA2
 G                      4.0*ALPHA2*ALPHA
 H                      12.0*ALPHA2
ENDATA
 ↑↑ ↑    ↑   ↑↑    ↑   ↑↑                                       ↑
 23 5    10  1415  20  2425                                     65
```

Figure 7.32: SGIF file for the nontrivial group types for the example of Section 1.2.3

The problem is again given the name DOC. The two types of nontrivial groups were assigned the names GTYPE1/2 by the previous SDIF file, each with group-type variables ALPHA. The function and derivatives values of the second group type, $g(\alpha) = \alpha^4$, all use some product of α^2, so an auxiliary variable is assigned to hold this value, the variable being local to the group type. Likewise, the derivatives of the first group type,$g(\alpha) = \alpha^2$ both use some product of 2.0, so another auxiliary variable is assigned to hold its value.

We gave a second example in Section 1.2.4. An SGIF file for this example is given in Figure 7.33 on page 240. The problem is again given the name DOC2. The single nontrivial group type was given the name SINE by the previous SDIF file, with the group-type variable ALPHA and the single parameter P. The function and second derivatives both depend on the product of the parameter with the sine of the group type variable, so an auxiliary variable is assigned to hold this value.

7.5 Free Form Input

So far, we have been quite specific in the format that we allow. In this section, we consider a second format which, though closely connected to the first, allows one to input problems in a less rigid fashion. Although we refer to this second

```
<> <---10---> <---10---> <-------------------41------------------->
F.1  Field 2    Field 3                 Field 7
GROUPS        DOC2
TEMPORARIES
 R  ISINA
 M  SIN
 M  COS
INDIVIDUALS
 T  SINE
 A  ISINA                P*SIN(ALPHA)
 F                       ISINA
 G                       P*COS(ALPHA)
 H                       -ISINA
ENDATA
 ↑↑ ↑    ↑   ↑↑    ↑   ↑↑                                         ↑
 2 3 5   10  1415  20  2425                                       65
```

Figure 7.33: SGIF file for the nontrivial group type for the example of Section 1.2.4

format as *free format*, the freedom really lies in how the data can be laid out in an input file, not in any extra enhancements to the content of a file.

The input style discussed in Sections 7.2–7.4 is known as fixed format. Each SDIF/SEIF/SGIF file is assumed to be in fixed format unless otherwise specified. A fixed format file has data arranged in specified fields of given length and normally does not allow for much data on a single card. A free form file, on the other hand, is one where considerable data may be conveyed on a single line. The data does not have to lie in prespecified fields. However, we shall insist that *any free form file can be translated to fixed format and interpreted correctly, in this format, in a single sequential pass through the file.*

We allow a further pair of indicator cards in any SDIF/SEIF/SGIF file. These cards, like those described in Section 7.2.1, Section 7.3.2 and Section 7.4.1.2, contain a single keyword starting in column 1. The new keywords are given in Figure 7.34.

Keyword	Presence
`FREE FORMAT`	optional
`FIXED FORMAT`	optional

Figure 7.34: Additional indicator cards

Any data that lies between a `FREE FORMAT` card and the next `FIXED FORMAT` or `ENDATA` card is considered to be in free format. Likewise, any data that lies between a `FIXED FORMAT` card and the next `FREE FORMAT` or `ENDATA` card is considered to be in fixed format. The file is considered to be in fixed format when the **NAME** (SDIF file), **ELEMENTS** (SEIF file) or **GROUPS** (SGIF file) card is first encountered and thus no initial **FIXED FORMAT** card is required.

Fixed format data is exactly as described in Sections 7.2–7.4. The data on a free format data card consists of a number of *strings* separated by *separators*. The characters "-", ";", "$" and " " (blank) are separators and should not therefore be used as significant characters within strings. For example, in free format, X1;2 will be interpreted as two strings X1 and 2. The separators have the following meanings:

(blank) indicates that the previous string has finished and that a new string will follow. One or more blanks is interpreted as a single blank.

; indicates that the previous string has finished and that a new string will follow. Moreover, if the file is translated into fixed format, the new string will appear on a new card.

_ indicates that the previous string has finished and that the next string is empty. Each - indicates a separate empty string so that ___ indicates three empty strings.

$ indicates that the previous string has finished and that the remainder of the card is to be considered as a comment (and thus ignored when the file is interpreted).

A free format card may contain up to 160 characters. On translation into fixed format, a free format card will be divided into one or more fixed format cards depending on how many card separators ";" are encountered. Each fixed format card may hold up to six strings; these strings are numbered 1 to 6.

String 1 is examined to see if the first 12 characters identify the card as an indicator. If so, these characters are placed in columns 1 to 12 on the card and the remaining strings discarded. Otherwise, the card is a data card and the first two-characters of string 1, together with the most recently identified indicator card are used to determine the structure of the remainder of the card; two character code must occur as field 1 in the indicated section of Sections 7.2–7.4 of this report. The first 2, 10, 10, 12 (41 on some SEIF/SGIF cards), 10, and 12 characters of strings 1–6, respectively, are extracted and placed on a single data card starting in columns 2, 5, 15, 25, 40, and 50, respectively. Left-over parts of strings are discarded. The assembled card is now in fixed format and may be interpreted as such. Thus although a free format card may appear to allow more flexibility, the requirement that the translated card conforms to the fixed input format places considerable responsibility on the user to specify the content of strings correctly.

As an example, a free format variant of the SDIF file given in Figure 7.20 might be:

```
NAME          DOC
FREE FORMAT
GROUPS;E GROUP1;E GROUP2;E GROUP3
VARIABLES;_X1 GROUP1 1.0;_X2 GROUP3 1.0;_X3
BOUNDS;FR BN1 X1;LO BN1 X2 -1.0;LO BN1 X3 1.0
       UP BN1 X2 1.0;UP BN1 X3 2.0
ELEMENT TYPE
EV ETYPE1 V1;EV ETYPE1 V2
EV ETYPE2 V1;EV ETYPE2 V2;EV ETYPE2 V3
IV ETYPE2 U1;IV ETYPE2 U2
ELEMENT USES
T G2E1 ETYPE1;V G2E1 V1_X2;V G2E1 V2_X3
T G3E1 ETYPE2;V G3E1 V1_X2;V G3E1 V2_X1;V G3E1 V3_X3
T G3E2 ETYPE1;V G2E1 V1_X1;V G2E1 V2_X3
GROUP TYPE;GV GTYPE1 ALPHA;GV GTYPE2 ALPHA
GROUP USES
T GROUP1 GTYPE1; GROUP2 GTYPE2
E GROUP2 G2E1;E GROUP3 G3E1;E GROUP3 G2E2
ENDATA
```

7.6 Other Standards and Proposals

There have been a number of other proposed standards for input. The most popular approaches use a high-level modelling language to specify problems. Typical examples are GAMS [2], AMPL [24] and OMP [17]. Such approaches are useful for specifying repetitious structures, but do not really attempt to cope with useful nonlinear structure (like invariant subspaces). Recent work by Fourer, Gay and Kernighan [25] hopes to overcome this disadvantage.

We have recently become aware of other suggestions for the input of large-scale structured problems. These proposals are based upon representing nonlinear functions in their factorable [40] or functional forms [42]. Such forms are the the logical extensions of (1.2.1) in which a function is decomposed completely into basic building blocks. The advantage of such schemes is the potential for the automatic calculation of derivatives, but this must be weighed against the difficulty of describing how the building blocks are assembled. We await further details of these interesting proposals.

7.7 Conclusions

We have made a proposal for a standard input format for the specification of (large-scale) nonlinear programming problems. In its full generality, the user needs to provide three input files. The first describes the structure of the problem and the decomposition of the problem into group and element functions. The second and third then specify the values and derivatives of these functions. It is anticipated that the first file will be used to provide input parameters for a user's optimization procedure, while the remaining two will be used to generate problem evaluation subprograms.

Chapter 8. The Specification of LANCELOT Subroutines

Although LANCELOT is a stand-alone Fortran package, it may sometimes be necessary for a user to call the main optimization programs independently. In this chapter we provide details of the Fortran argument lists for the primary minimization subroutines.

8.1 SBMIN

8.1.1 Summary

SBMIN is a ANSI Fortran 77 subroutine to minimize an objective function consisting of a sum of "group" functions. Each group function may be a linear or nonlinear functional whose argument is the sum of "finite element" functions; each finite element function is assumed to involve only a few variables or have a second derivative matrix with low rank for other reasons. Bounds on the variables and known values may be specified. The routine is especially effective on problems involving a large number of variables.

The objective function has the form

$$F(x) = \sum_{i=1}^{n_g} w_i g_i \left(\sum_{j \in J_i} w_{ij} f_j(\bar{x}_j) + a_i^T x - b_i \right), \qquad x = (x_1, x_2, \ldots, x_n)^T,$$

where each J_i is a subset of $J = \{1, \ldots, n_e\}$, the list of indices of nonlinear element functions f_j and where the w_i and w_{ij} are weights. Furthermore, the $\bar{x}_j$ are either small subsets of x or such that the rank of the second derivative matrix of the nonlinear element f_j is small for some other reason, and the vectors a_i of the linear elements are sparse. The least value of the objective function within the "box" region

$$l_i \leq x_i \leq u_i, \quad 1 \leq i \leq n,$$

is sought. Either bound on each variable may be infinite.

The method used is iterative. At each iteration, a quadratic model of the objective function is constructed. An approximation to the minimizer of this model within a trust-region is calculated. The trust region can be either a "box" or a hypersphere of specified radius, centered at the current best estimate of the minimizer. If there is an accurate agreement between the model and the true objective function at the new approximation to the minimizer, this approximation becomes the new best estimate. Otherwise, the radius of the trust region is reduced and a new approximate minimizer sought. The algorithm also allows the trust-region radius to increase when necessary. The minimization of the model function can be carried out by using either a direct-matrix or an iterative approach.

If there is a linear transformation of variables such that one or more of the element functions depend on fewer transformed "internal" variables than the original number of variables $\bar{x}_i$, this may be specified. This can lead to a substantial reduction in computing time.

There are facilities for the user to provide a preconditioner for the conjugate gradient solution of the internal linear systems (a default is provided) and to provide (optional) second derivatives for the element functions. The subroutine returns control to the user to calculate function and derivative values of the group functions g_i and element functions f_j.

If general constraints are present, the alternative subroutine `AUGLG` should be used instead.

ATTRIBUTES — Versions: `*SBMIN`, where `*` may be `S` (single precision) or `D` (double precision). **Origin:** A. R. Conn, IBM T. J. Watson Research Center, Yorktown Heights, Nick Gould, Rutherford Appleton Laboratory, M. Lescrenier and Ph. L. Toint, FUNDP, Namur.

8.1.2 How to Use the Routine

8.1.2.1 The Argument List

The single precision version:

```
CALL SSBMIN( N,  NG, NEL   , IELING, LELING, ISTADG, LSTADG,
*            IELVAR, LELVAR, ISTAEV, LSTAEV, INTVAR, LNTVAR,
*            ISTADH, LSTADH, ICNA  , LICNA,  ISTADA, LSTADA,
*            A , LA, B , LB, BL    , LBL   , BU    , LBU   ,
*            GSCALE, LGSCAL, ESCALE, LESCAL, VSCALE, LVSCAL,
*            GXEQX , LGXEQX, INTREP, LINTRE, RANGES, INFORM,
*            F , X , LX    , GVALS , LGVALS, FT    , LFT   ,
*            FUVALS, LFUVAL, XT    , LXT   , ICALCF, LCALCF,
*            NCALCF, ICALCG, LCALCG, NCALCG, IVAR  , LIVAR ,
*            NVAR  , Q , LQ, DGRAD , LDGRAD, ICHOSE, ITER  ,
*            MAXIT , QUADRT, STOPG , FLOWER, IWK   , LIWK  ,
*            WK    , LWK   , IPRINT, IOUT  )
```

The double precision version:

```
CALL DSBMIN( N,  NG, NEL   , IELING, LELING, ISTADG, LSTADG,
*            IELVAR, LELVAR, ISTAEV, LSTAEV, INTVAR, LNTVAR,
*            ISTADH, LSTADH, ICNA  , LICNA,  ISTADA, LSTADA,
*            A , LA, B , LB, BL    , LBL   , BU    , LBU   ,
*            GSCALE, LGSCAL, ESCALE, LESCAL, VSCALE, LVSCAL,
*            GXEQX , LGXEQX, INTREP, LINTRE, RANGES, INFORM,
*            F , X , LX    , GVALS , LGVALS, FT    , LFT   ,
*            FUVALS, LFUVAL, XT    , LXT   , ICALCF, LCALCF,
```

```
*              NCALCF, ICALCG, LCALCG, NCALCG, IVAR   , LIVAR ,
*              NVAR  , Q , LQ, DGRAD , LDGRAD, ICHOSE, ITER  ,
*              MAXIT , QUADRT, STOPG , FLOWER, IWK    , LIWK  ,
*              WK    , LWK   , IPRINT, IOUT  )
```

N is an INTEGER variable, that must be set by the user to n, the number of variables. It is not altered by the routine.

Restriction: N > 0.

NG is an INTEGER variable, that must be set by the user to n_g, the number of group functions. It is not altered by the routine.

Restriction: NG > 0.

NEL is an INTEGER variable, that must be set by the user to n_e, the number of nonlinear element functions. It is not altered by the routine.

Restriction: NEL ≥ 0.

IELING is an INTEGER array that must be set by the user to contain the indices of the nonlinear elements J_i used by each group. The indices for group i must immediately precede those for group $i+1$ and each group's indices must appear in a contiguous list. See Section 8.1.5 for an example. The contents of the array are not altered by the routine.

LELING is an INTEGER variable that must be set to the actual length of IELING in the calling program. It is not altered by the routine.

Restriction: LELING $\geq$ ISTADG(NG $+$ 1) $- 1$.

ISTADG is an INTEGER array of length at least NG+1, whose i-th value gives the position in IELING of the first nonlinear element in group function i. In addition, ISTADG(NG+1) should be equal to the position in IELING of the last nonlinear element in group NG plus one. See Table 8.1 and Section 8.1.5 for an example. It is not altered by the routine.

LSTADG is an INTEGER variable that must be set to the actual length of ISTADG in the calling program. It is not altered by the routine.

Restriction: LSTADG $\geq$ NG+1.

IELVAR is an INTEGER array containing the indices of the variables in the first nonlinear element f_1, followed by those in the second nonlinear element $f_2, \ldots$. See Section 8.1.5 for an example. It is not altered by the routine.

LELVAR is an INTEGER variable that must be set to the actual length of IELVAR in the calling program. It is not altered by the routine.

Restriction: LELVAR $\geq$ ISTAEV(NEL+1)-1.

elements in g_1	elements in g_2	...	elements in g_n		IELING
weights of elements in g_1	weights of elements in g_2	...	weights of elements in g_n		ESCALE
↑ ISTADG(1)	↑ ISTADG(2)	↑ ISTADG(3)	↑ ISTADG(n_g)	↑ ISTADG($n_g + 1$)	

Table 8.1: Contents of the arrays IELING, ESCALE and ISTADG

ISTAEV is an INTEGER array whose k-th value is the position of the first variable of the k-th nonlinear element function, in the list IELVAR. In addition, ISTAEV(n_e+1) must be equal to the position of the last variable of element n_e in IELVAR plus one. See Table 8.2 and Section 8.1.5 for an example. It is not altered by the routine.

variables in $\bar{x}_1$	variables in $\bar{x}_2$	... ...	variables in $\bar{x}_n$		IELVAR
↑ ISTAEV(1)	↑ ISTAEV(2)	↑ ISTAEV(3)	↑ ISTAEV(n_e)	↑ ISTAEV($n_e + 1$)	

Table 8.2: Contents of the arrays IELVAR and ISTAEV

LSTAEV is an INTEGER variable that must be set to the actual length of ISTAEV in the calling program. It is not altered by the routine.

Restriction: LSTAEV $\geq$ NEL+1.

INTVAR is an INTEGER array whose i-th value must be set to the number of internal variables required for the i-th nonlinear element function f_i on initial entry. See Section 8.1.2.2 for the precise definition of internal variables and Section 8.1.5 for an example. It will subsequently be reset so that its i-th value gives the position in the array FUVALS of the first component of the gradient of the i-th nonlinear element function, *with respect to its internal variables* (see FUVALS).

LNTVAR is an INTEGER variable that must be set to the actual length of INTVAR in the calling program. It is not altered by the routine.

Restriction: LNTVAR $\geq$ NEL+1.

ISTADH is an INTEGER array which need not be set on entry. The array is set in *SBMIN so that its i-th value ($1 \leq i \leq n_e$) gives the position in the array FUVALS of the first component of the Hessian matrix of the i-th nonlinear element function f_i, *with respect to its internal variables.* Only the upper triangular part of each Hessian matrix is stored and the storage is by columns (see Section 8.1.2.3 for details). The element ISTADH($n_e + 1$) gives the position in FUVALS of the first component of the gradient of the objective function (see FUVALS).

LSTADH is an INTEGER variable that must be set to the actual length of ISTADH in the calling program. It is not altered by the routine.

Restriction: LSTADH $\geq$ NEL+1.

ICNA is an INTEGER array containing the indices of the nonzero components of a_1, the gradient of the first linear element, in any order, followed by those in $a_2, \ldots$. See Table 8.3 and Section 8.1.5 for an example. It is not altered by the routine.

LICNA is an INTEGER variable that must be set to the actual length of ICNA in the calling program. It is not altered by the routine.

ISTADA is an INTEGER array whose i-th value is the position of the first nonzero component of the i-th linear element gradient, a_i, in the list ICNA. In addition, ISTADA(NG+1) must be equal to the position of the last nonzero component of a_{n_g} in ICNA plus one. See Table 8.3 and Section 8.1.5 for an example. It is not altered by the routine.

LSTADA is an INTEGER variable that must be set to the actual length of ISTADA in the calling program. It is not altered by the routine.

Restriction: LSTADA $\geq$ NG+1.

A is a REAL (DOUBLE PRECISION in the D version) array containing the values of the nonzero components of the gradients of the linear element functions, a_i, $i = 1, \ldots, n_g$. The values must appear in the same order as their indices appear in ICNA, i.e., the nonzero from element i, whose index is, say, ICNA(k) will have value A(k). See Section 8.1.5 for an example. A is not altered by the routine.

LA is an INTEGER variable that must be set to the actual length of A in the calling program. It is not altered by the routine.

B is a REAL (DOUBLE PRECISION in the D version) array whose i-th entry must be set to the value of the constant b_i for each group. B is not altered by the routine.

LB is an INTEGER variable that must be set to the actual length of B in the calling program. It is not altered by the routine.

Restriction: LB $\geq$ NG.

nonzeros in a_1	nonzeros in a_2	...	nonzeros in a_{n_g}	A
variables in a_1	variables in a_2	...	variables in a_{n_g}	ICNA
↑ ISTADA(1)	↑ ISTADA(2)	↑ ISTADA(3)	↑ ISTADA(n_g)	↑ ISTADA($n_g + 1$)

Table 8.3: Contents of the arrays A, ICNA and ISTADA

BL is a REAL (DOUBLE PRECISION in the D version) array whose i-th entry must be set to the value of the lower bound l_i on the i-th variable. If the i-th variable has no lower bound, BL(i) should be set to a large negative number. The array is not altered by the routine.

LBL is an INTEGER variable that must be set to the actual length of BL in the calling program. It is not altered by the routine.

Restriction: LBL $\geq$ N.

BU is a REAL (DOUBLE PRECISION in the D version) array whose i-th entry must be set to the value of the upper bound u_i on the i-th variable. If the i-th variable has no upper bound, BU(i) should be set to a large positive number. The array is not altered by the routine.

LBU is an INTEGER variable that must be set to the actual length of BU in the calling program. It is not altered by the routine.

Restriction: LBU $\geq$ N.

GSCALE is a REAL (DOUBLE PRECISION in the D version) array whose i-th entry must be set to the value of i-th group weight w_i. The array is not altered by the routine.

LGSCAL is an INTEGER variable that must be set to the actual length of GSCALE in the calling program. It is not altered by the routine.

Restriction: LGSCAL $\geq$ NG.

ESCALE is a REAL (DOUBLE PRECISION in the D version) array whose entries must be set to the values of element weights w_{ij}. The weights must occur in the same order as the indices of the elements assigned to each group in IELING, with the weights for the elements in group i preceding those in group $i+1$, $i = 1, \ldots,$ NG $- 1$. See Table 8.1 and Section 8.1.5 for an example. The array is not altered by the routine.

LESCAL is an INTEGER variable that must be set to the actual length of ESCALE in the calling program. It is not altered by the routine.

Restriction: LESCAL $\geq$ ISTADG(NG+1)−1.

VSCALE is a REAL (DOUBLE PRECISION in the D version) array whose entries must be set to suitable positive scale factors for the problem variables x. The i-th variable x_i will implicitly be divided by VSCALE(i) within *SBMIN. The scale factors should ideally be chosen so that the rescaled variables are of order one at the solution to the minimization problem. If the user does not know suitable scalings, each component of VSCALE should be set to 1.0. Good variable scalings can result in considerable savings in computing times. The array is not altered by the routine.

LVSCAL is an INTEGER variable that must be set to the actual length of VSCALE in the calling program. It is not altered by the routine.

Restriction: LVSCAL $\geq$ N.

GXEQX is a LOGICAL array whose i-th entry must be set .TRUE. if the i-th group function is the trivial function $g(x) = x$ and .FALSE. otherwise. It is not altered by the routine.

LGXEQX is an INTEGER variable that must be set to the actual length of GXEQX in the calling program. It is not altered by the routine.

Restriction: LGXEQX $\geq$ NG.

INTREP is a LOGICAL array whose i-th entry must be set .TRUE. if the i-th nonlinear element function has a useful transformation between elemental and internal variables and .FALSE. otherwise (see Section 8.1.2.2). It is not altered by the routine.

LINTRE is an INTEGER variable that must be set to the actual length of INTREP in the calling program. It is not altered by the routine.

Restriction: LINTRE $\geq$ NEL.

RANGES is a user supplied subroutine whose purpose is to define the linear transformation of variables for those nonlinear elements which have different elemental and internal variables. See Section 8.1.2.2 for details. RANGES must be declared EXTERNAL in the calling program.

INFORM is an INTEGER variable that is set within *SBMIN and passed back to the user to prompt further action (INFORM < 0), to indicate errors in the input parameters (INFORM > 0) and to record a successful minimization (INFORM $= 0$). On initial entry, INFORM must be set to 0. Possible nonzero values of INFORM and their consequences are discussed in Section 8.1.2.3 and Section 8.1.2.4.

FOBJ is a REAL (DOUBLE PRECISION in the D version) variable which need not be set on initial entry. On final exit, FOBJ contains the value of the objective function at the point X.

X is a REAL (DOUBLE PRECISION in the D version) array which must be set by the user to the value of the variables at the starting point. On exit, it contains the values of the variables at the best point found in the minimization (usually the solution).

LX is an INTEGER variable that must be set to the actual length of X in the calling program. It is not altered by the routine.

Restriction: LX $\geq$ N.

GVALS is a REAL (DOUBLE PRECISION in the D version) array of dimension (LGVALS,3) which is used to store function and derivative information for the group functions. The user is asked to provide values for these functions and/or derivatives, evaluated at the argument FT (see FT) when control is returned to the calling program with a negative value of the variable INFORM. This information needs to be stored by the user in specified locations within GVALS. Details of the required information are given in Section 8.1.2.3.

LGVALS is an INTEGER variable that must be set to the actual length of GVALS in the calling program. It is not altered by the routine.

Restriction: LGVALS $\geq$ NG.

FT is a REAL (DOUBLE PRECISION in the D version) array whose i-th entry is set within *SBMIN to a trial value of the argument of the i-th group function at which the user is required to evaluate the values and/or derivatives of that function. Precisely what group function information is required at FT is under the control of the variable INFORM and details are given in Section 8.1.2.3.

LFT is an INTEGER variable that must be set to the actual length of FT in the calling program. It is not altered by the routine.

Restriction: LFT $\geq$ NG.

FUVALS is a REAL (DOUBLE PRECISION in the D version) array which is used to store function and derivative information for the nonlinear element functions. The user is asked to provide values for these functions and/or derivatives, evaluated at the argument XT (see XT), at specified locations within FUVALS, when control is returned to the calling program with a negative value of the variable INFORM. Details of the required information are given in Section 8.1.2.3. The layout of FUVALS is indicated in Table 8.4.

The first segment of FUVALS contains the values of the nonlinear element functions; the next two segments contain their gradients and Hessian matrices, taken with respect to their internal variables, as indicated in Section 8.1.2.3. The remaining two used segments contain the gradient of

← n_e →			← n →	← n →		
element values	element gradients	element Hessians	objective gradient	Hessian diagonal	work-space	
	↑ INTVAR(1)	↑ ISTADH(1)	↑ ISTADH(n_e+1)			↑ LFUVALS

Table 8.4: Partitioning of the workspace array FUVALS

the objective function and the diagonal elements of the second derivative approximation, respectively. At the solution, the components of the gradient of the objective function corresponding to variables which lie on one of their bounds are of particular interest in many applications areas. In particular they are often called shadow prices and are used to assess the sensitivity of the solution to variations in the bounds on the variables.

LFUVAL is an INTEGER variable that must be set to the actual length of FUVALS in the calling program. It is not altered by the routine.

XT is a REAL (DOUBLE PRECISION in the D version) array which is set within *SBMIN to a trial value of the variables x at which the user is required to evaluate the values and/or derivatives of the nonlinear elements functions. Precisely what element function information is required at XT is under the control of the variable INFORM and details are given in Section 8.1.2.3.

LXT is an INTEGER variable that must be set to the actual length of XT in the calling program. It is not altered by the routine.

Restriction: LXT $\geq$ N.

ICALCF is an INTEGER array which need not be set on entry. If the value of INFORM on return from *SBMIN indicates that values of the nonlinear element functions are required prior to a re-entry, the first NCALCF components of ICALCF give the indices of the nonlinear element functions which need to be recalculated at XT. Precisely what element function information is required is under the control of the variable INFORM and details are given in Section 8.1.2.3.

LCALCF is an INTEGER variable that must be set to the actual length of ICALCF in the calling program. It is not altered by the routine.

Restriction: LCALCF $\geq$ NEL.

NCALCF is an INTEGER variable which need not be set on entry. If the value of INFORM on return from *SBMIN indicates that values of the nonlinear element functions are required prior to a re-entry, NCALCF values need to be calculated and they correspond to nonlinear elements `ICALCF(`i`)`, $i = 1, \ldots,$ `NCALCF`.

`ICALCG` is an `INTEGER` array which need not be set on entry. If the value of `INFORM` on return from `*SBMIN` indicates that further values of the group functions are required prior to a re-entry, the first `NCALCG` components of `ICALCG` give the indices of the group functions which need to be recalculated at `FT`. Precisely what group function information is required is under the control of the variable `INFORM` and details are given in Section 8.1.2.3.

`LCALCG` is an `INTEGER` variable that must be set to the actual length of `ICALCG` in the calling program. It is not altered by the routine.

Restriction: `LCALCG` $\geq$ `NG`.

`NCALCG` is an `INTEGER` variable which need not be set on entry. If the value of `INFORM` on return from `*SBMIN` indicates that values of the group functions are required prior to a re-entry, `NCALCG` values need to be calculated and they correspond to groups `ICALCG(`i`)`, $i = 1, \ldots,$ `NCALCG`.

`IVAR` is an `INTEGER` array which is required when the user is providing a special preconditioner for the conjugate gradient inner iteration (see Section 8.1.2.3).

`LIVAR` is an `INTEGER` variable that must be set to the actual length of `IVAR` in the calling program. It is not altered by the routine.

Restriction: `LIVAR` $\geq$ `N`.

`NVAR` is an `INTEGER` variable which need not be set on entry. `NVAR` is required when the user is providing a special preconditioner for the conjugate gradient inner iteration (see Section 8.1.2.3).

`Q` is a `REAL` (`DOUBLE PRECISION` in the D version) array which is required when the user is providing a special preconditioner for the conjugate gradient inner iteration (see Section 8.1.2.3).

`LQ` is an `INTEGER` variable that must be set to the actual length of `Q` in the calling program. It is not altered by the routine.

Restriction: `LQ` $\geq$ `N`.

`DGRAD` is a `REAL` (`DOUBLE PRECISION` in the D version) array which is required when the user is providing a special preconditioner for the conjugate gradient inner iteration (see Section 8.1.2.3).

`LDGRAD` is an `INTEGER` variable that must be set to the actual length of `LDGRAD` in the calling program. It is not altered by the routine.

Restriction: `LDGRAD` $\geq$ `N`.

`ICHOSE` is an `INTEGER` array of length 6 which allows the user to specify a number of the characteristics of the minimization algorithm used. The three entries and the information required are:

`ICHOSE(1)` must be set to 1 if a two-norm (hyperspherical) trust region is required, and any other value if an infinity-norm (box) trust region is to be used.

`ICHOSE(2)` is used to specify the method used to solve the linear systems of equations which arise at each iteration of the minimization algorithm. Possible values are

`1`: The conjugate gradients method will be used without preconditioning.

`2`: A preconditioned conjugate gradients method will be used with a diagonal preconditioner.

`3`: A preconditioned conjugate gradients method will be used with a user supplied preconditioner. Control will be passed back to the user to construct the product of the preconditioner with a given vector prior to a reentry to `*SBMIN` (see Section 8.1.2.3).

`4`: A preconditioned conjugate gradients method will be used with an expanding band incomplete Cholesky preconditioner.

`5`: A preconditioned conjugate gradients method will be used with Munksgaard's preconditioner.

`6`: A preconditioned conjugate gradients method will be used with a modified Cholesky preconditioner, in which small or negative diagonals are made sensibly positive during the factorization.

`7`: A preconditioned conjugate gradients method will be used with a modified Cholesky preconditioner, in which an indefinite factorization is altered to give a positive definite one.

`8`: A preconditioned conjugate gradients method will be used with a band preconditioner. The semi-bandwidth is given by the variable `NSEMIB` appearing in the common block `*COMSB` (see Section 8.1.2.6). The default semi-bandwidth is 5.

`11`: A multifrontal factorization method will be used.

`12`: A modified Cholesky factorization method will be used.

Any other value of `ICHOSE(2)` will be reset to 1.

`ICHOSE(3)` specifies what sort of first derivative approximation is to be used. If the user is able to provide analytical first derivatives for each nonlinear element function, `ICHOSE(3)` must be set to 0. If the user is unable to provide first derivatives, `ICHOSE(3)` should be set to 1 so that finite-difference approximations may be formed.

`ICHOSE(4)` specifies what sort of second derivative approximation is to be used. If the user is able to provide analytical second derivatives for each nonlinear element function, ICHOSE(3) must be set to 0. If the user is unable to provide second derivatives, these derivatives will be approximated using one of four secant approximation formulae. If ICHOSE(4) is set to 1, the BFGS formula is used; if it is set to 2, the DFP formula is used; if it is set to 3, the PSB formula is used; and if it is set to 4, the symmetric rank-one formula is used. The user is strongly advised to use exact second derivatives if at all possible as this often significantly improves the convergence of the method.

ICHOSE(5) specifies whether the exact generalized Cauchy point, the first estimate of the minimizer of the quadratic model within the box, is required (ICHOSE(5) = 1), or whether an approximation suffices (any other value of ICHOSE(5)).

ICHOSE(6) specifies whether an accurate minimizer of the quadratic model within the box is required (ICHOSE(6) = 1), or whether an approximation suffices (any other value of ICHOSE(6)).

ITER is an INTEGER variable which gives the number of iterations of the algorithm performed.

MAXIT is an INTEGER variable which must be set by the user to the maximium number of iterations that the algorithm will be allowed to perform. It is not altered by the routine.

QUADRT is a LOGICAL variable which must be set .TRUE. by the user if all the nonlinear element functions are quadratics and .FALSE.. otherwise. This will cut down on the number of derivative evaluations required for quadratic functions. It is not altered by the routine.

STOPG is a REAL (DOUBLE PRECISION in the D version) variable, that must be set by the user to a measure of the accuracy (in the sense of the infinity norm of the projected gradient of the objective function) required to stop the minimization procedure. If $0 \leq$ INFORM ≤ 3 on exit from *SBMIN, STOPG holds the value of the norm of the projected gradient of the objective function at the best point found in the minimization.

FLOWER is a REAL (DOUBLE PRECISION in the D version) variable, that must be set to a lower bound on the value of the objective function at the solution. If no lower bound is known, FLOWER should be set to a large negative value. It is not altered by the routine.

IWK is a INTEGER array of length LIWK used as workspace by *SBMIN.

LIWK is an INTEGER variable which must be set by the user to the length of the working array IWK and is not altered.

WK is a REAL (DOUBLE PRECISION in the D version) array of length LWK used as workspace by *SBMIN.

LWK is an INTEGER variable which must be set by the user to the length of the working array WK and is not altered.

IPRINT is an INTEGER variable which determines the amount of intermediate output produced by *SBMIN and must be set by the user. Possible values of IPRINT and the output obtained are described in Section 8.1.2.5. It is not altered by the routine.

IOUT is an INTEGER variable which must be set by the user to the Fortran unit number for intermediate output. It is not altered by the routine.

8.1.2.2 Internal Variables and Subroutine RANGES

A nonlinear element function f_j is assumed to be a function of the variables $\bar{x}_j$, a subset of the problem variables x. Suppose that $\bar{x}_j$ has n_j elements. Then another way of saying this is that we have an element function $f_j(v_1, \ldots, v_{n_j})$, where in our special case, we choose $v_1 = (\bar{x}_j)_1, \ldots, v_{n_j} = (\bar{x}_j)_{n_j}$. The *elemental* variables for the element function f_j are the variables $\bar{v}$ and, while we need to associate the particular values $\bar{x}_j$ with $\bar{v}$ (using the array IELVAR), it is the elemental variables which are important in defining the nonlinear element functions.

As an example, the seventh nonlinear element function for a particular problem might be

$$f_7(v_1, v_2, v_3) = (v_1 + v_2)e^{v_1 - v_3}, \tag{8.1.1}$$

where for our example $v_1 = x_{29}$, $v_2 = x_3$ and $v_3 = x_{17}$. For this example, there are three elemental variables. However, the example illustrates an additional point. Although f_7 is a function of three variables, the function itself is really only composed of *two* independent parts; the product of $(v_1 + v_2)$ with $e^{v_1 - v_3}$, or, if we write $u_1 = v_1 + v_2$ and $u_2 = v_1 - v_3$, the product of u_1 with e^{u_2}. The variables u_1 and u_2 are known as *internal* variables for the element function. They are obtained as *linear combinations* of the elemental variables. The important feature as far as *SBMIN is concerned is that each nonlinear function involves as few variables as possible, as this allows for compact storage and more efficient derivative approximation. By representing the function in terms of its internal variables, this goal is achieved. *SBMIN only stores derivatives of the element functions with respect to internal variables, so it pays to use an internal representation in this case. It frequently happens, however, that a function does not have useful internal variables. For instance, another element function might be

$$f_9(v_1, v_2) = v_1 \sin v_2, \tag{8.1.2}$$

where for example $v_1 = x_6$ and $v_2 = x_{12}$. Here, we have broken f_9 down into as few pieces as possible. Although there are internal variables, $u_1 = v_1$ and

$u_2 = v_2$, they are the same in this case as the elemental variables and there is no virtue in exploiting them. Moreover it can happen that although there are special internal variables, there are just as many internal as elemental variables and it therefore doesn't particularly help to exploit them. For instance, if

$$f_{14}(v_1, v_2) = (v_1 + v_2)\log(v_1 - v_2), \tag{8.1.3}$$

where for example $v_1 = x_{12}$ and $v_2 = x_2$, the function can be formed as $u_1 \log(u_2)$ where $u_1 = v_1 + v_2$ and $u_2 = v_1 - v_2$. But as there are just as many internal variables as elementals, it will not normally be advantageous to use this internal representation.

Finally, although an element function may have useful internal variables, the user need not bother with them. `*SBMIN` will still work, but at the expense of extra storage and computational effort. The user decides on input to `*SBMIN` which elements have useful transformations by setting the appropriate elements of the array `INTREP` to `.TRUE.`.

In general, there will be a linear transformation from the elemental variables to the internal ones. For example (8.1.1), we have

$$\begin{pmatrix} u_1 \\ u_2 \end{pmatrix} = \begin{pmatrix} 1 & 1 & 0 \\ 1 & 0 & -1 \end{pmatrix} \begin{pmatrix} v_1 \\ v_2 \\ v_3 \end{pmatrix},$$

while in (8.1.2), we have

$$\begin{pmatrix} u_1 \\ u_2 \end{pmatrix} = \begin{pmatrix} 1 & 1 \\ 1 & 0 \end{pmatrix} \begin{pmatrix} v_1 \\ v_2 \end{pmatrix}.$$

In general the transformation will be of the form

$$u = Wv, \tag{8.1.4}$$

and this transformation is *useful* if the matrix W has fewer rows than columns.

The purpose of the `RANGES` routine is to define the transformation between internal and elemental variables for nonlinear elements with useful internal representations. The routine has the following specification.

The single precision version:

```
CALL RANGES( IELEMN, TRANSP, W1, W2, NELVAR, NINVAR )
```

The double precision version:

```
CALL RANGES( IELEMN, TRANSP, W1, W2, NELVAR, NINVAR )
```

`IELEMN` is an `INTEGER` variable which gives the index of the nonlinear element whose transformation is required by `*SBMIN`. It must not be altered by the routine.

TRANSP is a LOGICAL variable. If TRANSP is .FALSE., the user must put the result of the transformation Wv in the array W2, where v is input in the array W1. Otherwise, the user must supply the result of the transposed transformation $W^T u$ in the array W2, where u is input in the array W1. It must not be altered by the routine.

W1 is a REAL (DOUBLE PRECISION in the D version) array whose length is the number of elemental variables if TRANSP is .FALSE., and the number of internal variables otherwise. It must not be altered by the routine.

W2 is a REAL (DOUBLE PRECISION in the D version) array, whose length is the number of internal variables if TRANSP is .FALSE., and the number of elemental variables otherwise. The result of the transformation of W1 or its transpose, as assigned by TRANSP, must be set in W2.

NELVAR is an INTEGER variable which gives the number of elemental variables for the element specified by IELEMN. It must not be altered by the routine.

NINVAR is an INTEGER variable which gives the number of internal variables for the element specified by IELEMN. It must not be altered by the routine.

The user will already have specified which elements have useful transformations in the array INTREP. RANGES will only be called for elements for which the corresponding component of INTREP is .TRUE..

8.1.2.3 Reverse Communication Information

When a return is made from *SBMIN with INFORM set negative, *SBMIN is asking the user for further information. The user should normally compute the required information and re-enter *SBMIN with INFORM unchanged (see INFORM $= -11$, below, for an exception).

Possible values of INFORM and the information required are

INFORM $= -1$: The user should compute the function and, if they are available, derivative values of the nonlinear element functions numbered ICALCF(i) for $i = 1, \ldots,$ NCALCF (see below).

INFORM $= -2$: The user should compute the function and derivative values of all non-trivial nonlinear group functions numbered ICALCG(i) for $i = 1, \ldots,$ NCALCG (see below).

INFORM $= -3$: The user should compute the function values of the nonlinear element functions numbered ICALCF(i) for $i = 1, \ldots,$ NCALCF (see below).

INFORM $= -4$: The user should compute the function values of the nontrivial nonlinear group functions numbered ICALCG(i) for $i = 1, \ldots,$ NCALCG (see below).

`INFORM` $= -5$: If exact derivatives are available, the user should compute the derivative values of all nonlinear element functions numbered `ICALCF(`i`)` for $i = 1, \ldots,$ `NCALCF`. The user should also compute the derivative values of all non-trivial nonlinear group functions numbered `ICALCG(`i`)` for $i = 1, \ldots,$ `NCALCG` (see below).

`INFORM` $= -6$: The user should compute the derivative values of all nonlinear element functions numbered `ICALCF(`i`)` for $i = 1, \ldots,$ `NCALCF` (see below).

`INFORM` $= -7$: The user should compute the function values of the nonlinear element functions numbered `ICALCF(`i`)` for $i = 1, \ldots,$ `NCALCF` (see below).

`INFORM` $= -8, -9, -10$: The user should compute the product of a preconditioning matrix M with the vector d and return the value in q.

When `INFORM` $= -1$, -3 or -7, the user must return the values of all of the nonlinear element functions $f_k(\bar{x}_k)$, $k =$`ICALCF(`i`)`, $i = 1, \ldots,$ `NCALCF`. The functions are to be evaluated at the point x given in the array `XT`. The k-th function value must be placed in `FUVALS`(k).

When `INFORM` $= -1, -5$ or -6 and first derivatives are available (that is when `ICHOSE(3)` $= 0$), the user must return the values of the gradients, *with respect to their internal variables*, of all the nonlinear element functions $f_k(\bar{x}_k)$, $k =$`ICALCF(`i`)`, $i = 1, \ldots,$ `NCALCF`. The gradients are to be evaluated at the point x given in the array `XT`. The gradient with respect to internal variable i of the k-th nonlinear element,

$$\frac{\partial f_k}{\partial u_i},$$

must be placed in `FUVALS(INTVAR(`k`)+`i`-1)`. If, in addition, exact second derivatives are to be provided (`ICHOSE(4)` $= 0$), the user must return the values of the Hessian matrices, *with respect to their internal variables*, of the same nonlinear element functions evaluated at x. Only the "upper triangular" part of the required Hessians should be specified. The component of the Hessian of the k-th nonlinear element with respect to internal variables i and j, $i \leq j$,

$$\frac{\partial^2 f_k}{\partial u_i \partial u_j},$$

must be placed in `FUVALS(ISTADH(`k`)+`$(j(j-1)/2)+i-1$`)`.

When `INFORM` $= -2$ or -4, the user must return the values of all the group functions g_k, $k =$`ICALCG(`i`)`, $i = 1, \ldots,$ `NCALCG`. The k-th such function should be evaluated with the argument `FT(`k`)` and the result placed in `GVALS(`k`,1)`.

When `INFORM` $= -2$ or -5, the user must return the values of the first and second derivatives of each of the group functions g_k, $k =$`ICALCG(`i`)`, $i = 1, \ldots,$ `NCALCG`, with respect to its argument. The derivatives of the k-th such function should be evaluated with the argument FT(k). The first derivative of the k-th group function should be placed in GVALS(k,2) and the corresponding second derivative returned in GVALS(k,3).

When INFORM $= -8, -9$ or -10, the user must return the values of the components Q(IVAR(i)) $= q_i$, $i = 1, \ldots,$ NVAR, such that $q = Md$ and where d_i =Q(IVAR(i)) $i = 1, \ldots,$ NVAR. Here M is a symmetric positive definite approximation to the inverse of the matrix whose i-th row and column are the IVAR(i)-th row and column of the Hessian matrix of $F(x)$, $i = 1, \ldots,$ NVAR. These values can only occur if ICHOSE(2) $= 3$.

If the user does not wish, or is unable, to compute an element or group function at a particular argument returned from *SBMIN, INFORM may be reset to -11 and *SBMIN re-entered. *SBMIN will treat such a re-entry as if the current iteration had been unsuccessful and reduce the trust-region radius. This facility is useful when, for instance, the user is asked to evaluate a function at a point outside its domain of definition.

8.1.2.4 Error Messages

If INFORM is positive on return from *SBMIN, an error has been detected. The user should correct the error and restart the minimization. Possible values of INFORM and their consequences are:

INFORM $= 1$: More than MAXIT iterations have been performed. This is often a symptom of incorrectly programmed derivatives or of the preconditioner used being insufficiently effective. Recheck the derivatives. Otherwise, increase MAXIT and re-enter *SBMIN at the best point found so far.

INFORM $= 2$: The trust-region radius has become too small. This is often a symptom of incorrectly programmed derivatives or of requesting more accuracy in the projected gradient than is reasonable on the user's machine. If the projected gradient is small, the minimization has probably succeeded. Otherwise, recheck the derivatives.

INFORM $= 3$: The step taken during the current iteration is so small that no difference will be observed in the function values. This sometimes occurs when too much accuracy is required of the final gradient. If the projected gradient is small, the minimization has probably succeeded.

INFORM $= 4$: One of the INTEGER arrays has been initialized with insufficient space. A message indicating which array is at fault and the required space will be printed on unit number IOUT.

INFORM $= 5$: One of the REAL (DOUBLE PRECISION in the D version) arrays has been initialized with insufficient space. A message indicating which array is at fault and the required space will be printed on unit number IOUT.

`INFORM` $= 6$: One of the `LOGICAL` arrays has been initialized with insufficient space. A message indicating which array is at fault and the required space will be printed on unit number `IOUT`.

8.1.2.5 Intermediate Printing

The user is able to control the amount of intermediate printing performed in the course of the minimization. Printing is under the control of the parameter `IPRINT` and output is sent to I/O unit number `IOUT`. Possible values of `IPRINT` and the levels of output produced are as follows.

`IPRINT` ≤ 0 : No printing, except warning messages, will be performed.

`IPRINT` ≥ 1 : Details of the minimization function will be output. This includes the number of variables, groups and nonlinear elements which are used and a list of the variables which occur in each of the linear and nonlinear elements in every group.

`IPRINT` $= 1$: A simple one line description of each iteration is given. This includes the iteration number, the number of derivative evaluations that have been made, the number of conjugate-gradient iterations that have been performed, the current function value, the (two-) norm of the projected gradient, the ratio ρ of the actual to predicted decrease in objective function value achieved, the current trust-region radius, the norm of the step taken, an indication of how the direct or iterative method ended, the number of variables which lie away from their bounds and the total time spent on the minimization.

`IPRINT` $= 2$: In addition to the information output with `IPRINT` $=$ 1, a short description of the approximate solution to the inner-iteration linear system is given. Before a successful (`INFORM` $=$ 0) exit, details of the estimate of the minimizer and the gradient of the objective function are given.

`IPRINT` $= 3$: A list of the current iteration number, function value, the number of derivative evaluations that have been made, the (two-) norm of the projected gradient, the number of conjugate-gradient iterations that have been performed and the current trust-region radius are given, followed by the current estimate of the minimizer. The values of the reduction in the model of the objective function and the actual reduction in the objective, together with their ratio, are also given. Before a successful (`INFORM` $=$ 0) exit, details of the estimate of the minimizer and the gradient of the objective function are given.

`IPRINT` $= 4$: In addition to the information output with `IPRINT` $=$ 3, the gradient of the objective function at the current estimate of the minimizer is given. Full details of the approximate solution to the inner-iteration linear system are also given. This level of output is intended as a debugging aid for the expert only.

IPRINT $= 5$: In addition to the information output with IPRINT $= 4$, the diagonal elements of the second derivative approximation are given.

IPRINT ≥ 6 : In addition to the information output with IPRINT $= 5$, the second derivative approximations (taken with respect to the internal variables) to each nonlinear element function are given.

8.1.2.6 Common

The variables in common block *COMSB are used as controlling parameters for the minimization and may provide the user with further information. These variables should not be altered by the user.
The single precision version:

```
      COMMON / SCOMSB / ACCCG , RATIO , RADIUS, RADMAX, FINDMX, CMA31,
     *                  ITERCG, ITCGMX, NGEVAL, ISKIP , IFIXED, NSEMIB
```

The double precision version:

```
      COMMON / DCOMSB / ACCCG , RATIO , RADIUS, RADMAX, FINDMX, CMA31,
     *                  ITERCG, ITCGMX, NGEVAL, ISKIP , IFIXED, NSEMIB
```

ACCCG is a REAL (DOUBLE PRECISION in the D version) variable which gives the size of the (two-) norm of the reduced gradient of the model function below which the conjugate-gradients iteration will terminate. This value will become small as the algorithm converges.

RATIO is a REAL (DOUBLE PRECISION in the D version) variable which gives the ratio of the actual reduction that has been made in the objective function value during the current iteration to that predicted by the model function. A value close to one is to be expected as the algorithm converges.

RADIUS is a REAL (DOUBLE PRECISION in the D version) variable which gives the current radius of the trust-region.

RADMAX is a REAL (DOUBLE PRECISION in the D version) variable which gives the maximum allowable trust-region radius.

FINDMX is a REAL (DOUBLE PRECISION in the D version) variable which allows the user to alter the printing of function and gradient information. Normally FINDMX should be set to 1.0. The printed values of the objective function and its derivatives actually correspond to the function FINDMX $F(x)$. This may be useful if, for instance, SBMIN is used to calculate the maximum value of the function $f(x)$ by minimizing $F(x) = -f(x)$. Then, values of f and its derivatives may be printed within SBMIN merely by setting FINDMX to -1.0.

CMA31 is a REAL (DOUBLE PRECISION in the D version) variable which gives the pivot tolerance used if Munksgaard's preconditioner is selected (that is when ICHOSE(2) = 5).

ITERCG is an INTEGER variable which gives the number of conjugate-gradient iterations that have been performed since the start of the minimization

ITCGMX is an INTEGER variable which gives the maximum number of conjugate-gradient iterations that will be allowed at each iteration of SBMIN.

NGEVAL is an INTEGER variable which gives the number of objective function gradient evaluations that have been made since the start of the minimization.

ISKIP is an INTEGER variable which gives the number of times that an approximation to the second derivatives of an element function has been rejected.

IFIXED is an INTEGER variable which gives the index of the variable that most recently encountered one of its bounds in the minimization process.

NSEMIB is an INTEGER variable which gives the bandwidth used if the band or expanding band preconditioner is selected (ICHOSE(2) = 4 or 8).

The variables in common block MACHNS hold machine dependent constants. These variables should not be altered by the user.

The single precision version:

```
COMMON / SMACHN / EPSMCH, EPSNEG, TINY, BIG
```

The double precision version:

```
COMMON / DMACHN / EPSMCH, EPSNEG, TINY, BIG
```

EPSMCH is a REAL (DOUBLE PRECISION in the D version) variable which gives the smallest positive machine number for which 1.0 + EPSMCH > 1.0.

EPSNEG is a REAL (DOUBLE PRECISION in the D version) variable which gives the smallest positive machine number for which 1.0 − EPSNEG < 1.0.

TINY is a REAL (DOUBLE PRECISION in the D version) variable which gives the smallest positive machine number.

BIG is a REAL (DOUBLE PRECISION in the D version) variable which gives the largest positive machine number.

There is a further named common block, INWKSP, which is used to pass information on partitioning of the workspace arrays between internal subroutines. The content of the common block should be of no interest to the general user.

8.1.3 General Information

Use of common: See Section 8.1.2.6.

Workspace: Provided by the user (see arguments IWK and WK).

Other routines called directly:

```
*CG,     *CAUCH, *ELGRD, *HSPRD, *FRNTL, *GTPGR, *PRECN, *PRGRA,
*SECNT, *CLCFG, *INXAC, *NRM2,  *INITW, *DOT,   *MACHR, *SCALH,
*FDGRD, CPUTIM.
```

Input/output: No input; output on device number IOUT.

Restrictions: N $>$ 0, NEL $>$ 0, NG $>$ 0, LELING $\geq$ ISTADG(NG + 1) − 1, LSTADG $\geq$ NG + 1, LELVAR $\geq$ ISTAEV(NEL + 1) − 1, LSTAEV $\geq$ NEL + 1, LNTVAR $\geq$ NEL + 1, LSTADH $\geq$ NEL + 1, LSTADA $\geq$ NG + 1, LB $\geq$ NG, LBL $\geq$ N, LBU $\geq$ N, LX $\geq$ N, LXT $\geq$ N, LIVAR $\geq$ N, LDGRAD $\geq$ N, LQ $\geq$ N, LGSCAL $\geq$ NG, LESCAL $\geq$ ISTADG(NG + 1) − 1, LVSCAL $\geq$ N, LGXEQX $\geq$ NG, LINTRE $\geq$ NEL, LGVALS $\geq$ NG, LFT $\geq$ NG, LCALCF $\geq$ NEL, LCALCG $\geq$ NG.

Portability: ANSI Fortran 77.

8.1.4 Method

The basic method that is implemented within subroutine SBMIN is described in detail by Conn, Gould and Toint (1988b). The concept of partial separability was first suggested by Griewank and Toint (1982). The extension to group partially separable functions is new. Also see Chapter 3 of this manual.

A step from the current estimate of the solution is determined using a trust-region approach. That is, a quadratic model of the objective function is approximately minimized within the intersection of the constraint "box" and another convex region, the trust-region. This minimization is carried out in two stages. Firstly, the so-called generalized Cauchy point for the quadratic subproblem is found. (The purpose of this point is to ensure that the algorithm converges and that the set of bounds which are satisfied as equations at the solution is rapidly identified.) Thereafter an improvement to the quadratic model is sought using either a direct-matrix or truncated conjugate gradients algorithm. The trust-region size is increased if the reduction obtained in the objective function is reasonable when compared with the reduction predicted by the model and reduced otherwise.

A central idea is that a collection of small matrices approximating the Hessian matrices of each f_k is used and updated at every iteration using one of a number of possible updating formulae. Objective function values and derivatives are assembled from these components as required.

The strategy for treating bound constraints is based on the usual projection device and is described in detail in Conn, Gould and Toint (1988a).

8.1.5 Example

We now consider the small example problem,

$$\text{minimize } F(x_1, x_2, x_3) = x_1^2 + x_2 \sin(x_1 + x_3) + 3x_2^4 x_3^4 + x_2.$$

subject to the bounds $-1 \le x_2 \le 1$ and $1 \le x_3 \le 2$. There are a number of ways of casting this problem in the framework required by `*SBMIN`. Here, we consider partitioning F into groups as Notice the following:

$(x_1)^2$	+	$(x_2 \sin(x_1 + x_3))$	+	$3(x_2 x_3)^4$	+	(x_2)
↑		↑		↑		↑
group 1		group 2		group 3		group 4

Table 8.5: An example of grouping

1. Group 1 uses the non-trivial group function $g_1(\alpha) = \alpha^2$. The group contains a single *linear* element; the element function is x_1.
2. Group 2 uses the trivial group function $g_2(\alpha) = \alpha$. The group contains a single *nonlinear* element; this element function is $x_2 \sin(x_1 + x_3)$. The element function has *three* elemental variables, v_1, v_2 and v_3, say, (with $v_1 = x_2$, $v_2 = x_1$ and $v_3 = x_3$), but may be expressed in terms of *two* internal variables u_1 and u_2, say, where $u_1 = v_1$ and $u_2 = v_2 + v_3$.
3. Group 3 uses the non-trivial group function $g_3(\alpha) = \alpha^4$, weighted by the factor 3. Again, the group contains a single *nonlinear* element; this element function is $x_2 x_3$. The element function has *two* elemental variables, v_1 and v_2, say, (with $v_1 = x_2$ and $v_2 = x_3$). This time, however, there is no useful transformation to internal variables.
4. Group 4 again uses the trivial group function $g_4(\alpha) = \alpha$. This time the group contains a single *linear* element x_2.

Thus we see that we can consider our objective function to be made up of four group functions; the first and third are non-trivial, so we need to provide function and derivative values for these when prompted by `*SBMIN`. There are two nonlinear elements, one each from groups two and three. Again this means that we need to provide function and derivative values for these when required by `*SBMIN`. Finally, one of these elements, the first, has a useful transformation from elemental to internal variables, so the routine `RANGES` must be set to provide this transformation.

The problem involves three variables, four groups and two nonlinear elements, so we set `N` = 3, `NG` = 4 and `NEL` = 2. As the first and third group functions are non-trivial, we set `GXEQX` as:

I	1	2	3	4
GXEQX(I)	.FALSE.	.TRUE.	.FALSE.	.TRUE.

The first nonlinear element occurs in group 2 and the second in group 3; both elements are unweighted. We thus set IELING, ESCALE and ISTADG as:

I	1	2
IELING(I)	1	2
ESCALE(I)	1.0	1.0

I	1	2	3	4	5
ISTADG(I)	1	1	2	3	3

The first nonlinear element function is not a quadratic function, of its internal variables so we set QUADRT=.FALSE.. Nonlinear element one assigns variables x_2, x_1 and x_3 to its elemental variables, while the second nonlinear element assigns variables x_2 and x_3 to its elemental variables. Hence IELVAR contains:

I	1	2	3	4	5
IELVAR(I)	2	1	3	2	3
	⟵	el. 1	⟶	← el. 2	→

In this vector, we now locate the position of the first variable of each nonlinear element, and build the vector ISTAEV as follows:

I	1	2	3
ISTAEV(I)	1	4	6

Both nonlinear elements have two internal variables — element two using its elemental variables as internals — so we set INTVAR as follows:

I	1	2	3
INTVAR(I)	2	2	-

As the first nonlinear element has a useful transformation between elemental and internal variables, while the second does not, INTREP contains:

I	1	2
INTREP(I)	.TRUE.	.FALSE.

Turning to the linear elements, groups one and four each have a linear element, each involving a single variable — x_1 for group 1 and x_2 for group 4. We thus set ISTADA as:

I	1	2	3	4	5
ISTADA(I)	1	2	2	2	3

and ICNA and A as

I	1	2
ICNA(I)	1	2
A(I)	1.0D+0	1.0D+0

There is no constant term for any group, so B is set to:

I	1	2	3	4
B(I)	0.0D+0	0.0D+0	0.0D+0	0.0D+0

The bounds for the problem are set in BL and BU. As the first variable is allowed to take any value, we specify lower and upper bounds of $\pm 10^{20}$. Thus we set

I	1	2	3
BL(I)	-1.0D+20	-1.0D+0	-1.0D+0
BU(I)	1.0D+20	1.0D+0	2.0D+0

The third group is scaled by 3.0, while the others are unscaled. Thus GSCALE is set to:

I	1	2	3	4
GSCALE(I)	1.0D+0	1.0D+0	3.0D+0	1.0D+0

Finally, it is unclear that the variables are badly scaled, so we set VSCALE to:

I	1	2	3
VSCALE(I)	1.0D+0	1.0D+0	1.0D+0

*SBMIN passes control back to the user to evaluate the function and derivative values of the group and nonlinear element functions. We need only to specify the group functions for the non-trivial groups, groups 1 and 3. Also note that the function and gradient values are only evaluated if the component is specified in ICALCG. For convenience, we evaluate the required values within the following subroutine:

```
SUBROUTINE GROUPS( GVALS, LGVALS, FT, ICALCG, NCALCG, DERIVS )
INTEGER LGVALS, ICALCG( * ), NCALCG, IG, JCALCG
LOGICAL DERIVS
DOUBLE PRECISION GVALS( LGVALS, 3 ), FT( * )
DOUBLE PRECISION ALPHA, ALPHA2
DO 100 JCALCG = 1, NCALCG
   IG        = ICALCG( JCALCG )
   IF ( IG .EQ. 1 ) THEN
      ALPHA = FT( 1 )
      IF ( .NOT. DERIVS ) THEN
         GVALS( 1, 1 ) = ALPHA * ALPHA
      ELSE
         GVALS( 1, 2 ) = 2.0D+0 * ALPHA
         GVALS( 1, 3 ) = 2.0D+0
      END IF
```

```
         END IF
         IF ( IG .EQ. 3 ) THEN
            ALPHA  = FT( 3 )
            ALPHA2 = ALPHA * ALPHA
            IF ( .NOT. DERIVS ) THEN
               GVALS( 3, 1 ) = ALPHA2 * ALPHA2
            ELSE
               GVALS( 3, 2 ) = 4.0D+0 * ALPHA2 * ALPHA
               GVALS( 3, 3 ) = 1.2D+1 * ALPHA2
            END IF
         END IF
  100 CONTINUE
      RETURN
      END
```

Here, the logical parameter DERIVS set .TRUE. specifies that the first and second derivatives are required; a .FALSE. value will return the function values.

To evaluate the values and derivatives of the nonlinear element functions we could use the following subroutine:

```
      SUBROUTINE ELFUNS( FUVALS, XT, ISTAEV, IELVAR, INTVAR,
     *                   ISTADH, ICALCF, NCALCF, DERIVS )
      INTEGER ISTAEV( * ), IELVAR( * ), INTVAR( * )
      INTEGER ISTADH( * ), ICALCF( * ), NCALCF
      DOUBLE PRECISION FUVALS( * ), XT( * )
      LOGICAL DERIVS
      INTEGER IEL, IGSTRT, IHSTRT, ILSTRT, JCALCF
      DOUBLE PRECISION V1, V2, U1, U2, U3
      DOUBLE PRECISION CS, SN
      INTRINSIC SIN, COS
      DO 100 JCALCF = 1, NCALCF
         IEL    = ICALCF(JCALCF)
         ILSTRT = ISTAEV( IEL ) - 1
         IGSTRT = INTVAR( IEL ) - 1
         IHSTRT = ISTADH( IEL ) - 1
         U1     = XT( IELVAR( ILSTRT + 1 ) )
         U2     = XT( IELVAR( ILSTRT + 2 ) )
         IF ( IEL .EQ. 1 ) THEN
            U3 = XT( IELVAR( ILSTRT + 3 ) )
            V1 = U1
            V2 = U2 + U3
            CS = COS( V2 )
            SN = SIN( V2 )
            IF ( .NOT. DERIVS ) THEN
               FUVALS( IEL ) = V1 * SN
            ELSE
               FUVALS( IGSTRT + 1 ) = SN
               FUVALS( IGSTRT + 2 ) = V1 * CS
               FUVALS( IHSTRT + 1 ) = 0.0D+0
               FUVALS( IHSTRT + 2 ) = CS
               FUVALS( IHSTRT + 3 ) = - V1 * SN
            END IF
         END IF
         IF ( IEL .EQ. 2 ) THEN
            IF ( .NOT. DERIVS ) THEN
               FUVALS( IEL ) = U1 * U2
            ELSE
               FUVALS( IGSTRT + 1 ) = U2
               FUVALS( IGSTRT + 2 ) = U1
               FUVALS( IHSTRT + 1 ) = 0.0D+0
               FUVALS( IHSTRT + 2 ) = 1.0D+0
               FUVALS( IHSTRT + 3 ) = 0.0D+0
            END IF
```

```
      END IF
100 CONTINUE
    RETURN
    END
```

Once again, the logical parameter DERIVS set .TRUE. specifies that the first and second derivatives are required; a .FALSE. value will return the function values. Notice that the derivatives are taken with respect to the internal variables. For the first nonlinear element function, this means that we must transform from elemental to internal variables before evaluating the derivatives. Thus, as this function may be written as $f(v_1, v_2) = v_1 \sin v_2$, where $v_1 = u_1$ and $v_2 = u_2 + u_3$, the gradient and Hessian matrix are

$$\begin{pmatrix} \sin v_2 \\ v_1 \cos v_2 \end{pmatrix} \quad \text{and} \quad \begin{pmatrix} 0 & \cos v_2 \\ \cos v_2 & -v_1 \sin v_2 \end{pmatrix},$$

respectively. Notice that as it is easy to specify the second derivatives for this example, we do so. Also note that the function and gradient values are only evaluated if the component is specified in ICALCF.

We must also specify the routine RANGES for our example. As we have observed, only the first element has a useful transformation. The transformation matrix is

$$\begin{pmatrix} 1 & 0 & 0 \\ 0 & 1 & 1 \end{pmatrix}$$

(see Section 8.1.2.2). As a consequence, the following routine RANGES is appropriate:

```
      SUBROUTINE RANGES( IELEMN, TRANSP, W1, W2, NELVAR, NINVAR )
      INTEGER IELEMN, NELVAR, NINVAR
      LOGICAL TRANSP
      DOUBLE PRECISION W1( * ), W2( * )
      IF ( IELEMN .NE. 1 ) RETURN
      IF ( TRANSP ) THEN
         W2( 1 ) = W1( 1 )
         W2( 2 ) = W1( 2 )
         W2( 3 ) = W1( 2 )
      ELSE
         W2( 1 ) = W1( 1 )
         W2( 2 ) = W1( 2 ) + W1( 3 )
      END IF
      RETURN
      END
```

This completes the supply of information to *SBMIN. The problem may now be solved using the following program. We choose to solve the model subproblem using a preconditioned conjugate gradient scheme, using diagonal preconditioning, a "box" shaped trust-region, calculating an exact Cauchy point at each iteration but not obtaining an accurate minimizer of the model within the box. Furthermore, we start from the initial estimate $x_1 = 0.0$, $x_2 = 0.0$ and $x_3 = 1.5$, observe that the objective function is bounded below by -2.0 within the feasible box and terminate when the norm of the projected gradient is smaller than 10^{-5}.

```
      PROGRAM MAIN
      INTEGER NMAX, NELMAX, NGMAX, LA, LELVAR, LFUVAL, MAXIT, IOUT
      INTEGER IPRINT, LWK, LIWK, NAMAX, NGM1, NELM1, LB, LICNA, LSTADA
      INTEGER LELING, LGVALS, LSTADG, LENGTH, LGXEQX, LSTAEV
      INTEGER LSTADH, LCALCF, LINTRE, LNTVAR, LFT
      INTEGER LCALCG, LGSCAL, LESCAL, LVSCAL, NCALCF, NCALCG
      INTEGER LIVAR , LX    , LBL   , LBU   , LQ    , LXT   , LDGRAD
      PARAMETER ( MAXIT = 100, IOUT = 6, IPRINT = 1 )
      PARAMETER ( NMAX  = 3, NGMAX = 4, NELMAX = 2, NAMAX = 2 )
      PARAMETER ( LWK = 2000, LIWK = 10000, LELVAR = 5, LFUVAL = 1000 )
      PARAMETER ( NGM1   = NGMAX + 1, NELM1 = NELMAX + 1 )
      PARAMETER ( LB     = NGMAX,  LICNA  = NAMAX,  LSTADA = NGM1   )
      PARAMETER ( LGVALS = NGMAX,  LSTADG = NGM1,   LENGTH = LELVAR )
      PARAMETER ( LGXEQX = NGMAX,  LSTAEV = NELM1,  LA     = NAMAX )
      PARAMETER ( LSTADH = NELM1,  LCALCF = NELMAX, LINTRE = NELMAX )
      PARAMETER ( LNTVAR = NELM1,  LFT    = NGMAX,  LELING = 2 )
      PARAMETER ( LGSCAL = NGMAX,  LESCAL = LELING, LVSCAL = NMAX )
      PARAMETER ( LCALCG = NGMAX,  LIVAR  = NMAX  , LX     = NMAX )
      PARAMETER ( LBL    = NMAX ,  LBU    = NMAX  , LQ     = NMAX )
      PARAMETER ( LXT    = NMAX ,  LDGRAD = NMAX   )
      INTEGER IELVAR( LELVAR ), ISTAEV( LSTAEV ), ICHOSE( 6 )
      INTEGER ISTADH( LSTADH ), ISTADG( LSTADG ), IVAR( LIVAR )
      INTEGER INTVAR( LNTVAR ), ICNA( LA ), IELING( LELING )
      INTEGER ISTADA( LSTADA ), IWK( LIWK )
      INTEGER ICALCF( LCALCF ), ICALCG( LCALCG )
      DOUBLE PRECISION X( LX ), F, FLOWER, WK( LWK )
      DOUBLE PRECISION BL( LBL ), BU( LBU ), GVALS( LGVALS, 3 )
      DOUBLE PRECISION A( LA ), DGRAD( LDGRAD ), Q( LQ ), XT( LXT )
      DOUBLE PRECISION FUVALS( LFUVAL ), B( LB ), FT( LFT )
      DOUBLE PRECISION GSCALE( LGSCAL ), ESCALE( LESCAL )
      DOUBLE PRECISION VSCALE( LVSCAL )
      LOGICAL GXEQX( LGXEQX ), INTREP( LINTRE )
      INTEGER I, INFORM, ITER, LGFX, N, NG, NEL, NVAR
      DOUBLE PRECISION STOPG
      LOGICAL QUADRT
      EXTERNAL RANGES, DSBMIN, ELFUNS, GROUPS
      DATA N  / 3 /, NG / 4 /, NEL / 2 /, ISTADG / 1,  1,  2,  3,  3 /
      DATA IELVAR / 2,  1,  3,  2,  3 /, ISTAEV / 1,  4,  6 /
      DATA INTVAR( 1 ), INTVAR( 2 ) / 2,  2 /, IELING / 1, 2 /
      DATA ICNA   / 1,  2 /, ISTADA / 1,  2,  2,  2,  3 /
      DATA ICHOSE / 0,  1, 0,  0,  0,  0 /
      DATA A / 1.0D+0, 1.0D+0 /, B / 0.0D+0, 0.0D+0, 0.0D+0, 0.0D+0 /
      DATA BL / -1.0D+20, -1.0D+0, 1.0D+0 /
      DATA BU / 1.0D+20, 1.0D+0, 2.0D+0 /, FLOWER / -2.0D+0 /
      DATA GSCALE / 1.0D+0, 1.0D+0, 3.0D+0, 1.0D+0 /
      DATA ESCALE / 1.0D+0, 1.0D+0 /
      DATA VSCALE / 1.0D+0, 1.0D+0, 1.0D+0 /
      DATA X  / 0.0D+0, 0.0D+0, 1.5D+0 /, STOPG / 1.0D-5 /
      DATA QUADRT / .FALSE. /, INTREP / .TRUE., .FALSE. /
      DATA GXEQX / .FALSE., .TRUE., .FALSE., .TRUE. /
      INFORM = 0
      WRITE( IOUT, 2010 )
  200 CONTINUE
      CALL DSBMIN( N,  NG, NEL   , IELING, LELING, ISTADG, LSTADG,
     *             IELVAR, LELVAR, ISTAEV, LSTAEV, INTVAR, LNTVAR,
     *             ISTADH, LSTADH, ICNA  , LICNA,  ISTADA, LSTADA,
     *             A     , LA    , B     , LB    , BL    , LBL   ,
     *             BU    , LBU   , GSCALE, LGSCAL, ESCALE, LESCAL,
     *             VSCALE, LVSCAL, GXEQX , LGXEQX, INTREP, LINTRE,
     *             RANGES, INFORM, F , X , LX    , GVALS , LGVALS,
     *             FT    , LFT   , FUVALS, LFUVAL, XT    , LXT   ,
     *             ICALCF, LCALCF, NCALCF, ICALCG, LCALCG, NCALCG,
     *             IVAR  , LIVAR , NVAR  , Q , LQ, DGRAD , LDGRAD,
     *             ICHOSE, ITER  , MAXIT , QUADRT, STOPG,  FLOWER,
     *             IWK   , LIWK  , WK    , LWK   , IPRINT, IOUT   )
      IF ( INFORM .LT. 0 ) THEN
         IF ( INFORM .EQ. - 1 .OR. INFORM .EQ. - 3 )
```

```
*        CALL ELFUNS( FUVALS, XT, ISTAEV, IELVAR, INTVAR,
*                     ISTADH, ICALCF, NCALCF, .FALSE. )
      IF ( INFORM .EQ. - 1 .OR. INFORM .EQ. - 5 .OR. INFORM
*          .EQ. - 6 ) CALL ELFUNS( FUVALS, XT, ISTAEV,
*                     IELVAR, INTVAR, ISTADH, ICALCF, NCALCF,
*                     .TRUE. )
      IF ( INFORM .EQ. - 2 .OR. INFORM .EQ. - 4 )
*        CALL GROUPS( GVALS, LGVALS, FT, ICALCG, NCALCG, .FALSE. )
      IF ( INFORM .EQ. - 2 .OR. INFORM .EQ. - 5 )
*        CALL GROUPS( GVALS, LGVALS, FT, ICALCG, NCALCG, .TRUE. )
      GO TO 200
   ELSE
      WRITE( IOUT, 2000 ) INFORM
      IF ( INFORM .EQ. 0 ) THEN
         LGFX = ISTADH( ISTADG( NG + 1 ) ) - 1
         WRITE( IOUT, 2020 )
         DO 300 I = 1, N
            WRITE( IOUT, 2030 ) I, X( I ), FUVALS( LGFX + I )
  300    CONTINUE
      END IF
      WRITE( IOUT, 2010 )
   END IF
   STOP
 2000 FORMAT( /, ' INFORM = ', I3, ' FROM DSBMIN ' )
 2010 FORMAT( /, ' ********** PROBLEM FOR SPEC-SHEET ************ ' )
 2020 FORMAT( /, ' VARIABLE NAME    VALUE    DUAL VALUE ',
*        /, ' -------------    -----    ---------- ' )
 2030 FORMAT( 7X, 'X', I1, 4X, 1P, 2D12.4 )
   END
```

This produces the following output.

```
********** PROBLEM FOR SPEC-SHEET ************

There are        3 variables
There are        4 groups
There are        2 nonlinear elements

 Iter #g.ev c.g.it      f      proj.g     rho     radius    step   cgend #free    time
   0     1     0  0.00D+00 1.0D+00      -        -         -        -     1       0.0
   1     2     1 -3.75D-01 8.0D-01  9.4D-01 2.0D-01 2.0D-01 BOUND    3       0.1
   2     3     2 -6.47D-01 6.0D-01  1.1D+00 4.0D-01 4.0D-01 BOUND    3       0.1
   3     4     5 -7.57D-01 2.8D-03  1.0D+00 8.0D-01 1.4D-01 CONVR    2       0.1
   4     5     7 -7.57D-01 1.4D-06  1.0D+00 8.0D-01 9.2D-04 CONVR    2       0.1

Iteration number               4  Function value          =  -7.56571572350D-01
No. derivative evaluations     5  Projected gradient norm =   1.39742721172D-06
C.G. iterations                7  Trust region radius     =   7.98997994642D-01
Number of updates skipped      0

There are      3 variables and       1 active bounds

Times for Cauchy, systems, products and updates    0.00   0.00   0.02   0.00

Approximate Cauchy step computed

Conjugate gradients without preconditioner used

One-norm trust region used

Exact second derivatives used

INFORM =   0 FROM DSBMIN

VARIABLE NAME    VALUE    DUAL VALUE
-------------    -----    ----------
      X1      1.1826D-01 -8.8383D-08
      X2     -5.4093D-01 -1.3974D-06
      X3      1.0000D+00  7.9089D-01

********** PROBLEM FOR SPEC-SHEET ************
```

8.2 AUGLG

8.2.1 Summary

AUGLG is a ANSI Fortran 77 subroutine to minimize a given objective function where the minimization variables are required to satisfy a set of auxiliary, possibly nonlinear, constraints. The objective function is represented as a sum of "group" functions. Each group function may be a linear or nonlinear functional whose argument is the sum of "finite element" functions; each finite element function is assumed to involve only a few variables or have a second derivative matrix with low rank for other reasons. The constraints are also represented as group functions. Bounds on the variables and known values may be specified. The routine is especially effective on problems involving a large number of variables.

The objective function has the form

$$F(x) = \sum_{i \in I_o} w_i g_i \left(\sum_{j \in J_i} w_{ij} f_j(\bar{x}_j) + a_i^T x - b_i \right), \qquad x = (x_1, x_2, \ldots, x_n)^T, \tag{8.2.5}$$

where I_o is a subset of $I = \{1, \ldots, n_g\}$, a list of indices of group functions g_i, each J_i is a subset of $J = \{1, \ldots, n_e\}$, the list of indices of nonlinear element functions f_j and, the w_i and w_{ij} are weights. Furthermor, the $\bar{x}_j$ are either small subsets of x or such that the rank of the second derivative matrix of the nonlinear element f_j is small for some other reason, and the vectors a_i of the linear elements are sparse. The least value of the objective function within the intersection of "box" region

$$l_i \leq x_i \leq u_i, \quad 1 \leq i \leq n,$$

and the region defined by the general constraints

$$c_i(x) = w_i g_i \left(\sum_{j \in J_i} w_{ij} f_j(\bar{x}_j) + a_i^T x - b_i \right) = 0, \qquad i \in I \setminus I_o, \tag{8.2.6}$$

is sought. Either bound on each variable may be infinite. If no general constraints are present, the alternative subroutine SBMIN should be used instead.

The method used is iterative. There are two levels of iteration. In the outer, a composite function, the augmented Lagrangian function, is formulated. The augmented Lagrangian function is a weighted combination of the objective function and general constraints. This function is (approximately) minimized within the feasible box using the auxiliary subroutine SBMIN.

SBMIN performs a series of inner iterations. At each inner iteration, a quadratic model of the augmented Lagrangian function is constructed. An approximation to the minimizer of this model within a trust-region is calculated. The trust region can be either a "box" or a hypersphere of specified radius, centered at the current best estimate of the minimizer. If there is an accurate agreement between the model and the true objective function at the new approximation to the minimizer, this approximation becomes the new best estimate. Otherwise, the radius of the trust region is reduced and a new approximate minimizer sought. The algorithm also allows the trust-region radius to increase when necessary. The minimization of the model function can be carried out by using either a direct-matrix or an iterative approach.

After an appropriate approximation to the minimizer of the augmented Lagrangian function is obtained, the weights are adjusted to ensure convergence of the outer iteration to the required solution of the constrained minimization problem.

If there is a linear transformation of variables such that one or more of the element functions depend on fewer transformed "internal" variables than the original number of variables $\bar{x}_i$, this may be specified. This can lead to a substantial reduction in computing time.

There are facilities for the user to provide a preconditioner for the conjugate gradient solution of the internal linear systems (a default is provided) and to provide (optional) second derivatives for the element functions. The subroutine returns control to the user to calculate function and derivative values of the group functions g_i and element functions f_j.

ATTRIBUTES — Versions: *AUGLG, where * may be S (single precision) or D (double precision). **Origin:** A. R. Conn, IBM T. J. Watson Research Center, Yorktown Heights, Nick Gould, Rutherford Appleton Laboratory, and Ph. L. Toint, FUNDP, Namur.

8.2.2 How to Use the Routine

8.2.2.1 The Argument List

The single precision version:

```
      CALL SAUGLG( N,  NG, NEL,    IELING, LELING, ISTADG, LSTADG,
     *             IELVAR, LELVAR, ISTAEV, LSTAEV, INTVAR, LNTVAR,
     *             ISTADH, LSTADH, ICNA  , LICNA , ISTADA, LSTADA,
     *             A , LA, B , LB, BL    , LBL   , BU    , LBU   ,
     *             GSCALE, LGSCAL, ESCALE, LESCAL, VSCALE, LVSCAL,
     *             GXEQX,  LGXEQX, INTREP, LINTRE, KNDOFC, LKNDOF,
     *             RANGES, INFORM, FOBJ  , X , LX, U , LU, GVALS ,
     *             LGVALS, FT    , LFT,    FUVALS, LFUVAL, XT    ,
     *             LXT   , ICALCF, LCALCF, NCALCF, ICALCG, LCALCG,
     *             NCALCG, IVAR  , LIVAR , NVAR  , Q , LQ, DGRAD ,
     *             LDGRAD, ICHOSE, ITER  , MAXIT , QUADRT, VNAMES,
     *             LVNAME, GNAMES, LGNAME, STOPG , STOPC , IWK   ,
     *             LIWK  , WK    , LWK   , IPRINT, IOUT  )
```

The double precision version:

```
      CALL DAUGLG( N,  NG, NEL,    IELING, LELING, ISTADG, LSTADG,
     *             IELVAR, LELVAR, ISTAEV, LSTAEV, INTVAR, LNTVAR,
     *             ISTADH, LSTADH, ICNA  , LICNA , ISTADA, LSTADA,
```

```
     *                A , LA, B , LB, BL    , LBL   , BU    , LBU   ,
     *                GSCALE, LGSCAL, ESCALE, LESCAL, VSCALE, LVSCAL,
     *                GXEQX,  LGXEQX, INTREP, LINTRE, KNDOFC, LKNDOF,
     *                RANGES, INFORM, FOBJ  , X , LX, U , LU, GVALS ,
     *                LGVALS, FT    , LFT,    FUVALS, LFUVAL, XT    ,
     *                LXT   , ICALCF, LCALCF, NCALCF, ICALCG, LCALCG,
     *                NCALCG, IVAR  , LIVAR , NVAR  , Q , LQ, DGRAD ,
     *                LDGRAD, ICHOSE, ITER  , MAXIT , QUADRT, VNAMES,
     *                LVNAME, GNAMES, LGNAME, STOPG , STOPC , IWK   ,
     *                LIWK  , WK    , LWK   , IPRINT, IOUT  )
```

`N` is an `INTEGER` variable, that must be set by the user to n, the number of variables. It is not altered by the routine.

Restriction: `N` > 0.

`NG` is an `INTEGER` variable, that must be set by the user to n_g, the number of group functions. It is not altered by the routine.

Restriction: `NG` > 0.

`NEL` is an `INTEGER` variable, that must be set by the user to n_e, the number of nonlinear element functions. It is not altered by the routine.

Restriction: `NEL` ≥ 0.

`IELING` is an `INTEGER` array that must be set by the user to contain the indices of the nonlinear elements J_i used by each group. The indices for group i must immediately precede those for group $i + 1$ and each group's indices must appear in a contiguous list. See Section 8.2.5 for an example. The contents of the array are not altered by the routine.

`LELING` is an `INTEGER` variable that must be set to the actual length of `IELING` in the calling program. It is not altered by the routine.

Restriction: `LELING` $\geq$ `ISTADG(NG + 1)` $- 1$.

`ISTADG` is an `INTEGER` array of length at least `NG+1`, whose i-th value gives the position in `IELING` of the first nonlinear element in group function i. In addition, `ISTADG(NG+1)` should be equal to the position in `IELING` of the last nonlinear element in group `NG` plus one. See Table 8.6 and Section 8.2.5 for an example. It is not altered by the routine.

`LSTADG` is an `INTEGER` variable that must be set to the actual length of `ISTADG` in the calling program. It is not altered by the routine.

Restriction: `LSTADG` $\geq$ `NG+1`.

`IELVAR` is an `INTEGER` array containing the indices of the variables in the first nonlinear element f_1, followed by those in the second nonlinear element $f_2, \ldots$. See Section 8.2.5 for an example. It is not altered by the routine.

elements in g_1	elements in g_2	...	elements in g_n		IELING
weights of elements in g_1	weights of elements in g_2	...	weights of elements in g_n		ESCALE
↑ ISTADG(1)	↑ ISTADG(2)	↑ ISTADG(3)	↑ ISTADG(n_g)	↑ ISTADG($n_g + 1$)	

Table 8.6: Contents of the arrays IELING, ESCALE and ISTADG

LELVAR is an INTEGER variable that must be set to the actual length of IELVAR in the calling program. It is not altered by the routine.

Restriction: LELVAR $\geq$ ISTAEV(NEL+1)-1.

ISTAEV is an INTEGER array, whose k-th value is the position of the first variable of the k-th nonlinear element function, in the list IELVAR. In addition, ISTAEV(n_e+1) must be equal to the position of the last variable of element n_e in IELVAR plus one. See Table 8.7 and Section 8.2.5 for an example. It is not altered by the routine.

variables in $\bar{x}_1$	variables in $\bar{x}_2$	...	variables in $\bar{x}_n$		IELVAR
↑ ISTAEV(1)	↑ ISTAEV(2)	↑ ISTAEV(3)	↑ ISTAEV(n_e)	↑ ISTAEV($n_e + 1$)	

Table 8.7: Contents of the arrays IELVAR and ISTAEV

LSTAEV is an INTEGER variable that must be set to the actual length of ISTAEV in the calling program. It is not altered by the routine.

Restriction: LSTAEV $\geq$ NEL+1.

INTVAR is an INTEGER array, whose i-th value must be set to the number of internal variables required for the i-th nonlinear element function f_i on initial entry. See Section 8.2.2.2 for the precise definition of internal variables and Section 8.2.5 for an example. It will subsequently be reset so that its i-th value gives the position in the array FUVALS of the first component of the gradient of the i-th nonlinear element function, *with respect to its internal variables* (see FUVALS).

LNTVAR is an INTEGER variable that must be set to the actual length of INTVAR in the calling program. It is not altered by the routine.

Restriction: LNTVAR $\geq$ NEL+1.

ISTADH is an INTEGER array The array is set in *AUGLG so that its i-th value ($1 \leq i \leq n_e$) gives the position in the array FUVALS of the first component of of the Hessian matrix of the i-th nonlinear element function f_i, *with respect to its internal variables.* Only the upper triangular part of each Hessian matrix is stored and the storage is by columns (see Section 8.2.2.3 for details). The element ISTADH($n_e + 1$) gives the position in FUVALS of the first component of the gradient of the objective function (see FUVALS).

LSTADH is an INTEGER variable that must be set to the actual length of ISTADH in the calling program. It is not altered by the routine.

Restriction: LSTADH $\geq$ NEL+1.

ICNA is an INTEGER array containing the indices of the nonzero components of a_1, the gradient of the first linear element, in any order, followed by those in a_2, etc. See Table 8.8 and Section 8.2.5 for an example. It is not altered by the routine.

LICNA is an INTEGER variable that must be set to the actual length of ICNA in the calling program. It is not altered by the routine.

ISTADA is an INTEGER array, whose i-th value is the position of the first nonzero component of the i-th linear element gradient, a_i, in the list ICNA. In addition, ISTADA(NG+1) must be equal to the position of the last nonzero component of a_{n_g} in ICNA plus one. See Table 8.8 and Section 8.2.5 for an example. It is not altered by the routine.

LSTADA is an INTEGER variable that must be set to the actual length of ISTADA in the calling program. It is not altered by the routine.

Restriction: LSTADA $\geq$ NG+1.

A is a REAL (DOUBLE PRECISION in the D version) array containing the values of the nonzero components of the gradients of the linear element functions, a_i, $i = 1, \ldots, n_g$. The values must appear in the same order as their indices appear in ICNA, i.e., the nonzero from element i, whose index is, say, ICNA(k) will have value A(k). See Section 8.2.5 for an example. A is not altered by the routine.

LA is an INTEGER variable that must be set to the actual length of A in the calling program. It is not altered by the routine.

B is a REAL (DOUBLE PRECISION in the D version) array whose i-th entry must be set to the value of the constant b_i for each group. B is not altered by the routine.

nonzeros in a_1	nonzeros in a_2	...	nonzeros in a_{n_g}		A
variables in a_1	variables in a_2	...	variables in a_{n_g}		ICNA
↑	↑	↑	↑	↑	
ISTADA(1)	ISTADA(2)	ISTADA(3)	ISTADA(n_g)	ISTADA($n_g + 1$)	

Table 8.8: Contents of the arrays A, ICNA and ISTADA

LB is an INTEGER variable that must be set to the actual length of B in the calling program. It is not altered by the routine.

Restriction: LB $\geq$ NG.

BL is a REAL (DOUBLE PRECISION in the D version) array whose i-th entry must be set to the value of the lower bound l_i on the i-th variable. If the i-th variable has no lower bound, BL(i) should be set to a large negative number. The array is not altered by the routine.

LBL is an INTEGER variable that must be set to the actual length of BL in the calling program. It is not altered by the routine.

Restriction: LBL $\geq$ N.

BU is a REAL (DOUBLE PRECISION in the D version) array whose i-th entry must be set to the value of the upper bound u_i on the i-th variable. If the i-th variable has no upper bound, BU(i) should be set to a large positive number. The array is not altered by the routine.

LBU is an INTEGER variable that must be set to the actual length of BU in the calling program. It is not altered by the routine.

Restriction: LBU $\geq$ N.

GSCALE is a REAL (DOUBLE PRECISION in the D version) array whose i-th entry must be set to the value of i-th group weight w_i. The array is not altered by the routine.

LGSCAL is an INTEGER variable that must be set to the actual length of GSCALE in the calling program. It is not altered by the routine.

Restriction: LGSCAL $\geq$ NG.

ESCALE is a REAL (DOUBLE PRECISION in the D version) array whose entries must be set to the values of element weights w_{ij}. The weights must occur in the same order as the indices of the elements assigned to each group in IELING, with the weights for the elements in group i preceding those in group $i+1$, $i = 1, \ldots,$ NG $-$ 1. See Table 8.6 and Section 8.2.5 for an example. The array is not altered by the routine.

LESCAL is an INTEGER variable that must be set to the actual length of ESCALE in the calling program. It is not altered by the routine.

Restriction: LESCAL $\geq$ ISTADG(NG+1)-1.

VSCALE is a REAL (DOUBLE PRECISION in the D version) array whose entries must be set to suitable positive scale factors for the problem variables x. The i-th variable x_i will implicitly be divided by VSCALE(i) within *AUGLG. The scale factors should ideally be chosen so that the rescaled variables are of order one at the solution to the minimization problem. If the user does not know suitable scalings, each component of VSCALE should be set to 1.0. Good variable scalings can result in considerable savings in computing times. The array is not altered by the routine.

LVSCAL is an INTEGER variable that must be set to the actual length of VSCALE in the calling program. It is not altered by the routine.

Restriction: LVSCAL $\geq$ N.

GXEQX is a LOGICAL array whose i-th entry, $1 \leq i \leq n_g$, must be set .TRUE. if the i-th group function is the trivial function $g(x) = x$ and .FALSE. otherwise. The entries GXEQX(i), for $n_g + 1 \leq i \leq 2n_g$ need not be set on entry and are subsequently used for workspace.

LGXEQX is an INTEGER variable that must be set to the actual length of GXEQX in the calling program. It is not altered by the routine.

Restriction: LGXEQX $\geq$ 2 * NG.

INTREP is a LOGICAL array whose i-th entry must be set .TRUE. if the i-th nonlinear element function has a useful transformation between elemental and internal variables and .FALSE. otherwise (see Section 8.2.2.2). It is not altered by the routine.

LINTRE is an INTEGER variable that must be set to the actual length of INTREP in the calling program. It is not altered by the routine.

Restriction: LINTRE $\geq$ NEL.

KNDOFC is an INTEGER array is used to indicate which of the groups are to be included in the objective function and which define equality constraints. Each of the first n_g entries of KNDOFC must be set to 1 or 2 on initial entry. If KNDOFC(i) has the value 1, i lies in the set I_o and the group will be included in the objective function (8.2.5). If, on the other hand, KNDOFC(i) has the value 2, the i-th group defines an equality constraint of the form (8.2.6). The array is not altered by the routine.

`LKNDOF` is an `INTEGER` variable that must be set to the actual length of `KNDOFC` in the calling program. It is not altered by the routine.

Restriction: `LKNDOF` $\geq$ `NG`.

`RANGES` is a user supplied subroutine whose purpose is to define the linear transformation of variables for those nonlinear elements which have different elemental and internal variables. See Section 8.2.2.2 for details. `RANGES` must be declared `EXTERNAL` in the calling program.

`INFORM` is an `INTEGER` variable that is set within `*AUGLG` and passed back to the user to prompt further action (`INFORM` < 0), to indicate errors in the input parameters (`INFORM` > 0) and to record a successful minimization (`INFORM` $= 0$). On initial entry, `INFORM` must be set to 0. Possible nonzero values of `INFORM` and their consequences are discussed in Section 8.2.2.3 and Section 8.2.2.4.

`FOBJ` is a `REAL` (`DOUBLE PRECISION` in the D version) variable which need not be set on initial entry. On final exit, `FOBJ` contains the value of the objective function at the point `X`.

`X` is a `REAL` (`DOUBLE PRECISION` in the D version) array which must be set by the user to the value of the variables at the starting point. On exit, it contains the values of the variables at the best point found in the minimization (usually the solution).

`LX` is an `INTEGER` variable that must be set to the actual length of `X` in the calling program. It is not altered by the routine.

Restriction: `LX` $\geq$ `N`.

`U` is a `REAL` (`DOUBLE PRECISION` in the D version) array which must be set by the user to suitable estimates of the Lagrange multipliers (see Section 8.2.4). If no reasonable estimates are known, zero is appropriate. If `KNDOFC(`i`)` $= 1$, `U(`i`)` need not be set. If `KNDOFC(`i`)` $= 2$, `U(`i`)` should contain the multiplier estimate for the constraint (8.2.6). On exit, `U` contains the values of the Lagrange multipliers at the best point found in the minimization (usually the solution).

`LU` is an `INTEGER` variable that must be set to the actual length of `U` in the calling program. It is not altered by the routine.

Restriction: `LU` $\geq$ `NG`.

`GVALS` is a `REAL` (`DOUBLE PRECISION` in the D version) array of dimension (`LGVALS`,3) which is used to store function and derivative information for the group functions. The user is asked to provide values for these functions and/or derivatives, evaluated at the argument `FT` (see `FT`) when control is returned to the calling program with a negative value of the variable INFORM. This information needs to be stored by the user in specified locations within GVALS. Details of the required information are given in Section 8.2.2.3.

LGVALS is an INTEGER variable that must be set to the actual length of GVALS in the calling program. It is not altered by the routine.

Restriction: LGVALS $\geq$ NG.

FT is a REAL (DOUBLE PRECISION in the D version) array whose i-th entry is set within *AUGLG to a trial value of the argument of the i-th group function at which the user is required to evaluate the values and/or derivatives of that function. Precisely what group function information is required at FT is under the control of the variable INFORM and details are given in Section 8.2.2.3.

LFT is an INTEGER variable that must be set to the actual length of FT in the calling program. It is not altered by the routine.

Restriction: LFT $\geq$ NG.

FUVALS is a REAL (DOUBLE PRECISION in the D version) array which is used to store function and derivative information for the nonlinear element functions. The user is asked to provide values for these functions and/or derivatives, evaluated at the argument XT (see XT), at specified locations within FUVALS, when control is returned to the calling program with a negative value of the variable INFORM. Details of the required information are given in Section 8.2.2.3. The layout of FUVALS is indicated in Table 8.9.

← n_e →			← n →	← n →		
element values	element gradients	element Hessians	objective gradient	Hessian diagonal	work-space	
	↑ INTVAR(1)	↑ ISTADH(1)	↑ ISTADH($n_e + 1$)			↑ LFUVALS

Table 8.9: Partitioning of the workspace array FUVALS

The first segment of FUVALS contains the values of the nonlinear element functions; the next two segments contain their gradients and Hessian matrices, taken with respect to their internal variables, as indicated in Section 8.2.2.3. The remaining two used segments contain the gradient of the objective function and the diagonal elements of the second derivative approximation, respectively. At the solution, the components of the gradient of the augmented Lagrangian function corresponding to variables which lie on one of their bounds are of particular interest in many applications areas. In particular they are often called shadow prices and are used to assess the sensitivity of the solution to variations in the bounds on the variables.

`LFUVAL` is an `INTEGER` variable that must be set to the actual length of `FUVALS` in the calling program. It is not altered by the routine.

`XT` is a `REAL` (`DOUBLE PRECISION` in the D version) array which is set within `*AUGLG` to a trial value of the variables x at which the user is required to evaluate the values and/or derivatives of the nonlinear elements functions. Precisely what element function information is required at `XT` is under the control of the variable `INFORM` and details are given in Section 8.2.2.3.

`LXT` is an `INTEGER` variable that must be set to the actual length of `XT` in the calling program. It is not altered by the routine.

Restriction: `LXT` $\geq$ `N`.

`ICALCF` is an `INTEGER` array which need not be set on entry. If the value of `INFORM` on return from `*AUGLG` indicates that values of the nonlinear element functions are required prior to a re-entry, the first `NCALCF` components of `ICALCF` give the indices of the nonlinear element functions which need to be recalculated at `XT`. Precisely what element function information is required is under the control of the variable `INFORM` and details are given in Section 8.2.2.3.

`LCALCF` is an `INTEGER` variable that must be set to the actual length of `ICALCF` in the calling program. It is not altered by the routine.

Restriction: `LCALCF` $\geq$ `NEL`.

`NCALCF` is an `INTEGER` variable which need not be set on entry. If the value of `INFORM` on return from `*AUGLG` indicates that values of the nonlinear element functions are required prior to a re-entry, `NCALCF` values need to be calculated and they correspond to nonlinear elements `ICALCF(`i`)`, $i = 1, \ldots,$ `NCALCF`.

`ICALCG` is an `INTEGER` array which need not be set on entry. If the value of `INFORM` on return from `*AUGLG` indicates that further values of the group functions are required prior to a re-entry, the first `NCALCG` components of `ICALCG` give the indices of the group functions which need to be recalculated at `FT`. Precisely what group function information is required is under the control of the variable `INFORM` and details are given in Section 8.2.2.3.

`LCALCG` is an `INTEGER` variable that must be set to the actual length of `ICALCG` in the calling program. It is not altered by the routine.

Restriction: `LCALCG` $\geq$ `NG`.

NCALCG is an INTEGER variable which need not be set on entry. If the value of INFORM on return from *AUGLG indicates that values of the group functions are required prior to a re-entry, NCALCG values need to be calculated and they correspond to groups ICALCG(i), $i = 1, \ldots,$ NCALCG.

IVAR is an INTEGER array which is required when the user is providing a special preconditioner for the conjugate gradient inner iteration (see Section 8.2.2.3).

LIVAR is an INTEGER variable that must be set to the actual length of IVAR in the calling program. It is not altered by the routine.

Restriction: LIVAR $\geq$ N.

NVAR is an INTEGER variable which need not be set on entry. NVAR is required when the user is providing a special preconditioner for the conjugate gradient inner iteration (see Section 8.2.2.3).

Q is a REAL (DOUBLE PRECISION in the D version) array which is required when the user is providing a special preconditioner for the conjugate gradient inner iteration (see Section 8.2.2.3).

LQ is an INTEGER variable that must be set to the actual length of Q in the calling program. It is not altered by the routine.

Restriction: LQ $\geq$ N.

DGRAD is a REAL (DOUBLE PRECISION in the D version) array which is required when the user is providing a special preconditioner for the conjugate gradient inner iteration (see Section 8.2.2.3).

LDGRAD is an INTEGER variable that must be set to the actual length of LDGRAD in the calling program. It is not altered by the routine.

Restriction: LDGRAD $\geq$ N.

ICHOSE is an INTEGER array of length 6 which allows the user to specify a number of the characteristics of the minimization algorithm used. The three entries and the information required are:

ICHOSE(1) must be set to 1 if a two-norm (hyperspherical) trust region is required, and any other value if an infinity-norm (box) trust region is to be used.

ICHOSE(2) is used to specify the method used to solve the linear systems of equations which arise at each iteration of the minimization algorithm. Possible values are

1: The conjugate gradient method will be used without preconditioning.

2: A preconditioned conjugate gradient method will be used with a diagonal preconditioner.

3: A preconditioned conjugate gradient method will be used with a user supplied preconditioner. Control will be passed back to the user to construct the product of the preconditioner with a given vector prior to a reentry to *AUGLG (see Section 8.2.2.3).

4: A preconditioned conjugate gradient method will be used with an expanding band incomplete Cholesky preconditioner.

5: A preconditioned conjugate gradient method will be used with Munksgaard's preconditioner.

6: A preconditioned conjugate gradient method will be used with a modified Cholesky preconditioner, in which small or negative diagonals are made sensibly positive during the factorization.

7: A preconditioned conjugate gradient method will be used with a modified Cholesky preconditioner, in which an indefinite factorization is altered to give a positive definite one.

8: A preconditioned conjugate gradient method will be used with a band preconditioner. The semi-bandwidth is given by the variable NSEMIB in the common block *COMSB (see Section 8.2.2.6). The default semi-bandwidth is 5.

11: A multifrontal factorization method will be used.

12: A modified Cholesky factorization method will be used.

Any other value of ICHOSE(2) will be reset to 1.

ICHOSE(3) specifies what sort of first derivative approximation is to be used. If the user is able to provide analytical first derivatives for each nonlinear element function, ICHOSE(3) must be set to 0. If the user is unable to provide first derivatives, ICHOSE(3) should be set to 1 so that finite-difference approximations may be formed.

ICHOSE(4) specifies what sort of second derivative approximation is to be used. If the user is able to provide analytical second derivatives for each nonlinear element function, ICHOSE(4) must be set to 0. If the user is unable to provide second derivatives, these derivatives will be approximated using one of four secant approximation formulae. If ICHOSE(4) is set to 1, the BFGS formula is used; if it is set to 2, the DFP formula is used; if it is set to 3, the PSB formula is used; and if it is set to 4, the symmetric rank-one formula is used. The user is strongly advised to use exact second derivatives if at all possible as this often significantly improves the convergence of the method.

ICHOSE(5) specifies whether the exact generalized Cauchy point, the first estimate of the minimizer of the quadratic model within the box, is required (ICHOSE(5) = 1), or whether an approximation suffices (any other value of ICHOSE(5)).

ICHOSE(6) specifies whether an accurate minimizer of the quadratic model within the box is required (ICHOSE(6) = 1), or whether an approximation suffices (any other value of ICHOSE(6)).

ITER is an INTEGER variable which gives the number of iterations of the algorithm performed.

MAXIT is an INTEGER variable which must be set by the user to the maximium number of iterations that the algorithm will be allowed to perform. It is not altered by the routine.

QUADRT is a LOGICAL variable which must be set .TRUE. by the user if all the nonlinear element functions are quadratics and .FALSE. otherwise. This will cut down on the number of derivative evaluations required for quadratic functions. It is not altered by the routine.

VNAMES is a CHARACTER * 10 array which must be initialized by the user on entry. A name of up to ten characters may be associated with each variable x_i and is printed at the solution to the minimization problem whenever IPRINT $\geq$ 2 (see Section 8.2.2.5). The name associated with the i-th variable should be placed in VNAMES(i). VNAMES is not altered by the routine.

LVNAME is an INTEGER variable that must be set to the actual length of VNAMES in the calling program. It is not altered by the routine.

Restriction: LVNAME $\geq$ N.

GNAMES is a CHARACTER * 10 array which must be initialized by the user on entry. A name of up to ten characters may be associated with each group g_i and is printed at the solution to the minimization problem whenever IPRINT $\geq$ 1 (see Section 8.2.2.5). The name associated with the i-th group should be placed in GNAMES(i). GNAMES is not altered by the routine.

LGNAME is an INTEGER variable that must be set to the actual length of GNAMES in the calling program. It is not altered by the routine.

Restriction: LGNAME $\geq$ NG.

STOPG is a REAL (DOUBLE PRECISION in the D version) variable, that must be set by the user to a measure of the accuracy (in the sense of the infinity norm of the projected gradient of the Lagrangian function) required to stop the minimization procedure. If 0 $\leq$ INFORM $\leq$ 3 or INFORM = 8 on exit from *AUGLG, STOPG holds the value of the norm of the projected gradient of the objective function at the best point found in the minimization.

STOPC is a REAL (DOUBLE PRECISION in the D version) variable, that must be set by the user to a measure of the accuracy (in the sense of the infinity norm of the general constraints) required to stop the minimization procedure. If $0 \leq$ INFORM ≤ 3 or INFORM $= 8$ on exit from *AUGLG, STOPC holds the value of the norm of the constraints at the best point found in the minimization.

IWK is a INTEGER array of length LIWK used as workspace by *AUGLG.

LIWK is an INTEGER variable which must be set by the user to the length of the working array IWK and is not altered.

WK is a REAL (DOUBLE PRECISION in the D version) array of length LWK used as workspace by *AUGLG.

LWK is an INTEGER variable which must be set by the user to the length of the working array WK and is not altered.

IPRINT is an INTEGER variable which determines the amount of intermediate output produced by *AUGLG and must be set by the user. Possible values of IPRINT and the output obtained are described in Section 8.2.2.5. It is not altered by the routine.

IOUT is an INTEGER variable which must be set by the user to the Fortran unit number for intermediate output. It is not altered by the routine.

8.2.2.2 Internal Variables and Subroutine RANGES

A nonlinear element function f_j is assumed to be a function of the variables $\bar{x}_j$, a subset of the problem variables x. Suppose that $\bar{x}_j$ has n_j elements. Then another way of saying this is that we have an element function $f_j(v_1, \ldots, v_{n_j})$, where in our special case, we choose $v_1 = (\bar{x}_j)_1, \ldots, v_{n_j} = (\bar{x}_j)_{n_j}$. The *elemental* variables for the element function f_j are the variables $\bar{v}$ and, while we need to associate the particular values $\bar{x}_j$ with $\bar{v}$ (using the array IELVAR), it is the elemental variables which are important in defining the nonlinear element functions.

As an example, the seventh nonlinear element function for a particular problem might be

$$f_7(v_1, v_2, v_3) = (v_1 + v_2)e^{v_1 - v_3}, \tag{8.2.7}$$

where for our example $v_1 = x_{29}$, $v_2 = x_3$ and $v_3 = x_{17}$. For this example, there are three elemental variables. However, the example illustrates an additional point. Although f_7 is a function of three variables, the function itself is really only composed of *two* independent parts; the product of $(v_1 + v_2)$ with $e^{v_1 - v_3}$, or, if we write $u_1 = v_1 + v_2$ and $u_2 = v_1 - v_3$, the product of u_1 with e^{u_2}. The variables u_1 and u_2 are known as *internal* variables for the element function. They are obtained as *linear combinations* of the elemental variables. The important feature as far as *AUGLG is concerned is that each nonlinear function involves as few variables as possible, as this allows for compact storage and more efficient derivative approximation. By representing the function in terms of its internal variables, this goal is achieved. *AUGLG only stores derivatives of the element functions with respect to internal variables, so it pays to use an internal representation in this case. It frequently happens, however, that a function does not have useful internal variables. For instance, another element function might be

$$f_9(v_1, v_2) = v_1 \sin v_2, \tag{8.2.8}$$

where for example $v_1 = x_6$ and $v_2 = x_{12}$. Here, we have broken f_9 down into as few pieces as possible. Although there are internal variables, $u_1 = v_1$ and $u_2 = v_2$, they are the same in this case as the elemental variables and there is no virtue in exploiting them. Moreover it can happen that although there are special internal variables, there are just as many internal as elemental variables and it therefore doesn't particularly help to exploit them. For instance, if

$$f_{14}(v_1, v_2) = (v_1 + v_2) \log(v_1 - v_2), \tag{8.2.9}$$

where for example $v_1 = x_{12}$ and $v_2 = x_2$, the function can be formed as $u_1 \log(u_2)$ where $u_1 = v_1 + v_2$ and $u_2 = v_1 - v_2$. But as there are just as many internal variables as elementals, it will not normally be advantageous to use this internal representation.

Finally, although an element function may have useful internal variables, the user need not bother with them. *AUGLG will still work, but at the expense of extra storage and computational effort. The user decides on input to *AUGLG which elements have useful transformations by setting the appropriate elements of the array INTREP to .TRUE. .

In general, there will be a linear transformation from the elemental variables to the internal ones. For example (8.2.7), we have

$$\begin{pmatrix} u_1 \\ u_2 \end{pmatrix} = \begin{pmatrix} 1 & 1 & 0 \\ 1 & 0 & -1 \end{pmatrix} \begin{pmatrix} v_1 \\ v_2 \\ v_3 \end{pmatrix},$$

while in (8.2.8), we have

$$\begin{pmatrix} u_1 \\ u_2 \end{pmatrix} = \begin{pmatrix} 1 & 1 \\ 1 & 0 \end{pmatrix} \begin{pmatrix} v_1 \\ v_2 \end{pmatrix}.$$

In general the transformation will be of the form

$$u = Wv, \tag{8.2.10}$$

and this transformation is *useful* if the matrix W has fewer rows than columns.

The purpose of the RANGES routine is to define the transformation between internal and elemental variables for nonlinear elements with useful internal representations. The routine has the following specification.

The single precision version:

```
CALL RANGES( IELEMN, TRANSP, W1, W2, NELVAR, NINVAR )
```

The double precision version:

```
CALL RANGES( IELEMN, TRANSP, W1, W2, NELVAR, NINVAR )
```

`IELEMN` is an `INTEGER` variable which gives the index of the nonlinear element whose transformation is required by `*AUGLG`. It must not be altered by the routine.

`TRANSP` is a `LOGICAL` variable. If `TRANSP` is `.FALSE.`, the user must put the result of the transformation Wv in the array `W2`, where v is input in the array `W1`. Otherwise, the user must supply the result of the transposed transformation $W^T u$ in the array `W2`, where u is input in the array `W1`. It must not be altered by the routine.

`W1` is a `REAL` (`DOUBLE PRECISION` in the D version) array whose length is the number of elemental variables if `TRANSP` is `.FALSE.`, and the number of internal variables otherwise. It must not be altered by the routine.

`W2` is a `REAL` (`DOUBLE PRECISION` in the D version) array whose length is the number of internal variables if `TRANSP` is `.FALSE.`, and the number of elemental variables otherwise. The result of the transformation of `W1` or its transpose, as assigned by `TRANSP`, must be set in `W2`.

`NELVAR` is an `INTEGER` variable which gives the number of elemental variables for the element specified by `IELEMN`. It must not be altered by the routine.

`NINVAR` is an `INTEGER` variable which gives the number of internal variables for the element specified by `IELEMN`. It must not be altered by the routine.

The user will already have specified which elements have useful transformations in the array `INTREP`. `RANGES` will only be called for elements for which the corresponding component of `INTREP` is `.TRUE.`.

8.2.2.3 Reverse Communication Information

When a return is made from `*AUGLG` with `INFORM` set negative, `*AUGLG` is asking the user for further information. The user should normally compute the required information and re-enter `*AUGLG` with `INFORM` unchanged (see `INFORM` $= -10$, below, for an exception).

Possible values of `INFORM` and the information required are

`INFORM` $= -1$: The user should compute the function and, if they are available, derivative values of the nonlinear element functions numbered `ICALCF(`i`)` for $i = 1, \ldots,$ `NCALCF` (see below).

INFORM $= -2$: The user should compute the function and derivative values of all non-trivial nonlinear group functions, numbered ICALCG(i) for $i = 1, \ldots,$ NCALCG (see below).

INFORM $= -3$: The user should compute the function values of the nonlinear element functions numbered ICALCF(i) for $i = 1, \ldots,$ NCALCF (see below).

INFORM $= -4$: The user should compute the function values of all non-trivial nonlinear group functions numbered ICALCG(i) for $i = 1, \ldots,$ NCALCG (see below).

INFORM $= -5$: If exact derivatives are available, the user should compute the derivative values of all nonlinear element functions numbered ICALCF(i) for $i = 1, \ldots,$ NCALCF. The user should also compute the derivative values of all non-trivial nonlinear group functions numbered ICALCG(i) for $i = 1, \ldots,$ NCALCG (see below).

INFORM $= -6$: The user should compute the derivative values of all nonlinear element functions numbered ICALCF(i) for $i = 1, \ldots,$ NCALCF (see below).

INFORM $= -7$: The user should compute the function values of the nonlinear element functions numbered ICALCF(i) for $i = 1, \ldots,$ NCALCF (see below).

INFORM $= -8, -9, -10$: The user should compute the product of a preconditioning matrix M with the vector d and return the value in q.

When INFORM $= -1$, -3 or -7, the user must return the values of all of the nonlinear element functions $f_k(\bar{x}_k)$, $k =$ICALCF(i), $i = 1, \ldots,$ NCALCF. The functions are to be evaluated at the point x given in the array XT. The k-th function value must be placed in FUVALS(k).

When INFORM $= -1, -5$ or -6 and first derivatives are available, (that is when ICHOSE(3) $= 0$), the user must return the values of the gradients, *with respect to their internal variables*, of all the nonlinear element functions $f_k(\bar{x}_k)$, $k =$ICALCF(i), $i = 1, \ldots,$ NCALCF. The gradients are to be evaluated at the point x given in the array XT. The gradient with respect to internal variable i of the k-th nonlinear element,

$$\frac{\partial f_k}{\partial u_i},$$

must be placed in FUVALS(INTVAR(k)+i-1). If, in addition, exact second derivatives are to be provided (ICHOSE(4) $= 0$), the user must return the values of the Hessian matrices, *with respect to their internal variables*, of the same nonlinear element functions evaluated at x. Only the "upper triangular" part of the required Hessians should be specified. The component of the Hessian of the k-th nonlinear element with respect to internal variables i and j, $i \leq j$,

$$\frac{\partial^2 f_k}{\partial u_i \partial u_j},$$

must be placed in `FUVALS(ISTADH(`k`)+`$(j(j-1)/2)+i-1$`)`.

When `INFORM` $= -2$ or -4, the user must return the values of all the group functions g_k, k =`ICALCG(`i`)`, $i = 1, \ldots,$ `NCALCG`. The k-th such function should be evaluated with the argument `FT(`k`)` and the result placed in `GVALS(`k`,1)`.

When `INFORM` $= -2$ or -5, the user must return the values of the first and second derivatives of each of the group functions g_k, k =`ICALCG(`i`)`, $i = 1, \ldots,$ `NCALCG`, with respect to its argument. The derivatives of the k-th such function should be evaluated with the argument `FT(`k`)`. The first derivative of the k-th group function should be placed in `GVALS(`k`,2)` and the corresponding second derivative returned in `GVALS(`k`,3)`.

When `INFORM` $= -8, -9$ or -10, the user must return the values of the components `Q(IVAR(`i`))` $= q_i$, $i = 1, \ldots,$ `NVAR`, such that $q = Md$ and where d_i =`Q(IVAR(`i`))` $i = 1, \ldots,$ `NVAR`. Here M is a symmetric positive definite approximation to the inverse of the matrix whose i-th row and column are the `IVAR(`i`)`-th row and column of the Hessian matrix of augmented Lagrangian function (see section 8.2.4), $i = 1, \ldots,$ `NVAR`. These values can only occur if `ICHOSE(2)` $= 3$.

If the user does not wish, or is unable, to compute an element or group function at a particular argument returned from `*AUGLG`, `INFORM` may be reset to -11 and `*AUGLG` re-entered. `*AUGLG` will treat such a re-entry as if the current iteration had been unsuccessful and reduce the trust-region radius. This facility is useful when, for instance, the user is asked to evaluate a function at a point outside its domain of definition.

8.2.2.4 Error Messages

If `INFORM` is positive on return from `*AUGLG`, an error has been detected. The user should correct the error and restart the minimization. Possible values of `INFORM` and their consequences are:

`INFORM` $= 1$: More than `MAXIT` iterations have been performed. This is often a symptom of incorrectly programmed derivatives or of the preconditioner used being insufficiently effective. Recheck the derivatives. Otherwise, increase `MAXIT` and re-enter `*AUGLG` at the best point found so far.

`INFORM` $= 2$: The trust-region radius has become too small. This is often a symptom of incorrectly programmed derivatives or of requesting more accuracy in the projected gradient than is reasonable on the user's machine. If the projected gradient is small, the minimization has probably succeeded. Otherwise, recheck the derivatives.

`INFORM` $= 3$: The step taken during the current iteration is so small that no difference will be observed in the function values. This sometimes occurs when too much accuracy is required of the final gradient. If the projected gradient is small, the minimization has probably succeeded.

`INFORM` $= 4$: One of the `INTEGER` arrays has been initialized with insufficient space. A message indicating which array is at fault and the required space will be printed on unit number `IOUT`.

`INFORM` $= 5$: One of the `REAL` (`DOUBLE PRECISION` in the D version) arrays has been initialized with insufficient space. A message indicating which array is at fault and the required space will be printed on unit number `IOUT`.

`INFORM` $= 6$: One of the `LOGICAL` arrays has been initialized with insufficient space. A message indicating which array is at fault and the required space will be printed on unit number `IOUT`.

`INFORM` $= 7$: One or more of the components of the array `KNDOFC` does not have the value 1 or 2.

`INFORM` $= 8$: The problem does not appear to have a feasible solution. Check the constraints and try starting with different initial values for x.

8.2.2.5 Intermediate Printing

The user is able to control the amount of intermediate printing performed in the course of the minimization. Printing is under the control of the parameter `IPRINT` and output is sent to I/O unit number `IOUT`. Possible values of `IPRINT` and the levels of output produced are as follows.

`IPRINT` ≤ 0 : No printing, except warning messages, will be performed.

`IPRINT` ≥ 1 : Details of the minimization function will be output. This includes the number of variables, groups and nonlinear elements which are used and a list of the variables which occur in each of the linear and nonlinear elements in every group.

If the current iterate provides an acceptable estimate of the minimizer of the augmented Lagrangian function, the two-norm of the general constraints and the current value of the penalty parameter are given.

`IPRINT` $= 1$: A simple one line description of each iteration is given. This includes the iteration number, the number of derivative evaluations that have been made, the number of conjugate-gradient iterations that have been performed, the current value of the augmented Lagrangian function, the (two-) norm of the projected gradient, the ratio ρ of the actual to predicted decrease in augmented Lagrangian function value achieved, the current trust-region radius, the norm of the step taken, an indication of how the direct or iterative method ended, the number of variables which lie away from their bounds and the total time spent on the minimization.

`IPRINT` = 2 : In addition to the information output with `IPRINT` = 1, a short description of the approximate solution to the inner-iteration linear system is given. Before a successful (`INFORM` = 0) exit, details of the estimate of the minimizer and the gradient of the augmented Lagrangian function are given.

`IPRINT` = 3 : A list of the current iteration number, the value of the augmented Lagrangian function, the number of derivative evaluations that have been made, the (two-) norm of the projected gradient, the number of conjugate gradients iterations that have been performed and the current trust-region radius are given, followed by the current estimate of the minimizer. The values of the reduction in the model of the augmented Lagrangian function and the actual reduction in this function, together with their ratio, are also given. Before a successful (`INFORM` = 0) exit, details of the estimate of the minimizer and the gradient of the augmented Lagrangian function are given.

If the current iterate also provides an acceptable estimate of the minimizer of the augmented Lagrangian function, values of the general constraints and estimates of the Lagrange multipliers are also given.

`IPRINT` = 4 : In addition to the information output with `IPRINT` = 3, the gradient of the augmented Lagrangian function at the current estimate of the minimizer is given. Full details of the approximate solution to the inner-iteration linear system are also given. This level of output is intended as a debugging aid for the expert only.

`IPRINT` = 5 : In addition to the information output with `IPRINT` = 4, the diagonal elements of the second derivative approximation are given.

`IPRINT` ≥ 6 . In addition to the information output with `IPRINT` = 5, the second derivative approximations (taken with respect to the internal variables) to each nonlinear element function are given.

8.2.2.6 Common

The variables in common block `*COMAL` are used as controlling parameters for the outer iteration and may provide the user with further information. These variables should not be altered by the user.

The single precision version:

```
COMMON / SCOMAL / RMU, RMUTOL, FIRSTC, FIRSTG, NEWSOL
```

The double precision version:

```
COMMON / DCOMAL / RMU, RMUTOL, FIRSTC, FIRSTG, NEWSOL
```

RMU is a REAL (DOUBLE PRECISION in the D version) variable which gives the current value of the penalty parameter.

RMUTOL is a REAL (DOUBLE PRECISION in the D version) variable which gives the value of the penalty parameter above which the algorithm will not attempt to update the estimates of the Lagrange multipliers.

FIRSTC is a REAL (DOUBLE PRECISION in the D version) variable which gives the required accuracy of the norm of the constraints at the end of the first major iteration.

FIRSTG is a REAL (DOUBLE PRECISION in the D version) variable which gives the required accuracy of the norm of the projected gradient at the end of the first major iteration.

NEWSOL is a REAL (DOUBLE PRECISION in the D version) variable which indicates whether a major iteration has just been concluded (NEWSOL = .TRUE.) or not (NEWSOL = .FALSE.).

The variables in common block *COMSB are used as controlling parameters for the inner minimization and may provide the user with further information. Once again, these variables should not be altered by the user.
The single precision version:

```
      COMMON / SCOMSB / ACCCG , RATIO , RADIUS, RADMAX, FINDMX, CMA31,
     *                  ITERCG, ITCGMX, NGEVAL, ISKIP , IFIXED, NSEMIB
```

The double precision version:

```
      COMMON / DCOMSB / ACCCG , RATIO , RADIUS, RADMAX, FINDMX, CMA31,
     *                  ITERCG, ITCGMX, NGEVAL, ISKIP , IFIXED, NSEMIB
```

ACCCG is a REAL (DOUBLE PRECISION in the D version) variable which gives the size of the (two-) norm of the reduced gradient of the model function below which the conjugate gradient iteration will terminate. This value will become small as the algorithm converges.

RATIO is a REAL (DOUBLE PRECISION in the D version) variable which gives the ratio of the actual reduction that has been made in the augmented Lagrangian function value during the current iteration to that predicted by the model function. A value close to one is to be expected as the algorithm converges.

`RADIUS` is a `REAL` (`DOUBLE PRECISION` in the D version) variable which gives the current radius of the trust-region.

`RADMAX` is a `REAL` (`DOUBLE PRECISION` in the D version) variable which gives the maximum allowable radius of the trust-region.

`FINDMX` is a `REAL` (`DOUBLE PRECISION` in the D version) variable which allows the user to alter the printing of function and gradient information. Normally `FINDMX` should be set to 1.0. The printed values of the objective function and its derivatives actually correspond to the function `FINDMX` $F(x)$. This may be useful if, for instance, `AUGLG` is used to calculate the maximum value of the function $f(x)$ by minimizing $F(x) = -f(x)$. Then, values of f and its derivatives may be printed within `AUGLG` merely by setting `FINDMX` to -1.0.

`CMA31` is a `REAL` (`DOUBLE PRECISION` in the D version) variable which gives the pivot tolerance used if Munksgaard's preconditioner is selected (that is when `ICHOSE(2)` = 5).

`ITERCG` is an `INTEGER` variable which gives the number of conjugate-gradient iterations that have been performed since the start of the minimization.

`ITCGMX` is an `INTEGER` variable which gives the maximum number of conjugate-gradient iterations that will be allowed at each iteration of the inner minimization.

`NGEVAL` is an `INTEGER` variable which gives the number of problem function gradient evaluations that have been made since the start of the minimization.

`ISKIP` is an `INTEGER` variable which gives the number of times that an approximation to the second derivatives of an element function has been rejected.

`IFIXED` is an `INTEGER` variable which gives the index of the variable that most recently encountered one of its bounds in the minimization process.

`NSEMIB` is an `INTEGER` variable which gives the bandwidth used if the band or expanding band preconditioner is selected (`ICHOSE(2)` = 4 or 8).

The variables in common block `MACHNS` hold machine dependent constants. These variables should not be altered by the user.

The single precision version:

```
COMMON / SMACHN / EPSMCH, EPSNEG, TINY, BIG
```

The double precision version:

```
COMMON / DMACHN / EPSMCH, EPSNEG, TINY, BIG
```

EPSMCH is a REAL (DOUBLE PRECISION in the D version) variable which gives the smallest positive machine number for which $1.0 + \text{EPSMCH} > 1.0$.

EPSNEG is a REAL (DOUBLE PRECISION in the D version) variable which gives the smallest positive machine number for which $1.0 - \text{EPSNEG} < 1.0$.

TINY is a REAL (DOUBLE PRECISION in the D version) variable which gives the smallest positive machine number.

BIG is a REAL (DOUBLE PRECISION in the D version) variable which gives the largest positive machine number.

There is a further named common block, INWKSP, which is used to pass information on partitioning of the workspace arrays between internal subroutines. The content of the common block should be of no interest to the general user.

8.2.3 General Information

Use of common: See Section 8.2.2.6.

Workspace: Provided by the user (see arguments IWK and WK).

Other routines called directly:

```
*SBMIN, *CLCFG.
```

Input/output: No input; output on device number IOUT.

Restrictions: N > 0, NEL > 0, NG > 0, LELING ≥ ISTADG(NG + 1) − 1, LSTADG ≥ NG + 1, LELVAR ≥ ISTAEV(NEL + 1) − 1, LSTAEV ≥ NEL + 1, LNTVAR ≥ NEL + 1, LSTADH ≥ NEL + 1, LSTADA ≥ NG + 1, LB ≥ NG, LBL ≥ N, LBU ≥ N, LX ≥ N, LU ≥ NG, LXT ≥ N, LIVAR ≥ N, LDGRAD ≥ N, LQ ≥ N, LGSCAL ≥ NG, LESCAL ≥ ISTADG(NG + 1) − 1, LVSCAL ≥ N, LGXEQX ≥ 2 ∗ NG, LINTRE ≥ NEL, LKNDOF ≥ NG, LGVALS ≥ NG, LFT ≥ NG, LCALCF ≥ NEL, LCALCG ≥ NG, LVNAME ≥ N, LGNAME ≥ NG.

Portability: ANSI Fortran 77.

8.2.4 Method

The basic method implemented within subroutine AUGLG is described in detail by Conn, Gould and Toint [11]. The method used to solve the inner iteration subproblem is described by Conn, Gould and Toint [9]. The concept of partial separability was first suggested by Griewank and Toint (1982). The extension to group partially separable functions is new. Also see Chapter 3 of this manual.

The Lagrangian function associated with objective function (8.2.5) and general constraints (8.2.6) is the composite function

$$l(x,u) = F(x) + \sum_{i \in I \setminus I_o} u_i c_i(x). \tag{8.2.11}$$

The scalars u_i are known as Lagrange multiplier estimates. At a solution $x^\star$ to the constrained minimization problem, there are Lagrange multipliers $u^\star$ for which the components of the gradient of the Lagrangian function $\partial l(x^\star, u^\star,)/\partial x_i = 0$ whenever the corresponding variable $x_i^\star$ lies strictly between its lower and upper bounds. It is useful, but not esential, for the initial Lagrange multiplier estimates to be close to $u^\star$.

The augmented Lagrangian function is the composite function

$$\phi(x,u,\mu) = l(x,u) + \frac{1}{2\mu} \sum_{i \in I \setminus I_o} (c_i(x))^2, \tag{8.2.12}$$

where μ is known as the penalty parameter. An inner iteration is used to find an approximate minimizer of (8.2.12) within the feasible box for fixed values of the penalty parameter and Lagrange multiplier estimates. The outer iteration of AUGLG automatically adjusts the penalty parameter and Lagrange multiplier estimates to ensure convergence of these approximate minimizers to a solution of the constrained optimization problem.

In the inner iteration, a step from the current estimate of the solution is determined using a trust-region approach. That is, a quadratic model of the augmented Lagrangian function is approximately minimized within the intersection of the constraint "box" and another convex region, the trust-region. This minimization is carried out in two stages. Firstly, the so-called generalized Cauchy point for the quadratic subproblem is found. (The purpose of this point is to ensure that the algorithm converges and that the set of bounds which are satisfied as equations at the solution is rapidly identified.) Thereafter an improvement to the quadratic model is sought using either a direct-matrix or truncated conjugate-gradient algorithm. The trust-region size is increased if the reduction obtained in the objective function is reasonable when compared with the reduction predicted by the model and reduced otherwise.

A central idea is that a collection of small matrices approximating the Hessian matrices of each f_k is used and updated at every iteration using one of a number of possible updating formulae. Augmented Lagrangian function values and derivatives are assembled from these components as required.

The strategy for treating bound constraints is based on the usual projection device and is described in detail in Conn, Gould and Toint (1988a).

8.2.5 Example

We now consider the small example problem,

$$\text{minimize } F(x_1, x_2, x_3) = x_1^2 + x_2 \sin(x_1 + x_3) + 3x_2^4 x_3^4 + x_2$$

subject to the general constraint

$$c(x_1, x_2, x_3) = \cos(x_1 + 2x_2 - 1) = 0$$

and the bounds $-1 \leq x_2 \leq 1$ and $1 \leq x_3 \leq 2$. There are a number of ways of casting this problem in the framework required by *AUGLG. Here, we consider partitioning F into groups as

$$\underset{\text{group 1}}{\underset{\uparrow}{(x_1)^2}} \quad + \quad \underset{\text{group 2}}{\underset{\uparrow}{(x_2 \sin(x_1 + x_3))}} \quad + \quad \underset{\text{group 3}}{\underset{\uparrow}{3(x_2 x_3)^4}} \quad + \quad \underset{\text{group 4}}{\underset{\uparrow}{(x_2)}}.$$

Similarly, we write c as

$$\underset{\text{group 5}}{\underset{\uparrow}{cos(x_1 + 2x_2 - 1)}}.$$

Notice the following:

1. Group 1 uses the non-trivial group function $g_1(\alpha) = \alpha^2$. The group contains a single *linear* element; the element function is x_1.

2. Group 2 uses the trivial group function $g_2(\alpha) = \alpha$. The group contains a single *nonlinear* element; this element function is $x_2 \sin(x_1 + x_3)$. The element function has *three* elemental variables, v_1, v_2 and v_3, say, (with $v_1 = x_2$, $v_2 = x_1$ and $v_3 = x_3$), but may be expressed in terms of *two* internal variables u_1 and u_2, say, where $u_1 = v_1$ and $u_2 = v_2 + v_3$.

3. Group 3 uses the non-trivial group function $g_3(\alpha) = \alpha^4$, weighted by the factor 3. Again, the group contains a single *nonlinear* element; this element function is $x_2 x_3$. The element function has *two* elemental variables, v_1 and v_2, say, (with $v_1 = x_2$ and $v_2 = x_3$). This time, however, there is no useful transformation to internal variables.

4. Group 4 again uses the trivial group function $g_4(\alpha) = \alpha$. This time the group contains a single *linear* element x_2.

5. Finally, group 5 uses the non-trivial group function $g_5(\alpha) = \cos(\alpha)$. The group contains a single linear element; the element function is $x_1 + 2x_2 - 1$.

Thus we see that we can consider our objective and constraint functions to be made up of five group functions; the first, third and fifth are non-trivial so we need to provide function and derivative values for these when prompted by `*AUGLG`. There are two nonlinear elements, one each from groups two and three. Again this means that we need to provide function and derivative values for these when required by `*AUGLG`. Finally, one of these elements, the first, has a useful transformation from elemental to internal variables so the routine `RANGES` must be set to provide this transformation.

The problem involves three variables, four groups and two nonlinear elements so we set `N` = 3, `NG` = 5 and `NEL` = 2. As the first and third group functions are non-trivial, we set `GXEQX` as:

I	1	2	3	4	5
GXEQX(I)	.FALSE.	.TRUE.	.FALSE.	.TRUE.	.FALSE.

The first four groups are associated with the objective function while the fifth defines a constraint so we set `KNDOFC` as:

I	1	2	3	4	5
KNDOFC(I)	1	1	1	1	2

The first nonlinear element occurs in group 2 and the second in group 3; both elements are unweighted. We thus set `IELING`, `ESCALE` and `ISTADG` as:

I	1	2
IELING(I)	1	2
ESCALE(I)	1.0	1.0

I	1	2	3	4	5	6
ISTADG(I)	1	1	2	3	3	3

The first nonlinear element function is not a quadratic function of its internal variables so we set `QUADRT=.FALSE.`. Nonlinear element one assigns variables x_2, x_1 and x_3 to its elemental variables, while the second nonlinear element assigns variables x_2 and x_3 to its elemental variables. Hence `IELVAR` contains:

I	1	2	3	4	5
IELVAR(I)	2	1	3	2	3
	⟵	el. 1	⟶	← el. 2	→

In this vector, we now locate the position of the first variable of each nonlinear element, and build the vector `ISTAEV` as follows:

I	1	2	3
ISTAEV(I)	1	4	6

Both nonlinear elements have two internal variables — element two using its elemental variables as internals — so we set **INTVAR** as follows:

I	1	2	3
INTVAR(I)	2	2	-

As the first nonlinear element has a useful transformation between elemental and internal variables, while the second does not, **INTREP** contains:

I	1	2
INTREP(I)	.TRUE.	.FALSE.

Turning to the linear elements, groups one, four and five each have a linear element. Those from groups one and four involve a single variable — x_1 for group one and x_2 for group four. That from group five involves the variables x_1 and x_3. All the coefficients, excepting the last, are one; the last is two. We thus set ISTADA as:

I	1	2	3	4	5	6
ISTADA(I)	1	2	2	2	3	5

and ICNA and A as

I	1	2	3	4
ICNA(I)	1	2	1	2
A(I)	1.0D+0	1.0D+0	1.0D+0	2.0D+0

Only the last group has a constant term so B is set to:

I	1	2	3	4	5
B(I)	0.0D+0	0.0D+0	0.0D+0	0.0D+0	1.0D+0

The bounds for the problem are set in BL and BU. As the first variable is allowed to take any value, we specify lower and upper bounds of $\pm 10^{20}$. Thus we set

I	1	2	3
BL(I)	-1.0D+20	-1.0D+0	-1.0D+0
BU(I)	1.0D+20	1.0D+0	2.0D+0

The 3rd group is scaled by 3.0 while the others are unscaled. Thus we set GSCALE to:

I	1	2	3	4	5
GSCALE(I)	1.0D+0	1.0D+0	3.0D+0	1.0D+0	1.0D+0

It is unclear that the variables are badly scaled so we set **VSCALE** to:

I	1	2	3
VSCALE(I)	1.0D+0	1.0D+0	1.0D+0

Finally, we choose to allocate the names `Obj 1` to `Obj 4` to the four groups associated with the objective function and call the constraint group `Constraint`. The variables are called `x 1` to `x 3`. We thus set `GNAMES` and `VNAMES` to:

I	1	2	3	4	5
GNAMES(I)	Obj 1	Obj 2	Obj 3	Obj 4	Constraint

and

I	1	2	3
VNAMES(I)	x 1	x 2	x 3

`*AUGLG` passes control back to the user to evaluate the function and derivative values of the group and nonlinear element functions. We need only to specify the group functions for the non-trivial groups, groups 1, 3 and 5. Also note that the function and gradient values are only evaluated if the component is specified in `ICALCG`. For convenience, we evaluate the required values within the following subroutine:

```
SUBROUTINE GROUPS( GVALS, LGVALS, FT, ICALCG, NCALCG, DERIVS )
INTEGER LGVALS, ICALCG( * ), NCALCG, IG, JCALCG
LOGICAL DERIVS
DOUBLE PRECISION GVALS( LGVALS, 3 ), FT( * )
DOUBLE PRECISION ALPHA, ALPHA2
INTRINSIC COS, SIN
DO 100 JCALCG = 1, NCALCG
   IG        = ICALCG( JCALCG )
   IF ( IG .EQ. 1 ) THEN
      ALPHA = FT( 1 )
      IF ( .NOT. DERIVS ) THEN
         GVALS( 1, 1 ) = ALPHA * ALPHA
      ELSE
         GVALS( 1, 2 ) = 2.0D+0 * ALPHA
         GVALS( 1, 3 ) = 2.0D+0
      END IF
   END IF
   IF ( IG .EQ. 3 ) THEN
      ALPHA  = FT( 3 )
      ALPHA2 = ALPHA * ALPHA
      IF ( .NOT. DERIVS ) THEN
         GVALS( 3, 1 ) = ALPHA2 * ALPHA2
      ELSE
         GVALS( 3, 2 ) = 4.0D+0 * ALPHA2 * ALPHA
         GVALS( 3, 3 ) = 1.2D+1 * ALPHA2
      END IF
   END IF
   IF ( IG .EQ. 5 ) THEN
      ALPHA  = FT( 5 )
      IF ( .NOT. DERIVS ) THEN
         GVALS( 5, 1 ) = COS( ALPHA )
      ELSE
         GVALS( 5, 2 ) = - SIN( ALPHA )
         GVALS( 5, 3 ) = - COS( ALPHA )
      END IF
```

```
         END IF
  100 CONTINUE
      RETURN
      END
```

Here, the logical parameter DERIVS set .TRUE. specifies that the first and second derivatives are required; a .FALSE. value will return the function values.

To evaluate the values and derivatives of the nonlinear element functions we could use the following subroutine:

```
      SUBROUTINE ELFUNS( FUVALS, XT, ISTAEV, IELVAR, INTVAR,
     *                   ISTADH, ICALCF, NCALCF, DERIVS )
      INTEGER ISTAEV( * ), IELVAR( * ), INTVAR( * )
      INTEGER ISTADH( * ), ICALCF( * ), NCALCF
      DOUBLE PRECISION FUVALS( * ), XT( * )
      LOGICAL DERIVS
      INTEGER IEL, IGSTRT, IHSTRT, ILSTRT, JCALCF
      DOUBLE PRECISION V1, V2, U1, U2, U3
      DOUBLE PRECISION CS, SN
      INTRINSIC SIN, COS
      DO 100 JCALCF = 1, NCALCF
         IEL    = ICALCF(JCALCF)
         ILSTRT = ISTAEV( IEL ) - 1
         IGSTRT = INTVAR( IEL ) - 1
         IHSTRT = ISTADH( IEL ) - 1
         U1     = XT( IELVAR( ILSTRT + 1 ) )
         U2     = XT( IELVAR( ILSTRT + 2 ) )
         IF ( IEL .EQ. 1 ) THEN
            U3 = XT( IELVAR( ILSTRT + 3 ) )
            V1 = U1
            V2 = U2 + U3
            CS = COS( V2 )
            SN = SIN( V2 )
            IF ( .NOT. DERIVS ) THEN
               FUVALS( IEL ) = V1 * SN
            ELSE
               FUVALS( IGSTRT + 1 ) = SN
               FUVALS( IGSTRT + 2 ) = V1 * CS
               FUVALS( IHSTRT + 1 ) = 0.0D+0
               FUVALS( IHSTRT + 2 ) = CS
               FUVALS( IHSTRT + 3 ) = - V1 * SN
            END IF
         END IF
         IF ( IEL .EQ. 2 ) THEN
            IF ( .NOT. DERIVS ) THEN
               FUVALS( IEL ) = U1 * U2
            ELSE
               FUVALS( IGSTRT + 1 ) = U2
               FUVALS( IGSTRT + 2 ) = U1
               FUVALS( IHSTRT + 1 ) = 0.0D+0
               FUVALS( IHSTRT + 2 ) = 1.0D+0
               FUVALS( IHSTRT + 3 ) = 0.0D+0
            END IF
         END IF
  100 CONTINUE
      RETURN
      END
```

Once again, the logical parameter DERIVS set .TRUE. specifies that the first and second derivatives are required; a .FALSE. value will return the function values. Notice that the derivatives are taken with respect to the internal variables. For the first nonlinear element function, this means that we must transform from elemental to internal variables before evaluating the derivatives. Thus, as this function may be written as $f(v_1, v_2) = v_1 \sin v_2$, where $v_1 = u_1$ and $v_2 = u_2 + u_3$, the gradient and Hessian matrix are

$$\begin{pmatrix} \sin v_2 \\ v_1 \cos v_2 \end{pmatrix} \quad \text{and} \quad \begin{pmatrix} 0 & \cos v_2 \\ \cos v_2 & -v_1 \sin v_2 \end{pmatrix},$$

respectively. Notice that it is easy to specify the second derivatives for this example so we do so. Also note that the function and gradient values are only evaluated if the component is specified in `ICALCF`.

We must also specify the routine `RANGES` for our example. As we have observed, only the first element has a useful transformation. The transformation matrix is

$$\begin{pmatrix} 1 & 0 & 0 \\ 0 & 1 & 1 \end{pmatrix}$$

(see Section 8.2.2.2). As a consequence, the following routine `RANGES` is appropriate:

```
      SUBROUTINE RANGES( IELEMN, TRANSP, W1, W2, NELVAR, NINVAR )
      INTEGER IELEMN, NELVAR, NINVAR
      LOGICAL TRANSP
      DOUBLE PRECISION W1( * ), W2( * )
      IF ( IELEMN .NE. 1 ) RETURN
      IF ( TRANSP ) THEN
         W2( 1 ) = W1( 1 )
         W2( 2 ) = W1( 2 )
         W2( 3 ) = W1( 2 )
      ELSE
         W2( 1 ) = W1( 1 )
         W2( 2 ) = W1( 2 ) + W1( 3 )
      END IF
      RETURN
      END
```

This completes the supply of information to `*AUGLG`. The problem may now be solved using the following program. We choose to solve the model subproblem using a preconditioned conjugate gradient scheme, using diagonal preconditioning, a "box" shaped trust-region, calculating an exact Cauchy point at each iteration but not obtaining an accurate minimizer of the model within the box. Furthermore, we start from the initial estimate $x_1 = 0.0$, $x_2 = 0.0$ and $x_3 = 1.5$. As we have no idea of a good estimate for the Lagrange multiplier associated with the constraint (group 5), we set $u_5 = 0$. We observe that the objective function is bounded below by -2.0 within the feasible box and terminate when the norms of the reduced gradient of the Lagrangian function and constraints are smaller than 10^{-5}.

```
      PROGRAM MAIN
      INTEGER NMAX  , NELMAX, NGMAX , LELVAR, LFUVAL, MAXIT  , LSTADA
      INTEGER IPRINT, LWK   , LIWK  , NAMAX , NGM1  , NELM1  , LICNA
      INTEGER LELING, LGVALS, LSTADG, LENGTH, LGXEQX, LSTAEV , LU, LB
```

```
      INTEGER LSTADH, LCALCF, LINTRE, LNTVAR, LFT   , LKNDOF , LVNAME
      INTEGER LCALCG, LGSCAL, LESCAL, LVSCAL, NCALCF, NCALCG , LGNAME
      INTEGER LIVAR , LA, LX, LBL   , LBU   , LQ    , LXT    , LDGRAD
      INTEGER IOUT
      PARAMETER ( MAXIT = 100, IOUT = 6, IPRINT = 1 )
      PARAMETER ( NMAX  = 3, NGMAX = 5, NELMAX = 2, NAMAX = 4 )
      PARAMETER ( LWK = 2000, LIWK = 10000, LELVAR = 5, LFUVAL = 1000 )
      PARAMETER ( NGM1   = NGMAX + 1, NELM1 = NELMAX + 1 )
      PARAMETER ( LB     = NGMAX,  LICNA  = NAMAX,  LSTADA = NGM1  )
      PARAMETER ( LGVALS = NGMAX,  LSTADG = NGM1,   LENGTH = LELVAR )
      PARAMETER ( LGXEQX = NGMAX + NGMAX, LSTAEV = NELM1, LA = NAMAX )
      PARAMETER ( LSTADH = NELM1,  LCALCF = NELMAX, LINTRE = NELMAX )
      PARAMETER ( LNTVAR = NELM1,  LFT    = NGMAX,  LELING = 2 )
      PARAMETER ( LGSCAL = NGMAX,  LESCAL = LELING, LVSCAL = NMAX )
      PARAMETER ( LCALCG = NGMAX,  LIVAR  = NMAX  , LX     = NMAX )
      PARAMETER ( LBL    = NMAX ,  LBU    = NMAX  , LQ     = NMAX )
      PARAMETER ( LXT    = NMAX ,  LDGRAD = NMAX  , LU     = NGMAX )
      PARAMETER ( LKNDOF = NGMAX,  LVNAME = NMAX ,  LGNAME = NGMAX )
      INTEGER IELVAR( LELVAR ), ISTAEV( LSTAEV ), ICHOSE( 6 )
      INTEGER ISTADH( LSTADH ), ISTADG( LSTADG ), IVAR( LIVAR )
      INTEGER INTVAR( LNTVAR ), ICNA( LA ), IELING( LELING )
      INTEGER ISTADA( LSTADA ), IWK( LIWK ), KNDOFC( LKNDOF )
      INTEGER ICALCF( LCALCF ), ICALCG( LCALCG )
      DOUBLE PRECISION X( LX ), FOBJ, WK( LWK ), U( LU )
      DOUBLE PRECISION BL( LBL ), BU( LBU ), GVALS( LGVALS, 3 )
      DOUBLE PRECISION A( LA ), DGRAD( LDGRAD ), Q( LQ ), XT( LXT )
      DOUBLE PRECISION FUVALS( LFUVAL ), B( LB ), FT( LFT )
      DOUBLE PRECISION GSCALE( LGSCAL ), ESCALE( LESCAL )
      DOUBLE PRECISION VSCALE( LVSCAL )
      LOGICAL GXEQX( LGXEQX ), INTREP( LINTRE )
      CHARACTER * 10 GNAMES( LGNAME ), VNAMES( LVNAME )
      INTEGER INFORM, ITER, N, NG, NEL, NVAR
      DOUBLE PRECISION STOPG, STOPC
      LOGICAL QUADRT
      EXTERNAL RANGES, DAUGLG, ELFUNS, GROUPS
      DATA N  / 3 /, NG / 5 /, NEL / 2 /
      DATA ISTADG / 1,  1,  2,  3,  3,  3 /
      DATA IELVAR / 2,  1,  3,  2,  3 /, ISTAEV / 1,  4,  6 /
      DATA INTVAR( 1 ), INTVAR( 2 ) / 2,  2 /, IELING / 1, 2 /
      DATA ICNA   / 1,  2,  1, 2 /, ISTADA / 1,  2,  2,  2,  3,  5 /
      DATA ICHOSE / 0,  1, 0,  0,  0,  0 /, KNDOFC / 1, 1, 1, 1, 2 /
      DATA A / 1.0D+0, 1.0D+0, 1.0D+0, 2.0D+0 /
      DATA B / 0.0D+0, 0.0D+0, 0.0D+0, 0.0D+0, 1.0D+0 /
      DATA BL / -1.0D+20, -1.0D+0, 1.0D+0 /
      DATA BU / 1.0D+20, 1.0D+0, 2.0D+0 /
      DATA GSCALE / 1.0D+0, 1.0D+0, 3.0D+0, 1.0D+0, 1.0D+0 /
      DATA ESCALE / 1.0D+0, 1.0D+0 /
      DATA VSCALE / 1.0D+0, 1.0D+0, 1.0D+0 /
      DATA X  / 0.0D+0, 0.0D+0, 1.5D+0 /, U( 5 ) / 0.0D+0 /
      DATA STOPG / 1.0D-5 /, STOPC / 1.0D-5 /
      DATA QUADRT / .FALSE. /, INTREP / .TRUE., .FALSE. /
      DATA GXEQX( 1 ) / .FALSE. / GXEQX( 2 ) / .TRUE. /
      DATA GXEQX( 3 ) / .FALSE. / GXEQX( 4 ) / .TRUE. /
      DATA GXEQX( 5 ) / .FALSE. /
      DATA GNAMES / 'Obj 1', 'Obj 2', 'Obj 3', 'Obj 4', 'Constraint' /
      DATA VNAMES / 'x 1', 'x 2', 'x 3' /
      INFORM = 0
      WRITE( IOUT, 2010 )
  200 CONTINUE
      CALL DAUGLG( N,  NG, NEL,    IELING, LELING, ISTADG, LSTADG,
     *             IELVAR, LELVAR, ISTAEV, LSTAEV, INTVAR, LNTVAR,
     *             ISTADH, LSTADH, ICNA  , LICNA , ISTADA, LSTADA,
     *             A , LA, B , LB, BL    , LBL   , BU    , LBU   ,
     *             GSCALE, LGSCAL, ESCALE, LESCAL, VSCALE, LVSCAL,
     *             GXEQX,  LGXEQX, INTREP, LINTRE, KNDOFC, LKNDOF,
     *             RANGES, INFORM, FOBJ  , X , LX, U , LU, GVALS ,
     *             LGVALS, FT    , LFT,    FUVALS, LFUVAL, XT    ,
```

```
     *             LXT   , ICALCF, LCALCF, NCALCF, ICALCG, LCALCG,
     *             NCALCG, IVAR  , LIVAR , NVAR  , Q , LQ, DGRAD ,
     *             LDGRAD, ICHOSE, ITER  , MAXIT , QUADRT, VNAMES,
     *             LVNAME, GNAMES, LGNAME, STOPG , STOPC , IWK   ,
     *             LIWK  , WK    , LWK   , IPRINT, IOUT  )
      IF ( INFORM .LT. 0 ) THEN
         IF ( INFORM .EQ. - 1 .OR. INFORM .EQ. - 3 )
     *      CALL ELFUNS( FUVALS, XT, ISTAEV, IELVAR, INTVAR,
     *                   ISTADH, ICALCF, NCALCF, .FALSE. )
         IF ( INFORM .EQ. - 1 .OR. INFORM .EQ. - 5 .OR. INFORM
     *        .EQ. - 6 ) CALL ELFUNS( FUVALS, XT, ISTAEV,
     *                   IELVAR, INTVAR, ISTADH, ICALCF, NCALCF,
     *                   .TRUE. )
         IF ( INFORM .EQ. - 2 .OR. INFORM .EQ. - 4 )
     *      CALL GROUPS( GVALS, LGVALS, FT, ICALCG, NCALCG, .FALSE. )
         IF ( INFORM .EQ. - 2 .OR. INFORM .EQ. - 5 )
     *      CALL GROUPS( GVALS, LGVALS, FT, ICALCG, NCALCG, .TRUE. )
         GO TO 200
      ELSE
         WRITE( IOUT, 2000 ) INFORM
         IF ( INFORM .EQ. 0 ) THEN
            LGFX = ISTADH( ISTADG( NG + 1 ) ) - 1
            WRITE( IOUT, 2020 )
            DO 300 I = 1, N
               WRITE( IOUT, 2030 ) I, X( I ), FUVALS( LGFX + I )
  300       CONTINUE
         END IF
         WRITE( IOUT, 2010 )
      END IF
      STOP
 2000 FORMAT( /, ' INFORM = ', I3, ' FROM DAUGLG ' )
 2010 FORMAT( /, ' ********** PROBLEM FOR SPEC-SHEET ************ ' )
      END
```

This produces the following output.

```
********** PROBLEM FOR SPEC-SHEET ************

*********  Starting optimization  ***************

Penalty parameter   1.0000D-01 Required projected gradient norm =   1.0000D-01
                               Required constraint         norm =   1.0000D-01

There are        3 variables
There are        5 groups
There are        2 nonlinear elements

 Iter #g.ev c.g.it       f     proj.g    rho    radius   step  cgend #free   time
    0     1     0  1.46D+00 4.5D+00      -        -       -      -      2    0.0
    1     1     2  1.46D+00 4.5D+00 -3.1D+00 1.2D+00 9.4D-01 CONVR     2    0.1
    2     2     3  4.32D-01 3.0D+00  9.7D-01 7.5D-02 7.5D-02 BOUND     3    0.1
    3     3     3 -3.77D-01 6.7D-01  8.8D-01 1.5D-01 1.5D-01 CONVR     3    0.1
    4     4     6 -5.46D-01 3.2D-01  1.0D+00 3.0D-01 3.0D-01 BOUND     3    0.1
    5     5    10 -6.67D-01 8.9D-02  9.5D-01 6.0D-01 3.2D-01 CONVR     2    0.1

Iteration number                5  Function value           =  -6.67315955161D-01
No. derivative evaluations      5  Projected gradient norm =   8.94271224455D-02
C.G. iterations                10  Trust region radius      =   5.99310131638D-01
Number of updates skipped       0

There are      3 variables and       1 active bounds

Times for Cauchy, systems, products and updates    0.01    0.00    0.02    0.00

Approximate Cauchy step computed

Conjugate gradients without preconditioner used

One-norm trust region used
```

```
Exact second derivatives used

Penalty parameter         =  1.0000D-01
Projected gradient norm =  8.9427D-02 Required gradient    norm =   1.0000D-01
Constraint          norm =  4.1924D-02 Required constraint norm =   1.0000D-01

******** Updating multiplier estimates **********

Penalty parameter    1.0000D-01 Required projected gradient norm =   1.0000D-02
                                Required constraint          norm =   1.2589D-02

 Iter #g.ev c.g.it       f     proj.g    rho     radius    step   cgend #free   time
    5     6     10 -6.41D-01 7.5D-01      -         -        -       -      2    0.2
    6     7     10 -6.48D-01 6.3D-02  1.0D+00 9.2D-02 1.3D-02 CONVR      2    0.2
    7     8     11 -6.49D-01 6.3D-03  9.9D-01 9.2D-02 2.0D-02 CONVR      2    0.2

Iteration number                 7  Function value            =  -6.48520519138D-01
No. derivative evaluations       8  Projected gradient norm =   6.27409734493D-03
C.G. iterations                 11  Trust region radius       =   9.21695380150D-02
Number of updates skipped        0

There are       3 variables and       1 active bounds

Times for Cauchy, systems, products and updates    0.01    0.00    0.02    0.00

Approximate Cauchy step computed

Conjugate gradients without preconditioner used

One-norm trust region used

Exact second derivatives used

Penalty parameter         =  1.0000D-01
Projected gradient norm =  6.2741D-03 Required gradient    norm =   1.0000D-02
Constraint          norm =  6.1445D-03 Required constraint norm =   1.2589D-02

******** Updating multiplier estimates **********

Penalty parameter    1.0000D-01 Required projected gradient norm =   1.0000D-03
                                Required constraint          norm =   1.5849D-03

 Iter #g.ev c.g.it       f     proj.g    rho     radius    step   cgend #free   time
    7     9     11 -6.45D-01 1.2D-01      -         -        -       -      2    0.2
    8    10     13 -6.46D-01 8.7D-06  1.0D+00 3.5D-02 4.2D-03 CONVR      2    0.2

Iteration number                 8  Function value            =  -6.45566615097D-01
No. derivative evaluations      10  Projected gradient norm =   8.68934084425D-06
C.G. iterations                 13  Trust region radius       =   3.46744957930D-02
Number of updates skipped        0

There are       3 variables and       1 active bounds

Times for Cauchy, systems, products and updates    0.01    0.00    0.03    0.00

Approximate Cauchy step computed

Conjugate gradients without preconditioner used

One-norm trust region used

Exact second derivatives used

Penalty parameter         =  1.0000D-01
Projected gradient norm =  8.6893D-06 Required gradient    norm =   1.0000D-03
Constraint          norm =  4.1780D-04 Required constraint norm =   1.5849D-03

******** Updating multiplier estimates **********

Penalty parameter    1.0000D-01 Required projected gradient norm =   1.0000D-04
                                Required constraint          norm =   1.9953D-04

 Iter #g.ev c.g.it       f     proj.g    rho     radius    step   cgend #free   time
    8    11     13 -6.45D-01 8.4D-03      -         -        -       -      2    0.2
    9    12     14 -6.45D-01 1.2D-04  1.0D+00 3.2D-02 1.5D-04 CONVR      2    0.2
   10    13     14 -6.45D-01 3.3D-06  1.0D+00 3.2D-02 2.1D-06 CONVR      2    0.2

Iteration number                10  Function value            =  -6.45363951028D-01
```

```
No. derivative evaluations    13  Projected gradient norm =   3.27797290367D-06
C.G. iterations               14  Trust region radius     =   3.17079795255D-02
Number of updates skipped      0

There are      3 variables and      1 active bounds

Times for Cauchy, systems, products and updates    0.01   0.00   0.03   0.00

Approximate Cauchy step computed

Conjugate gradients without preconditioner used

One-norm trust region used

Exact second derivatives used

Penalty parameter       =   1.0000D-01
Projected gradient norm =   3.2780D-06 Required gradient    norm =   1.0000D-04
Constraint         norm =   4.0766D-05 Required constraint norm =   1.9953D-04

******** Updating multiplier estimates **********

Penalty parameter   1.0000D-01 Required projected gradient norm =   1.0000D-05
                               Required constraint          norm =   2.5119D-05

 Iter #g.ev c.g.it      f     proj.g    rho     radius    step  cgend #free   time
   10    14     14 -6.45D-01 8.1D-04     -        -        -      -      2    0.3
   11    15     16 -6.45D-01 1.2D-09  1.0D+00 3.2D-02 1.6D-05 CONVR     2    0.3

Iteration number              11  Function value          =  -6.45344167788D-01
No. derivative evaluations    15  Projected gradient norm =   1.15821718882D-09
C.G. iterations               16  Trust region radius     =   3.16588216695D-02
Number of updates skipped      0

There are      3 variables and      1 active bounds

Times for Cauchy, systems, products and updates    0.02   0.00   0.03   0.00

Approximate Cauchy step computed

Conjugate gradients without preconditioner used

One-norm trust region used

Exact second derivatives used

Penalty parameter       =   1.0000D-01
Projected gradient norm =   1.1582D-09 Required gradient    norm =   1.0000D-05
Constraint         norm =   3.8509D-06 Required constraint norm =   2.5119D-05

Variable name Number Status     Value    Lower bound Upper bound  |  Dual value
------------- ------ ------     -----    ----------- -----------  |  ----------
 x 1             1     FREE  3.0078D-01 -1.0000D+20  1.0000D+20  |  1.8499D-10
 x 2             2     FREE -4.3579D-01 -1.0000D+00  1.0000D+00  | -1.1582D-09
 x 3             3    LOWER  1.0000D+00  1.0000D+00  2.0000D+00  |  3.1656D-01

Constraint name Number      Value     Scale factor | Lagrange multiplier
--------------- ------      -----     ----- ------ | -------------------
   Constraint      5    -3.8509D-06  1.0000D+00    |     -4.8531D-01

Objective function value  -6.57120542661544D-01

There are        3 variables in total.
There are        1 equality constraints.
Of these         0 are primal degenerate.
There are        1 variables on their bounds.
Of these         0 are dual degenerate.

INFORM =    0 FROM DAUGLG

********** PROBLEM FOR SPEC-SHEET ************
```

Chapter 9. Coda

We hope this book has given the reader a flavour of the scope of the first release of the LANCELOT optimization package. The only way to really judge the effectiveness of the package is for the reader to try examples for his or herself. We believe that our approach is sound and have demonstrated this, at least to ourselves, by solving a large and diverse set of test problems.

Nonetheless, we realise that the package is not perfect, that a package with the general scope of LANCELOT is unlikely to be the most appropriate in all situations and that it needs to evolve to overcome current deficiencies. We are aware of some shortcomings. In particular, we are considering improved facilities for handling linear constraints, especially network constraints, alternative merit functions for inequality constraints (for example, shifted barrier functions) and the exploitation of group partial separability in more fundamental ways. Nevertheless, we are most interested in learning of others' experiences with LANCELOT as this will undoubtedly be helpful in the future evolution of the package.

Appendix A. Conditions of Use

The LANCELOT package has been developed by A.R. Conn, Nick Gould and Ph.L. Toint as a research tool and service for the scientific community.

The conditions under which the package and its constituent parts can be used are separated into

- general conditions, valid for all users,
- more specific additional conditions for 'academic' users.

Conditions for "commercial" use of the LANCELOT package or any of its constituent parts have to be separately negotiated with Nick Gould or Philippe Toint.

A.1 General Conditions for all Users

1. The LANCELOT package and the LANCELOT, SBMIN and AUGLG names are a copyright of the authors.

2. The package comes with no guarantee, expressed or implied, that it is suitable for any specific purpose nor that it is free of error. It should not be relied on as the basis to solve a problem whose incorrect solution could result in injury to person or property. If the package or any of its constituent parts is employed in such a manner, it is at the users' own risk and the authors disclaim all liability for such misuse.

3. The use of the package, or of any of its parts, in which the principal or exclusive purpose is as a tool for research into, or the production of, military weapons, or weapon systems, is prohibited.

4. The package is distributed "as is". In particular, no maintenance, support, troubleshooting or subsequent upgrade is implied.

5. The use of LANCELOT, SBMIN or AUGLG must be acknowledged, in any publication which contains results obtained with the package or any of its parts. A citation of the present book is suitable.

A.2 Additional Conditions for "Academic" Use

This package is distributed at low cost under the following additional conditions.

1. An "academic" licence for the package is granted to a user (hereafter called the licensee) on condition that one of the authors has received a copy of subsections A.1 and A.2 of the present "Conditions of use" signed and dated by the licensee.

2. Licences are issued for a fixed term. Unless the licence has been renewed, neither the package nor its constituent parts may be used beyond the expiration date of the licence.

3. The package, or any modification thereof, may not be embedded within other software products which are subsequently sold, or supplied, to an unlicenced third party.

4. It is the responsibility of the licensee to send a *representative* problem, of the type he or she has been solving, to the authors at most 11 months after the date given in Condition 1 in section A.2. It is acceptable to do this via electronic mail. A representative problem is one which arises in a field of the licensee's work or research and is typical in structure and size of the licensee's application. Problem data should be representative of the true application but may be simulated to protect the confidentiality of the licensee.

   The authors reserve the right to decline to renew the licence if they have difficulty obtaining from the licensee what they consider to be a representative problem. They also reserve the right to pass all problems so obtained to any third party.

   The purpose of the above requirement is to gather information to enhance the current version of the package and also to collect test problems from the general scientific community. It is indeed the authors' intention to make selected test problems obtained in this fashion publicly available.

5. A separate licence is required for each machine on which the package is available.

6. It is the responsibility of the licensee to ensure that each additional user of the package on a particular machine is aware of, and agrees to abide by, all the conditions given above, with the exception of Condition 4 in Section A.2.

A.3 Authors' Present Addresses

Dr A. R. Conn

Mathematical Sciences Department
IBM T. J. Watson Research Center
P.O. Box 218
Yorktown Heights, NY 10598, USA
email: arconn@watson.ibm.com

Dr Nick Gould

Central Computing Department
Rutherford Appleton Laboratory
Chilton, Oxfordshire, OX11 0QX, England
email: nimg@ib.rl.ac.uk

Pr Ph. L. Toint

Department of Mathematics,
Facultés Universitaires ND de la Paix,
61, rue de Bruxelles,
B-5000 Namur, Belgium
email: pht@math.fundp.ac.be

Appendix B. Trademarks

- Macintosh is a trademark of Apple Computers Inc.
- Unix is a trademark of AT&T Bell Laboratories.
- VAX, Ultrix and VMS are trademarks of Digital Corporation.
- AIX, PS/2, VM and CMS are trademarks of IBM.
- UNICOS is a trademark of Cray Inc.
- CM2 is a trademark of Thinking Machines Corporation.

Bibliography

[1] J. Bisschop and A. Meeraus. Matrix augmentation and structure preservation in linearly constrained control problems. *Mathematical Programming*, 18:7–15, 1977.

[2] A. Brooke, D. Kendrick, and A. Meeraus. *GAMS: a User's Guide*. The Scientific Press, Redwood City, USA., 1988.

[3] A. G. Buckley. Test functions for unconstrained minimization. Technical Report CS-3, Computing Science Division, Dalhousie University, Dalhousie, Canada, 1989.

[4] J. V. Burke and J. J. Moré. On the identification of active constraints. *SIAM Journal on Numerical Analysis*, 25:1197–1211, 1988.

[5] J. V. Burke, J. J. Moré, and G. Toraldo. Convergence properties of trust region methods for linear and convex constraints. *Mathematical Programming, Series A*, 47(3):305–336, 1990.

[6] P. H. Calamai and J. J. Moré. Projected gradient methods for linearly constrained problems. *Mathematical Programming*, 39:93–116, 1987.

[7] A. R. Conn, N. I. M. Gould, M. Lescrenier, and Ph. L. Toint. Performance of a multifrontal scheme for partially separable optimization. In *Advances in numerical partial differential equations and optimization, Proceedings of the sixth Mexico-United States Workshop*, Philadelphia, USA, 1992. SIAM.

[8] A. R. Conn, N. I. M. Gould, and Ph. L. Toint. Global convergence of a class of trust region algorithms for optimization with simple bounds. *SIAM Journal on Numerical Analysis*, 25:433–460, 1988. See also same journal 26:764–767, 1989.

[9] A. R. Conn, N. I. M. Gould, and Ph. L. Toint. Testing a class of methods for solving minimization problems with simple bounds on the variables. *Mathematics of Computation*, 50:399–430, 1988.

[10] A. R. Conn, N. I. M. Gould, and Ph. L. Toint. An introduction to the structure of large scale nonlinear optimization problems and the LANCELOT project. In R. Glowinski and A. Lichnewsky, editors, *Computing Methods in Applied Sciences and Engineering*, pages 42–54, Philadelphia, USA, 1990. SIAM.

[11] A. R. Conn, N. I. M. Gould, and Ph. L. Toint. Convergence of quasi-Newton matrices generated by the symmetric rank one update. *Mathematical Programming*, 50(2):177–196, 1991.

[12] A. R. Conn, N. I. M. Gould, and Ph. L. Toint. A globally convergent augmented Lagrangian algorithm for optimization with general constraints and simple bounds. *SIAM Journal on Numerical Analysis*, 28(2):545–572, 1991.

[13] A. R. Conn, N. I. M. Gould, and Ph. L. Toint. On the number of inner iterations per outer iteration of a globally convergent algorithm for optimization with general nonlinear equality constraints and simple bounds. In D.F Griffiths and G.A. Watson, editors, *Proceedings of the 14th Biennal Numerical Analysis Conference Dundee 1991*. Longmans, 1992.

[14] International Business Machine Corporation. Mathematical programming system/360 version 2, linear and separable programming-user's manual. Technical Report H20-0476-2, IBM Corporation, 1969. MPS Standard.

[15] A. R. Curtis and J. K. Reid. On the automatic scaling of matrices for Gaussian elimination. *Journal of the Institute of Mathematics and its Applications*, 10:118–124, 1972.

[16] G. B. Dantzig. *Linear Programming and Extensions.* Princeton University Press, Princeton, USA, 1963.

[17] D. Decker, F. Louveaux, G. Mortier, G. Schepens, and A. V. Looveren. *Linear and Mixed Integer Programming with OMP*. Beyers and Partners, Brasschaat, Belgium, 1987.

[18] J. E. Dennis and R. B. Schnabel. *Numerical methods for unconstrained optimization and nonlinear equations.* Prentice-Hall, Englewood Cliffs, USA, 1983.

[19] I. D. Duff, R. G. Grimes, and J. G. Lewis. Sparse matrix test problems. *ACM Transactions on Mathematical Software*, 15:1–14, 1989.

[20] I. S. Duff, A. M. Erisman, and J. K. Reid. *Direct methods for sparse matrices.* Clarendon Press, Oxford, UK, 1986.

[21] I. S. Duff and J. K. Reid. MA27: A set of Fortran subroutines for solving sparse symmetric sets of linear equations. Report R-10533, AERE Harwell Laboratory, Harwell, UK, 1982.

[22] I. S. Duff and J. K. Reid. The multifrontal solution of indefinite sparse symmetric linear equations. *ACM Transactions on Mathematical Software*, 9(3):302–325, 1983.

[23] R. Fletcher. *Practical Methods of Optimization.* J. Wiley and Sons, Chichester, second edition, 1987.

[24] R. Fourer, D. M. Gay, and B. W. Kernighan. AMPL: A mathematical programming language. Computer science technical report, AT&T Bell Laboratories, Murray Hill, USA, 1987.

[25] R. Fourer, D. M. Gay, and B. W. Kernighan. A high-level language would make a good standard form for nonlinear programming problems. talk at the CORS/TIMS/ORSA Meeting, Vanvouver, 1989.

[26] D. M. Gay. Electronic mail distribution of linear programming test problems. Mathematical Programming Society COAL Newsletter, December 1985.

[27] A. George and J. W.-H. Liu. *Computer solution of large sparse positive definite systems.* Prentice-Hall, Englewood Cliffs, USA, 1981.

[28] P. E. Gill and W. Murray. Newton-type methods for unconstrained and linearly constrained optimization. *Mathematical Programming*, 28:311–350, 1974.

[29] P. E. Gill, W. Murray, M. A. Saunders, and M. H. Wright. Sparse matrix methods in optimization. *SIAM Journal on Scientific and Statistical Computing*, 5:562–589, 1984.

[30] P. E. Gill, W. Murray, M. A. Saunders, and M. H. Wright. A Schur complement method for sparse quadratic programming. In M. G. Cox and S. Hammarling, editors, *Reliable Numerical Computation*, pages 113–138, Oxford, 1990. Clarendon Press.

[31] P. E. Gill, W. Murray, and M. H. Wright. *Practical Optimization.* Academic Press, London and New York, 1981.

[32] G. H. Golub and C. F. Van Loan. *Matrix Computations.* Johns Hopkins University Press, Baltimore, second edition, 1989.

[33] N. I. M. Gould. On growth in Gaussian elimination with complete pivoting. *SIAM Journal on Matrix Analysis*, 12(2):354–361, 1991.

[34] A. Griewank and Ph. L. Toint. On the unconstained optimization of partially separable functions. In M. J. D. Powell, editor, *Nonlinear Optimization 1981*, pages 301–312, London and New York, 1982. Academic Press.

[35] A. Griewank and Ph. L. Toint. Partitioned variable metric updates for large structured optimization problems. *Numerische Mathematik*, 39:429–448, 1982.

[36] Harwell Subroutine Library. *A catalogue of subroutines (release 10)*. Advanced Computing Department, Harwell Laboratory, Harwell, UK, 1990.

[37] Haverley Systems, Inc. *OMNI linear programming system: User and operating manual.* Haverley Systems, Denvile, USA, 1976.

[38] W. Hock and K. Schittkowski. *Test Examples for Nonlinear Programming Codes.* Springer Verlag, Berlin, 1981. Lectures Notes in Economics and Mathematical Systems 187.

[39] P. J. Huber. *Robust Statistics.* J. Wiley and Sons, New York, 1981.

[40] M. Lenard. Standardizing the interface with nonlinear optimizers. talk at the CORS/TIMS/ORSA Meeting, Vancouver, 1989.

[41] M. Lescrenier. Convergence of trust region algorithms for optimization with bounds when strict complementarity does not hold. *SIAM Journal on Numerical Analysis*, 28(2):476–495, 1991.

[42] G. P. McCormick and P. Rahnavard. Representation of unconstrained optimization. talk at the CORS/TIMS/ORSA Meeting, Vancouver, 1989.

[43] J. J. Moré. Trust regions and projected gradients. In M. Iri and K. Yajima, editors, *System Modelling and Optimization*, volume 113, pages 1–13, Berlin, 1988. Springer Verlag. Lecture Notes in Control and Information Sciences.

[44] J. J. Moré. A collection of nonlinear model problems. Technical Report ANL/MCS-P60-0289, Argonne National Laboratory, Argonne, USA, 1989.

[45] J. J. Moré, B. S. Garbow, and K. E. Hillstrom. Testing unconstrained optimization software. *ACM Transactions on Mathematical Software*, 7(1):17–41, 1981.

[46] J. J. Moré and G. Toraldo. Algorithms for bound constrained quadratic programming problems. *SIAM Journal on Optimization*, 1(1):93–113, 1991.

[47] N. Munksgaard. Solving sparse symmetric systems of linear equations by preconditioned conjugate gradients. *ACM Transactions on Mathematical Software*, 6:206–219, 1980.

[48] B. A. Murtagh and M. A. Saunders. MINOS 5.1 USER'S GUIDE. Technical Report SOL83-20R, Department of Operations Research, Stanford University, Stanford, USA, 1987.

[49] Oxford University Press. *Dictionnary of Computing.* Oxford University Press, Oxford, UK, 1983.

[50] K. Schittkowski. *More Test Examples for Nonlinear Programming Codes.* Springer Verlag, Berlin, 1987. Lecture notes in economics and mathematical systems, volume 282.

[51] R. B. Schnabel and E. Eskow. A new modified Cholesky factorization. *SIAM Journal on Scientific and Statistical Computing*, 11:1136–1158, 1991.

[52] Ph. L. Toint. Test problems for partially separable optimization and results for the routine pspmin. Technical Report 83/4, Department of Mathematics, FUNDP, Namur, Belgium, 1983.

[53] Ph. L. Toint. Global convergence of a class of trust region methods for nonconvex minimization in Hilbert space. *IMA Journal of Numerical Analysis*, 8:231–252, 1988.

[54] J. W. J. Williams. Algorithm 232, Heapsort. *Communications of the ACM*, 7:347–348, 1964.

Index